Chemical Thermodynamics

BASIC THEORY AND METHODS

PRENTICE-HALL CHEMISTRY SERIES

WENDELL M. LATIMER, *Editor*

Chemical Thermodynamics

BASIC THEORY AND METHODS

by

IRVING M. KLOTZ, Ph.D.

PROFESSOR OF CHEMISTRY
NORTHWESTERN UNIVERSITY

With Advice and Suggestions from

THOMAS FRASER YOUNG, Ph.D.

UNIVERSITY OF CHICAGO

Englewood Cliffs, N. J.
PRENTICE-HALL, INC.

First printing .... September, 1950
Second printing .... July, 1953
Third printing .... January, 1955
Fourth printing .... May, 1956
Fifth printing .... June, 1957

PRINTED IN THE UNITED STATES OF AMERICA
12885

Preface

As its title implies, this book is a text on thermodynamics designed primarily for chemists. It is, furthermore, an introductory volume on this subject. As such, it has two prime objectives: first, to present to the student the logical foundations and inter-relationships of the theory of thermodynamics; second, to teach the student the methods by which the theoretical principles may be applied to practical problems.

In the treatment of the theoretical principles, the author has adopted the classical, or phenomenological, approach to thermodynamics and has excluded entirely the statistical viewpoint. This attitude has several pedagogical advantages. First, it permits the maintenance of a logical unity throughout the book. In addition, it offers an opportunity to stress the "operational" approach to abstract concepts. Furthermore, it makes some contribution toward freeing the student from a perpetual yearning for a mechanical analogue for every new concept to which he is introduced. Finally, and perhaps most important, it avoids the promulgation of an all-too-common point of view toward statistical mechanics as an appendage which can be conveniently grafted on to the body of thermodynamics. A logical development of the statistical approach should probably be based on a previous introduction to the fundamental quantum-mechanical concept of energy-level and should emphasize the much broader scope of the phenomena which it can treat. An effective presentation of statistical theory should be, therefore, an independent and complementary one to phenomenological thermodynamics.

A great deal of attention is also paid in this text to training the student in the application of the theory of thermodynamics to problems which are commonly encountered by the chemist. The mathematical tools which are necessary for this purpose are considered in more detail than is usual. In addition, computational techniques, both graphical and analytical, are described fully and are used frequently, both in illustrative and in assigned problems. Furthermore, exercises have been designed to simulate more closely than in most texts, the type of problem that may actually be encountered by the practicing chemist. Short, unrelated exercises are thus kept to a minimum, whereas series of computations or derivations, illustrating a technique or principle of general applicability, are emphasized.

A definite effort has been made also to keep this volume within limits which may be covered in a course of lectures extending over a period of twelve to fifteen weeks. Too often, a textbook which attempts to be exhaustive in its coverage merely serves to overwhelm the student. On the

v

other hand, if a student can be guided to a sound grasp of the fundamental principles and shown how these may be applied to a few typical practical problems, he will be capable of examining other special topics either by himself or with the aid of one of the excellent comprehensive treatises which are available.

Another feature of this book which may be a little unusual is the extensive use of sub-headings in outline form to indicate the position of a given topic in the general sequence of presentation. In using this method of arrangement, the author has been influenced strongly by the viewpoint expressed so aptly by Poincaré: "The order in which these elements are placed is much more important than the elements themselves. If I have the feeling . . . of this order, so as to perceive at a glance the reasoning as a whole, I need no longer fear lest I forget one of the elements, for each of them will take its allotted place in the array, and that without any effort of memory on my part."[1] It is a universal experience of teachers, that students are able to retain a body of information much more effectively if they are aware of the place of the parts in the whole.

The preparation of this volume was facilitated greatly by the assistance rendered by several colleagues and students, particularly Professors Malcolm Dole and Ralph Pearson, and Doctors Robert Dobres and A. C. English, who read large portions of the manuscript critically. A special debt of gratitude is due to Professor T. F. Young who contributed so generously of his time and experience in reading the entire manuscript and in making a host of suggestions to improve the clarity and rigor of the presentation. Many of the exercises have also been taken from problems developed originally by Professor Young. I am also greatly indebted to my wife, Themis Askounis Klotz, for her many services above and beyond the call of duty. The assistance of Mrs. Jean Urquhart Dunham in the preparation of the drawings is also gratefully acknowledged.

IRVING M. KLOTZ

Evanston, Illinois

---

[1] H. Poincaré, *The Foundations of Science*, translated by G. B. Halsted, Science Press, 1913.

Contents

Thermodynamics of Systems of Variable Composition

CHAPTER 1

Introduction

I. OBJECTIVES OF CHEMICAL THERMODYNAMICS

Thermodynamics is distinguished from other general fields in the natural sciences by its concern with temperature and thermal manifestations. In its most general applications, thermodynamic analysis may involve electromagnetic and surface factors as well as mechanical and chemical parameters, but thermal variables always occupy a dominant position.

The concepts of thermodynamics have been powerful tools for the engineer, the physicist, and the chemist. For the mechanical engineer their great use lies in their ability to predict the maximum efficiency of various types of heat engine and the maximum work obtainable from a given fuel—in other words, in problems of combustion and power. For the physicist and chemist these problems are of secondary interest. Although no clear division can be made between the fields of primary interest to the physicist and chemist, there is generally greater emphasis by the physicist on phenomena involving interactions with radiation or with electric and magnetic fields. In contrast, the chemist is concerned primarily with physical and chemical transformations under conditions in which electromagnetic phenomena exert no significant influences.

The primary objective of *chemical* thermodynamics is the establishment of a criterion for the determination of the feasibility or spontaneity of a given transformation. This transformation may be either physical or chemical. For example, one may be interested in a criterion for determining the possibility of a spontaneous transformation from one phase to another, as from one crystalline form to another, or one may wish to know whether a reaction suggested for the preparation of some substance has a reasonable chance of success or is utterly hopeless. Based on empirical principles, chemical thermodynamics has developed theoretical concepts and mathematical functions which form a framework for answering these questions.

Having determined that a reaction is feasible, the chemist wishes to know further the maximum yields he may hope to obtain in a particular chemical transformation. Thermodynamic methods provide a basis for the formulation of the mathematical relations which are necessary to estimate such equilibrium yields.

Although the analysis of problems of chemical transformations is prob-

1

ably the main objective of chemical thermodynamics, the theory is capable also of application to many other problems encountered in chemical operations. As an illustration, the study of phase equilibria, for ideal and non-ideal systems, is basic to the intelligent use of the techniques of extraction, fractional distillation, and crystallization. Similarly, one may be interested in the energy changes—either in the form of heat or of work—which may accompany a chemical or physical transformation. Thermodynamic concepts and methods provide a powerful approach to the solution of these and related chemical problems.

II. LIMITATIONS OF CLASSICAL THERMODYNAMICS

Though the theory of thermodynamics provides the foundation for the solution of many chemical problems, the answers obtained are generally not definitive. In the language of the mathematician, we might say that classical thermodynamics is capable of formulating necessary conditions but not sufficient conditions. Thus, a thermodynamic analysis may rule out a given reaction for the synthesis of some substance by indicating that such a transformation cannot proceed spontaneously under any set of available conditions. In such a case we have a definitive answer. On the other hand, if the analysis indicates that a reaction may proceed spontaneously, no statement can be made from classical thermodynamics alone that it will do so in any finite time. For example, benzene is unstable with respect to its elements, according to thermodynamic calculations. In other words, according to thermodynamic theory, benzene *may* decompose spontaneously into carbon and hydrogen even at room temperature. Yet every chemist is aware of the practical stability of benzene standing in a room. This is not a contradiction of the theory, because thermodynamics claims no ability to predict the *time* required for a reaction. Much experience has indicated that when suitable catalysts are available, substances which apparently are very stable can be made to undergo transformations which thermodynamic calculation indicates are theoretically spontaneous.

An analogous situation is encountered in considerations of the *extent* to which a given chemical reaction may proceed. No statement can be made about the actual yield obtainable under a given set of conditions; only the maximum yield, the equilibrium yield, may be predicted. This limitation, too, is a consequence of the inability of classical thermodynamics to make any statement about the rate of a reaction. There is no relation between the kinetics of a reaction and the thermodynamic energy changes accompanying the reaction. The problem of kinetics is one which classical thermodynamic analysis has been unable to solve. Only since the introduction of statistical methods and the concepts of the kinetic-molecular theory has a promising approach been made.

Similarly, in connection with the *work* obtainable from a chemical or physical transformation, only limiting values may be calculated. Thermodynamic functions predict the work which may be obtained if the reaction is carried out with infinite slowness, in a so-called "reversible" manner. However, it is impossible to specify the actual work obtained in a real or natural process in which the time interval is finite, except for the statement that the real work will be less than the quantity obtainable in a reversible situation.

Thus classical thermodynamic methods can treat only limiting cases.[1] Nevertheless, such a restriction is not nearly as severe as it may seem at first glance, since in many cases it is possible to approach equilibrium conditions very closely, and the thermodynamic quantities coincide with actual values, within experimental error. In other situations, thermodynamic analysis may rule out certain reactions under *any* conditions, and a great deal of time and effort may be saved. Even in their most constrained applications, thermodynamic methods, by limiting solutions within certain boundary values, can reduce materially the amount of experimental work necessary to yield a definitive answer to a particular problem.

---

[1] Classical thermodynamics also encounters difficulties in treating fluctuation phenomena, such as the "shot effect" and Brownian motion. These difficulties, however, can be traced back to the nature of the postulations of thermodynamic theory and are different in character from the limitations inherent in the framework of the analytical method.

CHAPTER 2

Mathematical Apparatus[1]

A N EXACT science must be founded on precise definitions. Precision in definition, as well as in reasoning, is difficult to obtain by verbalization. Mathematics, on the other hand, offers a precise mode of expression as well as a rigorous logical procedure. Mathematics, therefore, has become the language of thermodynamics. Indeed, it has been aptly said that chemical thermodynamics is essentially a mathematical framework built on a foundation of three basic postulates: the first, the second, and the third laws of thermodynamics. It is evident, therefore, that a thorough appreciation of the power of thermodynamic methods depends on an intensive familiarity with its mathematical tools.

The most elegant and formal approach to thermodynamics, that of Caratheodory,[2] depends on a familiarity with a special type of differential equation (Pfaff equation), with which the usual student of chemistry is unacquainted. An introductory presentation follows best along historical lines of development, for which only the elementary principles of calculus are necessary. Nevertheless, since many concepts and derivations can be presented in a much more satisfying and precise manner if based on the use of exact differentials and of Euler's theorem, we shall introduce these propositions also, after a review of some elementary principles.

I. VARIABLES OF THERMODYNAMICS

A. Extensive and intensive quantities. It is convenient to distinguish two kinds of thermodynamic variable, *extensive* and *intensive*. The values of extensive variables are proportional to the quantity of matter which is under consideration. Volume and heat capacity are typical examples of such variables. Intensive variables, on the other hand, are independent of the amount of matter under consideration, and are exemplified by quantities such as temperature, pressure, concentration, and molal heat capacity.

In general, both extensive and intensive variables are handled by the

---

[1] For convenience in writing, the author has assembled in this single chapter all the mathematical methods to be described in this textbook. In actual classroom presentation, however, only the first third of this chapter is discussed at this point. The remaining portions are introduced as they are needed further along in the course.

[2] C. Caratheodory, *Math. Ann.*, **61**, 355 (1909); P. Frank, *Thermodynamics*, Brown University, Providence, R. I., 1945.

same mathematical techniques. Nevertheless, as we shall see in Chapter 13, extensive and intensive variables differ in their degree of homogeneity.

It is perhaps pertinent to point out also that some variables encountered in thermodynamic problems are neither extensive nor intensive. An example is the square root of the volume. The appearance of such variables produces no formal difficulties, however.

B. Units and conversion factors. The numerical value which is assigned to a particular variable in a given problem depends, of course, on the units in which it is expressed. Critical evaluations of conversion factors between units are available in standard reference works. The authoritative values, chosen by the National Bureau of Standards, for the units and constants which occur in thermodynamic calculations are assembled in Table 1. In accordance with international agreement, units of energy should be expressed, as of January 1, 1948, in terms of the *absolute* joule.

TABLE 1

SOME DEFINITIONS AND UNITS*

$$
\begin{aligned}
\text{1 absolute volt} &= 0.99967 \text{ international volt} \\
\text{1 absolute ampere} &= 1.00017 \text{ international amperes} \\
\text{1 absolute coulomb} &= 1.00017 \text{ international coulombs} \\
\text{1 absolute joule} &= 0.99984 \text{ international joule} \\
\text{1 (defined) calorie} &= 4.1840 \text{ absolute joules} \\
&= 4.1833 \text{ international joules} \\
&= 0.041292 \text{ liter-atmosphere} \\
\text{Absolute temperature at } 0\,^\circ\text{C} &= 273.16\,^\circ\text{K} \\
R, \text{ gas constant} &= 8.3144 \text{ absolute joules deg}^{-1}\text{ mole}^{-1} \\
&= 1.9872 \text{ cal deg}^{-1}\text{ mole}^{-1} \\
&= 0.082054 \text{ liter-atm deg}^{-1}\text{ mole}^{-1} \\
R\,(298.16)\ln x &= 1364.28 \log_{10} x \text{ cal mole}^{-1} \\
N, \text{ Avogadro number} &= 6.023 \times 10^{23}\text{ mole}^{-1} \\
\mathfrak{F}, \text{ the Faraday constant} &= 96,485 \text{ abs coul (gram-equivalent)}^{-1} \\
&= 23,060 \text{ cal (abs volt)}^{-1}\text{(gram-equivalent)}^{-1} \\
&= 23,068 \text{ cal (int volt)}^{-1}\text{(gram-equivalent)}^{-1}
\end{aligned}
$$

* *Selected Values of Chemical Thermodynamic Properties,* Values of Constants, pages 1–2, National Bureau of Standards, Washington, D. C., December 31, 1947.

II. THEORETICAL METHODS

A. Partial differentiation.

1. *Equation for the total differential.* Since the state of a given thermodynamic system is generally a function of more than one independent variable, it is necessary to consider the mathematical techniques of handling these polyvariable relations. Since many thermodynamic problems involve only two independent variables and since the extension to more variables is generally obvious, we shall limit our illustrations to functions of two variables.

Let us consider a specific example such as the volume of a pure substance.

This volume is a function of the temperature and pressure of the substance, and the relation may be written in general notation as

$$V = f(P,T). \tag{2.1}$$

Utilizing the principles of calculus,[3] we may write for the total differential

$$dV = \left(\frac{\partial V}{\partial P}\right)_T dP + \left(\frac{\partial V}{\partial T}\right)_P dT. \tag{2.2}$$

For the special case of one mole of an ideal gas, equation (2.1) is

$$V = \frac{RT}{P} = R(T)\left(\frac{1}{P}\right). \tag{2.3}$$

Since the partial derivatives are given by the expressions

$$\left(\frac{\partial V}{\partial P}\right)_T = -\frac{RT}{P^2} \tag{2.4}$$

and

$$\left(\frac{\partial V}{\partial T}\right)_P = \frac{R}{P}, \tag{2.5}$$

it is obvious that the total differential for the special case of the ideal gas may be obtained by substitution into equation (2.2) and is given by the relation

$$dV = -\frac{RT}{P^2} dP + \frac{R}{P} dT. \tag{2.6}$$

We shall have frequent occasion to make use of this expression.

2. *Transformation formulas.* It frequently happens in thermodynamic problems that a given partial derivative, for example $(\partial V/\partial T)_P$, is necessary for the solution of an equation or numerical problem, but that there is no convenient experimental method of evaluating this derivative. If an expression were available which related this derivative to other partial derivatives which are known or readily obtainable, we could solve our problem without difficulty. For this purpose we must be able to transform a partial derivative into some alternative form.

We shall illustrate the procedure in deriving such transformation formulas with the example of the preceding section, the volume function.

a. We can derive the first transformation formula readily[4] by referring to equation (2.2) and imposing the restriction that V shall be a constant.

[3] W. A. Granville, P. F. Smith, and W. R. Longley, *Elements of the Differential and Integral Calculus*, Ginn and Company, Boston, 1941.

J. W. Mellor, *Higher Mathematics for Students of Chemistry and Physics*, Dover Publications, New York, 1946, pages 68–75.

[4] For a more rigorous derivation of these transformation formulas, see H. Margenau and G. M. Murphy, *The Mathematics of Physics and Chemistry*, D. Van Nostrand and Company, New York, 1943, pages 6–8.

Keeping in mind, then, that $dV = 0$, we may obtain

$$\frac{dV}{dT} = 0 = \left(\frac{\partial V}{\partial P}\right)_T \frac{dP}{dT} + \left(\frac{\partial V}{\partial T}\right)_P. \tag{2.7}$$

If we now indicate explicitly for the second factor of the first term that V is constant, and if we rearrange terms, we obtain

$$\left(\frac{\partial V}{\partial T}\right)_P = -\left(\frac{\partial V}{\partial P}\right)_T \left(\frac{\partial P}{\partial T}\right)_V. \tag{2.8}$$

Thus if in some situation we needed $(\partial V / \partial T)_P$ but had no method of direct evaluation, we could establish its value if $(\partial V / \partial P)_T$ and $(\partial P / \partial T)_V$ were available.

It may be desirable to verify the validity of equation (2.8) for an ideal gas by evaluating both sides explicitly and showing that the equality holds. The values of the partial derivatives can be determined readily by reference to equation (2.3), and the following deductions can be made:

$$\frac{R}{P} = -\left(-\frac{RT}{P^2}\right)\left(\frac{R}{V}\right) = \frac{R^2 T}{P^2 V} = \frac{RT}{PV}\frac{R}{P} = \frac{R}{P}. \tag{2.9}$$

b. The second transformation formula is obtained by a procedure analogous to the preceding one in that V is held constant. After suitable rearrangement we obtain

$$\left(\frac{\partial V}{\partial P}\right)_T = -\left(\frac{\partial V}{\partial T}\right)_P \left(\frac{\partial T}{\partial P}\right)_V. \tag{2.10}$$

c. A third formula is obtainable by rearranging equation (2.10) to

$$\left(\frac{\partial V}{\partial T}\right)_P = -\frac{(\partial V / \partial P)_T}{(\partial T / \partial P)_V} \tag{2.11}$$

and setting the right-hand side of this expression equal to the right-hand side of equation (2.8):

$$-\left(\frac{\partial V}{\partial P}\right)_T \left(\frac{\partial P}{\partial T}\right)_V = -\left(\frac{\partial V}{\partial P}\right)_T \frac{1}{(\partial T / \partial P)_V}. \tag{2.12}$$

Therefore
$$\left(\frac{\partial P}{\partial T}\right)_V = \frac{1}{(\partial T / \partial P)_V}. \tag{2.13}$$

Thus, within limits, the derivatives may be handled formally as if they were fractions.

d. Another important relation must be obtained for use in a problem in which a new but not independent variable is introduced. For example, we might consider the energy, E, of a pure substance as a function of pressure and temperature:

$$E = g(P,T). \tag{2.14}$$

We may then wish to evaluate the partial derivative $(\partial V/\partial P)_E$, that is, the change of volume with change in pressure at constant energy. A suitable expression for this derivative in terms of other partial derivatives may be obtained readily from equation (2.2) by putting in explicitly the restriction that E is to be held constant. The result obtained is the relation

$$\left(\frac{\partial V}{\partial P}\right)_E = \left(\frac{\partial V}{\partial P}\right)_T + \left(\frac{\partial V}{\partial T}\right)_P \left(\frac{\partial T}{\partial P}\right)_E. \tag{2.15}$$

e. A fifth transformation formula, for use in certain situations where a new variable $x(P,T)$ is to be introduced, is derived conveniently (though not rigorously) as follows:

$$\left(\frac{\partial V}{\partial T}\right)_P = \left(\frac{\partial V}{\partial T}\frac{\partial x}{\partial x}\right)_P = \left(\frac{\partial V}{\partial x}\frac{\partial x}{\partial T}\right)_P$$

$$= \left(\frac{\partial V}{\partial x}\right)_P \left(\frac{\partial x}{\partial T}\right)_P. \tag{2.16}$$

These illustrations, based on the example of the volume function, are typical of the type of transformation which is required so frequently in thermodynamic manipulations. Greater facility in their use may be acquired by solution of the appropriate exercises at the end of this chapter.

B. Exact differentials. Many thermodynamic relations can be derived with such ease by use of the properties of the exact differential that it is highly desirable for the student to become familiar with this type of function. As an introduction to the characteristics of exact differentials, we shall consider the properties of certain simple functions used in connection with a gravitational field.

1. *Example of the gravitational field.* Let us compare the change in potential energy and the work done, respectively, in moving a large boulder up a hill, that is, against the force of gravity. From our familiarity with elementary physics, it is easy to see that these two quantities, ΔE and W, differ in the following respects.

a. The change in potential energy depends only on the initial and final heights of the stone, whereas the work done depends upon the path used. That is, the quantity of work expended if we use a pulley and tackle to raise the boulder directly will be much less than if we have to bring the object up the hill by pushing it over a long, muddy, and tortuous road. On the other hand, the change in potential energy is the same for both paths, so long as they have the same starting point and the same end point.

Independence of the path is a characteristic of an exact differential; dependence, of an inexact differential. Therefore

$$dE \quad = \quad \text{exact differential,}$$
$$dW = DW = \text{inexact differential.}$$

In general we shall use the capital letter D to indicate inexactness.

b. There is an explicit expression for the potential energy, E, and this function can be differentiated to give dE, whereas no explicit expression leading to DW can be obtained. The function for the potential energy, E, is a particularly simple one for the gravitational field, in that two of the space coordinates drop out and only the height, h, remains:

$$E = \text{constant} + mgh. \tag{2.17}$$

The symbols m and g have the usual significance, mass and acceleration due to gravity, respectively. This characteristic, the existence of a function which leads to the differential expression, is another property of the exact differential, in this case dE, as contrasted to DW.

c. A third difference between ΔE and W lies in the values obtained if one uses a cyclic path, as in moving the boulder up the hill and then back down again to the initial point. For such a cyclic or closed path, the net change in potential energy is zero, since the final and initial points are identical:

$$\oint dE = 0. \tag{2.18}$$

On the other hand, the value of W is not fixed at all. It depends upon the path used. If one uses each of the same two paths described above to bring the stone uphill and then, in each case, allows the stone to fall freely through the air back to its starting point, the net works done in the two cyclic paths are neither zero nor equal.

2. *General formulation.* With the gravitational example as a guide to the nature of exact differentials, we may set up the properties of these functions in general terms for use in thermodynamic analysis. To understand the notation which is generally adopted, we shall rewrite equation (2.2) in the following form, which makes explicit recognition of the fact that the partial derivatives and the total differential are functions (indicated by M and N) of the independent variables P and T:

$$dV(P,T) = M(P,T) \, dP + N(P,T) \, dT. \tag{2.19}$$

It is obvious, then, that a formulation for a general case with two independent variables, x and y, could take the form

$$dL(x,y) = M(x,y) \, dx + N(x,y) \, dy. \tag{2.20}$$

Using this expression, we may summarize the characteristics of an exact differential as follows.

a. A (linear) differential expression containing two variables, of the form of equation (2.20), is an exact differential if there exists a function $f(x,y)$ such that

$$df(x,y) = dL(x,y). \tag{2.21}$$

b. If dL is an exact differential, the line integral (that is, the integral over some path), $\int dL(x,y)$, depends only on the initial and final states and not on the path between them.

c. If dL is an exact differential, the line integral over a closed path is zero:

$$\oint dL(x,y) = 0. \tag{2.22}$$

It is this characteristic which is most frequently used in testing thermodynamic functions for exactness. If the differential of a thermodynamic function, dG, *is exact*, then G *is called a thermodynamic property*.

3. *Reciprocity characteristic.* A common test of exactness of a differential expression $dL(x,y)$ is to see whether the following relation holds:

$$\frac{\partial}{\partial y} M(x,y) = \frac{\partial}{\partial x} N(x,y). \tag{2.23}$$

We can see readily that this relation must be true if dL is exact, since in that case there exists a function $f(x,y)$ such that

$$df(x,y) = \left(\frac{\partial f}{\partial x}\right)_y dx + \left(\frac{\partial f}{\partial y}\right)_x dy = dL(x,y). \tag{2.24}$$

But for the function $f(x,y)$ we know from the principles of calculus that

$$\frac{\partial}{\partial y}\left(\frac{\partial f}{\partial x}\right)_y = \frac{\partial^2 f}{\partial y\,\partial x} = \frac{\partial}{\partial x}\left(\frac{\partial f}{\partial y}\right)_x. \tag{2.25}$$

Since it follows from equations (2.24) and (2.20) that

$$M(x,y) = \left(\frac{\partial f}{\partial x}\right)_y \tag{2.26}$$

and

$$N(x,y) = \left(\frac{\partial f}{\partial y}\right)_x, \tag{2.27}$$

it is obvious that if dL is exact,

$$\frac{\partial}{\partial y} M(x,y) = \frac{\partial}{\partial x} N(x,y). \tag{2.23}$$

To apply this criterion of exactness to a simple example, let us assume that we know only the expression for the total differential of the volume of an ideal gas, equation (2.6), and do not know whether this differential is exact or not. Applying the procedure of (2.23) to equation (2.6), we obtain

$$\frac{\partial}{\partial P}\left(\frac{R}{P}\right) = -\frac{R}{P^2} = \frac{\partial}{\partial T}\left(-\frac{RT}{P^2}\right). \tag{2.28}$$

Thus we would know that the volume of an ideal gas is a thermodynamic property, even if we had not been aware previously of an explicit function for V.

C. Homogeneous functions. In connection with the development of the thermodynamic concept of partial molal quantities it will be desirable to be familiar with a mathematical transformation known as *Euler's theorem.* Since this theorem is stated with reference to "homogeneous" functions, we shall consider briefly the nature of these functions.

1. *Definition.* Let us consider, as a simple example, the function

$$u = ax^2 + bxy + cy^2. \tag{2.29}$$

If we replace the variables x and y by kx and ky, where k is a parameter, it is obvious that

$$\left.\begin{aligned}
u^* &= a(kx)^2 + b(kx)(ky) + c(ky)^2\\
&= k^2ax^2 + k^2bxy + k^2cy^2\\
&= k^2(ax^2 + bxy + cy^2)\\
&= k^2u.
\end{aligned}\right\} \tag{2.30}$$

The net result of multiplying each independent variable by the parameter k has been merely to multiply the function by k^2. Since the k can be factored out, the function is called *homogeneous;* since the exponent of the factor k^2 is 2, the function is of the second degree.

We turn now to an example of experimental significance. If we mix certain quantities of benzene and toluene, the total volume, V, will be given by the expression

$$V = v_b n_b + v_t n_t, \tag{2.31}$$

where n_b is the number of moles of benzene, v_b is the volume of one mole of pure benzene, n_t is the number of moles of toluene, and v_t is the volume of one mole of pure toluene. Suppose, now, that we increase the quantity of each of the independent variables, n_b and n_t, by the same factor, say 2. We know from experience that the volume of the mixture will be doubled. Similarly, if the factor were k, the volume of the new mixture would be k times that of the original. In terms of equation (2.31) we can see also that if we replace n_b by kn_b and n_t by kn_t, the new volume V^* will be given by

$$\left.\begin{aligned}
V^* &= v_b kn_b + v_t kn_t\\
&= k(v_b n_b + v_t n_t)\\
&= kV,
\end{aligned}\right\} \tag{2.32}$$

The volume function then is homogeneous of the first degree since the parameter k which factors out occurs to the first power.

Proceeding to a general definition, we may say that a function $f(x,y,z, \ldots)$ is homogeneous of degree n if upon replacement of each independent variable by an arbitrary parameter, k, times the variable, the function is merely multiplied by k^n, that is, if

$$f(kx,ky,kz, \ldots) = k^n f(x,y,z, \ldots).$$ (2.33)

2. *Euler's theorem.* We shall state and prove Euler's theorem only for a function of two variables, $f(x,y)$. The extension to more variables will be obvious.

The statement of the theorem may be made as follows: if $f(x,y)$ is a homogeneous function of degree n, then

$$x\,\frac{\partial f}{\partial x} + y\,\frac{\partial f}{\partial y} = nf(x,y).$$ (2.34)

The proof may be carried out by the following steps. Let us represent the variables x^* and y^* by

$$x^* = kx$$ (2.35)

and

$$y^* = ky.$$ (2.36)

Then since $f(x,y)$ is homogeneous,

$$f^* = f(x^*,y^*) = f(kx,ky) = k^n f(x,y).$$ (2.37)

The total differential, df^*, is given by

$$df^* = \frac{\partial f^*}{\partial x^*}\,dx^* + \frac{\partial f^*}{\partial y^*}\,dy^*.$$ (2.38)

Hence

$$\frac{df^*}{dk} = \frac{\partial f^*}{\partial x^*}\frac{dx^*}{dk} + \frac{\partial f^*}{\partial y^*}\frac{dy^*}{dk}.$$ (2.39)

From equations (2.35) and (2.36) it is clear that

$$\frac{dx^*}{dk} = x$$ (2.40)

and

$$\frac{dy^*}{dk} = y.$$ (2.41)

Consequently, equation (2.39) may be transformed into

$$\frac{df^*}{dk} = \frac{\partial f^*}{\partial x^*}\,x + \frac{\partial f^*}{\partial y^*}\,y.$$ (2.42)

Making use of the equalities in (2.37), we can obtain

$$\frac{df^*}{dk} = \frac{df^*(x^*,y^*)}{dk} = \frac{d[k^n f(x,y)]}{dk}$$

$$= nk^{n-1}f(x,y). \tag{2.43}$$

Equating (2.42) and (2.43), we obtain

$$x\frac{\partial f^*}{\partial x^*} + y\frac{\partial f^*}{\partial y^*} = nk^{n-1}f(x,y). \tag{2.44}$$

Since k is an arbitrary parameter, equation (2.44) must hold for any particular value. It must be true, then, for $k = 1$. In such an instance, equation (2.44) reduces to

$$x\frac{\partial f}{\partial x} + y\frac{\partial f}{\partial y} = nf(x,y). \tag{2.34}$$

This equation is Euler's theorem.

As one example of the application of Euler's theorem, we may refer again to the volume of a two-component system. Evidently the total volume is a function of the number of moles of each component:

$$V = f(n_1,n_2). \tag{2.45}$$

As we have seen previously, the volume function is known from experience to be homogeneous of the first degree; that is, if we double the number of moles of each component, we also double the total volume. Applying Euler's theorem, then, we obtain the relation

$$n_1\frac{\partial V}{\partial n_1} + n_2\frac{\partial V}{\partial n_2} = V. \tag{2.46}$$

Although the significance of equation (2.46) may not be clear at this point, when we discuss partial molal quantities it will become apparent that this expression is basic to the entire subsequent development.

III. PRACTICAL TECHNIQUES

Throughout all our discussions we shall emphasize the application of thermodynamic methods to chemical problems. Successful solutions of such problems depend upon a familiarity with practical graphical and analytical techniques, as well as with the theoretical methods of mathematics. We shall consider these techniques at this point, therefore, so that they may be available for use as we approach specific problems.

A. Graphical methods. Experimental data of thermodynamic importance may be represented either graphically or in terms of an analytical equation. Often these data do not fit into a simple pattern which can be transcribed into a convenient equation. Consequently, graphical techniques, particularly for differentiation and integration, have assumed an important position among methods of treating thermodynamic data.

1. *Graphical differentiation.* Numerous procedures have been developed for graphical differentiation. A particularly convenient one,[5] which we may call the *chord-area method*, may be illustrated by the following example.

Let us consider a set of experimental determinations of the standard potential, $\varepsilon°$, at a series of temperatures, such as is listed in Table 2. A graph of these data (Fig. 1) shows that the slope varies slowly but uni-

TABLE 2

Standard Potentials* for Reaction: $\frac{1}{2}H_2 + AgCl = Ag + HCl$

t, °C	$\varepsilon°$, volt
0	0.23634
5	0.23392
10	0.23126
15	0.22847
20	0.22551
25	0.22239
30	0.21912
35	0.21563
40	0.21200
45	0.20821
50	0.20437
55	0.20035
60	0.19620

* H. S. Harned and R. W. Ehlers, *J. Am. Chem. Soc.*, **55**, 2179 (1933).

formly over the entire range of temperature. For thermodynamic purposes, such as in the calculation of the heat of reaction in the transformation

$$\frac{1}{2}H_2 + AgCl = Ag + HCl,$$

it is necessary to calculate precise values of the slope, $\partial\varepsilon°/\partial t$. It is clear from Fig. 1 that if we choose a sufficiently small interval of temperature, then the slope will be given approximately by $\Delta\varepsilon°/\Delta t$. In the example we are considering, an interval of 5° is sufficiently small for the average slope within this region to be given by $\Delta\varepsilon°/\Delta t$. We proceed then to tabulate values of $\Delta\varepsilon°/\Delta t$ from 0° on, as is illustrated in Table 3 for the first few data. Note that values of $\Delta\varepsilon°$ are placed between the values of $\varepsilon°$ to which they refer, and the temperature intervals, 5°, are indicated between their extremities. Similarly, since $\Delta\varepsilon°/\Delta t$ is an average value

[5] T. R. Running, *Graphical Mathematics*, John Wiley and Sons, Inc., New York, 1927, pages 65–66.

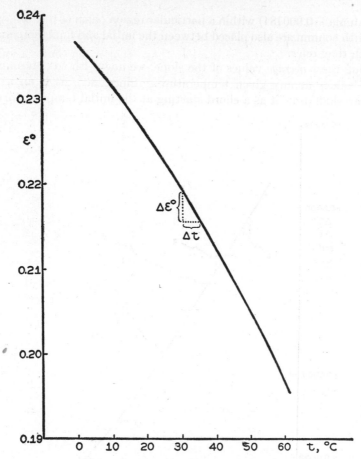

Fig. 1.   Standard electrode potentials for the reaction
$\tfrac{1}{2}H_2$ (g) + AgCl (s) = Ag (s) + HCl (aqueous).

TABLE 3

Tabulation for Graphical Differentiation

1 t	2 $\varepsilon°$	3 $\Delta\varepsilon°$	4 Δt	5 $\Delta\varepsilon°/\Delta t$	6 $\partial\varepsilon°/\partial t$
0	0.23634				−0.000476
		−0.00242	5	−0.000484	
5	0.23392				−0.000509
		−0.00266	5	−0.000532	
10	0.23126				−0.000543
		−0.00279	5	−0.000558	
15	0.22847				−0.000576
		−0.00296	5	−0.000592	
20	0.22551				0.000610

(for example -0.000484) within a particular region (such as 0 to 5°), values in the fifth column are also placed between the initial and final temperatures to which they refer.

Having these *average* values of the slope, we now wish to determine the *specific* values at any given temperature. Since $\Delta\mathscr{E}°/\Delta t$ is an average value, we shall draw it as a chord starting at the initial temperature of the

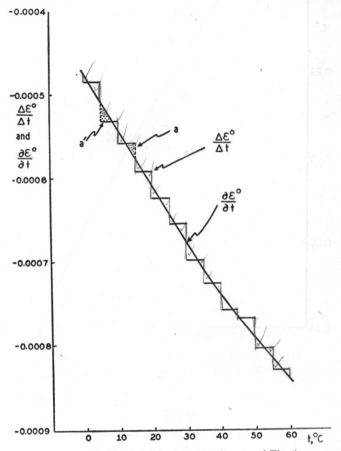

Fig. 2. Chord-area plot of slopes of curve of Fig. 1.

interval and terminating at the final temperature. A graph of these chords over the entire temperature region from 0 to 60°C is illustrated in Fig. 2. To find the slope, $\partial\mathscr{E}°/\partial t$, a curve is drawn through these chords in such a manner that the sum of the areas of the triangles, such as a, of which the chords form the upper sides, shall be equal to the sum of the areas of the triangles, such as a', of which the chords form the lower sides. This smooth curve gives $\partial\mathscr{E}°/\partial t$ as a function of the temperature. Some values at various temperatures have been entered in Column 6 of Table 3.

In the example given above, the chords have been taken for equal intervals, since the curve changes slope only gradually and the data are given at rounded temperatures at equal intervals. In many cases, the intervals will not be equal, nor will they occur at rounded numbers. Nevertheless, the chord-area method of differentiation may be used in substantially the same manner, though a little more care is necessary to avoid numerical errors in calculations.

2. *Graphical integration.* The procedure for integration is rather closely analogous to that for differentiation. Again we shall cite an example of use in thermodynamic problems, the integration of heat-capacity data. Let us consider the heat-capacity data for solid *n*-heptane listed in Table

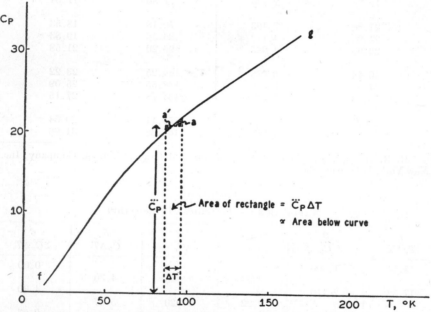

Fig. 3. Graphical integration of heat-capacity curve.

4. A graph of these data (Fig. 3) shows an *S*-shaped curve, for which it may not be convenient to use an analytical equation. Nevertheless, in connection with determinations of certain thermodynamic functions it may be desirable to evaluate the integral $\int C_p \, dT$. It is most convenient, therefore, to use a graphical method.

Once again we consider small intervals of the independent variable, T, as is indicated in Fig. 3. At the mid-point of this interval we have an average value of the heat capacity, $\overline{C_p}$, indicated by the broken horizontal line in the figure. It is clear that the area of the rectangle below this dotted line is $\overline{C_p} \, \Delta T$. If the interval chosen is so small that the section of the curve which has been cut is practically linear, then it is also evident that the area below this section of the curve is essentially the same as that of the rectangle, since the area a' is practically equal to a. Hence it follows that the area under the curve between the limits f and g is given very closely by the sum of the areas of the rectangles taken over sufficiently short tem-

TABLE 4

HEAT CAPACITIES OF SOLID n-HEPTANE*

T, °K	C_p cal mole^{-1} deg^{-1}	T, °K	C_p cal mole^{-1} deg^{-1}
15.14	1.500	53.18	12.80
17.52	2.110	65.25	15.69
19.74	2.730	71.86	17.04
21.80	3.403	79.18	18.53
24.00	4.112	86.56	19.83
26.68	4.935	96.20	21.58
30.44	6.078	106.25	23.22
34.34	7.370	118.55	25.09
38.43	8.731	134.28	27.15
42.96	10.02	151.11	29.54
47.87	11.36	167.38	31.96

* R. R. Wenner, *Thermochemical Calculations*, McGraw-Hill Book Company, Inc., New York, 1941, page 356.

TABLE 5

TABULATION FOR GRAPHICAL INTEGRATION

1 T, °K	2 C_p	3 ΔT	4 $\bar{C}_p$	5 $\bar{C}_p \Delta T$	6 $\Sigma \bar{C}_p \Delta T$
15.14	1.500				0.00
		2.38	1.805	4.30	
17.52	2.110				4.30
		2.22	2.420	5.37	
19.74	2.730				9.67
		2.06	3.067	6.32	
21.80	3.403				15.99
		2.20	3.758	8.27	
24.00	4.112				24.26

perature intervals. Since the area under the curve corresponds to the integral $\int C_p \, dT$, it follows that

$$\sum_f^g \bar{C}_p \, \Delta T = \int_f^g C_p \, dT. \tag{2.47}$$

Since the first few data in Table 4 are given at closely successive temperatures, we may use these values to form the temperature intervals. For $\bar{C}_p$ between any two temperatures we may take the arithmetic mean between the listed experimental values. The values of ΔT and $\overline{C_p}$ are then tabulated conveniently as in Columns 3 and 4 of Table 5. Column 5 lists the area for the given interval. Finally, to obtain the area between any two of the temperatures listed in Column 1, the sums of the areas of the intervals from 15.14°K are tabulated in Column 6.

It is evident that if we wish to obtain the value of the integral at some temperature not listed in Table 5, we may plot the figures in Column 6 as a function of T and read off values of the integral at the desired upper limit.

B. Analytical methods. In many cases it is possible to summarize data in terms of a convenient algebraic expression. Such an equation is highly desirable, whenever it can be used with sufficient precision, because it is a very concise summary of much information. We shall consider two common methods of fitting an algebraic expression to a set of experimental data. In each case we shall use a simple quadratic equation as an example. The extension to power series with terms of higher or lower degree will be obvious. Methods of fitting other types of equation to experimental data are described in appropriate mathematical treatises.[6]

The two approaches we shall consider are the method of averages and the method of least squares. In both cases we shall assume that we have a series of data, such as equilibrium constants as a function of pressure, to which we wish to fit a quadratic equation,

$$y = a + bx + cx^2. \tag{2.48}$$

Assuming that a quadratic equation can be used, we wish to obtain the best values of the constants, a, b, and c. The two methods differ as to the criterion of a "best" equation. In both, however, it is assumed implicitly that all of the error lies in the dependent variable, y, and none in the independent variable, x.[7]

1. *The method of averages.* Let us suppose that we have a series of numerical values of equilibrium constants and corresponding pressures such as are given in Table 6 and that we wish to obtain a quadratic expression for K_p as a function of the total pressure, P. Let us define the residual, r, as the difference between the experimentally determined K_p and the value which may be calculated from the analytical expression; that is,

$$r = K_p - (a + bP + cP^2). \tag{2.49}$$

There will be a value of r for each of the six experiments listed in Table 6. In obtaining an equation by the method of averages, we assume that the best values of the constants a, b, and c are those for which the sum of all (six) residuals is zero. Using this assumption,

---

[6] A. G. Worthing and J. Geffner, *Treatment of Experimental Data*, J. Wiley and Sons, Inc., New York, 1943.

[7] The theory has also been developed for the case where a significant error may appear in x. See Worthing and Geffner, *op. cit.*, pages 258–260.

$$\sum r = 0, \tag{2.50}$$

we obtain the following condition for evaluating the constants:

$$\sum r = [K_{p1} - (a + bP_1 + cP_1^2)] + [K_{p_2} - (a + bP_2 + cP_2^2)] + \cdots$$

$$= [K_{p1} + K_{p_2} + \cdots] - [a + a + \cdots] - b[P_1 + P_2 + \cdots] - c[P_1^2 + P_2^2 + \cdots]$$

$$= 0 = \sum_{i=1}^{n} K_{p_i} - na - b\sum_{i=1}^{n} P_i - c\sum_{i=1}^{n} P_i^2, \tag{2.51}$$

where n represents the number of experimental measurements. Rearrangement of the last expression gives

$$\sum K_{p_i} = na + b\sum P_i + c\sum P_i^2. \tag{2.52}$$

TABLE 6

EQUILIBRIUM CONSTANTS FOR REACTION:* $\frac{1}{2}N_2$ (g) $+ \frac{3}{2}H_2$ (g) $= NH_3$ (g)

P (atm)	K_p (at 500°C)
10	0.00381
30	0.00386
50	0.00388
100	0.00402
300	0.00498
600	0.00651

* A. T. Larson and R. L. Dodge, *J. Am. Chem. Soc.*, **45**, 2918 (1923); **46**, 367 (1924).

We now have one condition but three unknowns, a, b, and c. To fix three unknowns, it is necessary to have expressions for three restrictions. The three equations may be obtained if we divide the data into three equal (or nearly equal) groups and carry out the summations indicated in equation (2.52). With the data listed in Table 6, we might obtain the following set of relations:

$$
\begin{aligned}
0.00381 &= a + b(10) + c(100)\\
0.00386 &= a + b(30) + c(900)\\
\sum = 0.00767 &= 2a + 40b + 1000c
\end{aligned}
\tag{2.53}
$$

$$
\begin{aligned}
0.00388 &= a + b(50) + c(2500)\\
0.00402 &= a + b(100) + c(10{,}000)\\
\sum = 0.00790 &= 2a + 150b + 12{,}500c
\end{aligned}
\tag{2.54}
$$

$$
\begin{aligned}
0.00498 &= a + b(300) + c(90{,}000)\\
0.00651 &= a + b(600) + c(360{,}000)\\
\sum = 0.01149 &= 2a + 900b + 450{,}000c
\end{aligned}
\tag{2.55}
$$

Equations (2.53), (2.54), and (2.55) may now be solved simultaneously. They lead to the following analytical expression for K_p:

$$K_p = 0.003805 + 1.500 \times 10^{-6}P + 5.63 \times 10^{-9}P^2. \tag{2.56}$$

For the general case we proceed as above to divide the experimental data into three equal (or nearly equal) groups. For each group we find the summations required by the expression

$$\sum y = na + b\sum x + c\sum x^2, \qquad (2.57)$$

where the y's and x's are the known values of the dependent and independent variables, respectively. From the three equations so obtained it is possible to solve for the constants a, b, and c and thereby to obtain an explicit relation for y as a function of x.

2. *The method of least squares.* With the method of least squares, we obtain three independent equations which the three constants of the quadratic equation must obey. The procedure follows from the assumption that the best expression is that for which the sum of the *squares* of the residuals is a minimum. If we define the residual for the general quadratic expression as

$$r = y - (a + bx + cx^2), \qquad (2.58)$$

where y and x refer to experimentally determined values, then according to the method of least squares we should obtain an equation for which

$$\sum r^2 = \text{a minimum.} \qquad (2.59)$$

This condition will be satisfied when the partial derivative of $\sum r^2$ with respect to each of the constants, a, b, and c, respectively, is zero. Let us consider first the partial derivative with respect to a:

$$\sum r^2 = [y_1 - (a + bx_1 + cx_1^2)]^2 + [y_2 - (a + bx_2 + cx_2^2)]^2 + \ldots, \quad (2.60)$$

$$\left(\frac{\partial}{\partial a}\sum r^2\right)_{b,c,x,y} = -2[y_1 - a - bx_1 - cx_1^2] - 2[y_2 - a - bx_2 - cx_2^2] - \ldots$$

$$= 0 = -2[y_1 + y_2 + \ldots] - 2[-na] -$$
$$2[-bx_1 - bx_2 - \ldots] - 2[-cx_1^2 - cx_2^2 - \ldots]. \quad (2.61)$$

Rearranging gives

$$\sum y = na + b\sum x + c\sum x^2. \qquad (2.62)$$

By a similar procedure we may obtain the following expression from the partial derivative of $\sum r^2$ with respect to the parameter b:

$$\sum yx = a\sum x + b\sum x^2 + c\sum x^3. \qquad (2.63)$$

Similarly, the differentiation with respect to c leads to an expression which can be reduced to

$$\sum yx^2 = a\sum x^2 + b\sum x^3 + c\sum x^4. \qquad (2.64)$$

The three simultaneous equations (2.62), (2.63), and (2.64) may be used to solve for the constants a, b, and c.

In order to obtain the sums required for the solution of the three simultaneous equations, it is convenient to set up a table of the form of Table 7. To take a specific example, let us use the data in Table 6 for the synthesis of ammonia. The calculations leading to the required sums are tabulated in Table 8.

TABLE 7
OUTLINE OF CALCULATIONS FOR LEAST SQUARE QUADRATIC EQUATION

y	x	x^2	x^3	x^4	yx	yx^2
.	.	.	.	.	.	.
.	.	.	.	.	.	.
.	.	.	.	.	.	.
.	.	.	.	.	.	.
.	.	.	.	.	.	.
.	.	.	.	.	.	.
.	.	.	.	.	.	.
.	.	.	.	.	.	.
.	.	.	.	.	.	.
.	.	.	.	.	.	.
.	.	.	.	.	.	.
.	.	.	.	.	.	.
Σy	Σx	Σx^2	Σx^3	Σx^4	Σyx	Σyx^2

TABLE 8
CALCULATIONS FOR THE LEAST SQUARE QUADRATIC EQUATION

K_p	P	P^2	P^3	P^4	K_pP	K_pP^2
0.00381	10	100	1,000	10,000	0.0381	0.381
0.00386	30	900	27,000	810,000	0.1158	3.474
0.00388	50	2,500	125,000	6,250,000	0.1940	9.700
0.00402	100	10,000	1,000,000	100,000,000	0.4020	40.200
0.00498	300	90,000	27,000,000	8,100,000,000	1.4940	448.200
0.00651	600	360,000	216,000,000	129,600,000,000	3.9060	2,343.600
0.02706	1090	463,500	244,153,000	137,807,070,000	6.1499	2,845.555

From the sums listed at the bottoms of the columns of Table 8, we can set up the following three specific simultaneous equations corresponding to the respective general equations (2.62), (2.63), and (2.64):

$$0.02706 = 6a + 1090b + 4.635 \times 10^5 c \qquad (2.65)$$

$$6.1499 = 1090a + 4.635 \times 10^5 b + 2.44153 \times 10^8 c \qquad (2.66)$$

$$2845.555 = 4.635 \times 10^5 a + 2.44153 \times 10^8 b + 1.3780707 \times 10^{11} c. \qquad (2.67)$$

Solution of these equations leads to the following least square expression for the equilibrium constant as a function of the pressure:

$$K_p = 0.003743 + 3.392 \times 10^{-6}P + 2.083 \times 10^{-9}P^2. \qquad (2.68)$$

. Thus we have two methods of obtaining an analytical expression for representing data in a concise form. In practice, the method of averages is much more rapid, but it is not so sound from a theoretical point of view. Where data of high precision are available, the method of least squares should be used.

Exercises

1. Find the conversion factor for changing liter-atmosphere to: (a) erg, (b) calorie.

2. Find the conversion factor for changing calorie to: (a) cubic foot-atmosphere, (b) volt-faraday.

3. The area, a, of a rectangle may be considered a function of the breadth, b, and the length, l:

$$a = bl.$$

b and l are the independent variables, a is the dependent one. Other possible dependent variables are the perimeter, p,

$$p = 2b + 2l$$

and the diagonal, d, $\qquad d = \sqrt{b^2 + l^2}.$

(a) Find the values of the following partial derivatives in terms of b, l, or a numerical answer:

$$\left(\frac{\partial a}{\partial l}\right)_b, \quad \left(\frac{\partial l}{\partial b}\right)_a, \quad \left(\frac{\partial p}{\partial l}\right)_b, \quad \left(\frac{\partial l}{\partial b}\right)_p,$$

$$\left(\frac{\partial d}{\partial b}\right)_l, \quad \left(\frac{\partial p}{\partial b}\right)_l, \quad \left(\frac{\partial a}{\partial b}\right)_l.$$

(b) Find suitable transformation expressions in terms of the partial derivatives given in (a) for each of the following derivatives, and then evaluate the results in terms of b and l:

$$\left(\frac{\partial a}{\partial b}\right)_d, \quad \left(\frac{\partial b}{\partial p}\right)_l, \quad \left(\frac{\partial a}{\partial b}\right)_p.$$

(c) Find suitable transformation expressions in terms of the preceding partial derivatives for each of the following derivatives, and then evaluate the result in terms of b and l:

$$\left(\frac{\partial p}{\partial b}\right)_d, \quad \left(\frac{\partial a}{\partial p}\right)_l, \quad \left(\frac{\partial b}{\partial p}\right)_d, \quad \left(\frac{\partial a}{\partial p}\right)_d.$$

4. In a right triangle, such as is illustrated in Fig. 4, the following relations are valid:

$$D^2 = H^2 + B^2,$$
$$P = H + B + D,$$
$$A = \tfrac{1}{2}BH.$$

(a) Given the special conditions

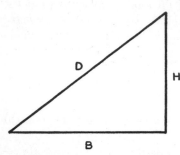

Fig. 4. A right triangle.

$$H = 1000 \text{ cm}, \qquad \left(\frac{\partial H}{\partial B}\right)_A = -2,$$

$$\left(\frac{\partial H}{\partial B}\right)_D = -0.5, \qquad \left(\frac{\partial B}{\partial H}\right)_P = -1.309,$$

compute the values of the following partial derivatives, using transformation relations if necessary:

$$\left(\frac{\partial A}{\partial B}\right)_H, \qquad \left(\frac{\partial A}{\partial H}\right)_B, \qquad \left(\frac{\partial A}{\partial B}\right)_D, \qquad \left(\frac{\partial A}{\partial H}\right)_P.$$

(b) Given the following different set of special conditions:

$$B = 4 \text{ cm}, \quad \left(\frac{\partial H}{\partial A}\right)_P = -0.310, \quad \left(\frac{\partial H}{\partial B}\right)_A = -2.0, \quad \left(\frac{\partial P}{\partial B}\right)_A = -2.341,$$

(1) compute the values of the following partial derivatives, using transformation relations if necessary:

$$\left(\frac{\partial H}{\partial A}\right)_B, \qquad \left(\frac{\partial B}{\partial A}\right)_P, \qquad \left(\frac{\partial P}{\partial A}\right)_B;$$

(2) compute A.

5. Considering E as a function of any two of the variables P, V, and T, prove that

$$\left(\frac{\partial E}{\partial T}\right)_P \left(\frac{\partial T}{\partial P}\right)_V = -\left(\frac{\partial E}{\partial V}\right)_P \left(\frac{\partial V}{\partial P}\right)_T.$$

6. Making use of the definition $H = E + PV$ and considering H (or E) as a function of any two of the variables P, V, and T, prove the following relationships:

(a) $\left(\dfrac{\partial H}{\partial T}\right)_P = \left(\dfrac{\partial E}{\partial T}\right)_V + \left[P + \left(\dfrac{\partial E}{\partial V}\right)_T\right]\left(\dfrac{\partial V}{\partial T}\right)_P.$

(b) $\left(\dfrac{\partial H}{\partial T}\right)_P = \left(\dfrac{\partial E}{\partial T}\right)_V + \left[V - \left(\dfrac{\partial H}{\partial P}\right)_T\right]\left(\dfrac{\partial P}{\partial T}\right)_V.$

(c) $\left(\dfrac{\partial E}{\partial T}\right)_V = \left(\dfrac{\partial H}{\partial T}\right)_P - \left[\left(\dfrac{\partial H}{\partial T}\right)_P\left(\dfrac{\partial T}{\partial P}\right)_H + V\right]\left(\dfrac{\partial P}{\partial T}\right)_V.$

7.[8] An ideal gas in State A (Fig. 5) is changed to State C. This transformation may be carried out by an infinite number of paths of which only two will be considered, one directly from A to C and the other from A to B to C.

---

[8] Adapted from H. Margenau and G. M. Murphy, *The Mathematics of Physics and Chemistry*, D. Van Nostrand and Company, New York, 1943, pages 8–11.

(a) Calculate and compare the changes in volume in going from A to C by each of the two paths, AC and ABC, respectively.

Proceed by integrating the differential equation

$$dV = \left(\frac{\partial V}{\partial T}\right)_P dT + \left(\frac{\partial V}{\partial P}\right)_T dP \qquad (2.2)$$

or

$$dV = \frac{R}{P}\, dT - \frac{RT}{P^2}\, dP. \qquad (2.6)$$

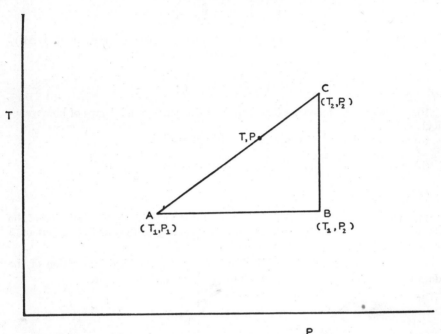

Fig. 5. Two paths for carrying an ideal gas from State A to State C.

Before the integration is carried out along the path AC, use the following relation to make the necessary substitutions:

$$\text{Slope of line } AC = \frac{T_2 - T_1}{P_2 - P_1} = \frac{T - T_1}{P - P_1}; \qquad (2.69)$$

therefore

$$T = T_1 + \frac{T_2 - T_1}{P_2 - P_1}\,(P - P_1) \qquad (2.70)$$

and

$$dT = \frac{T_2 - T_1}{P_2 - P_1}\, dP. \qquad (2.71)$$

(b) Applying the reciprocity test to equation (2.6), show that dV is an exact differential.

(c) Calculate and compare the works done in going from A to C by each of the two paths. Make use of the relation

$$dW = P \, dV = R \, dT - \frac{RT}{P} dP \qquad\qquad (2.72)$$

and the substitution suggested by equation (2.70).

(d)   Applying the reciprocity test to equation (2.72), show that dW is an inexact differential.

8.   The Gibbs free energy, F, is a thermodynamic property. If $(\partial F/\partial P)_T = V$ and $(\partial F/\partial T)_P = -S$, prove the following relation:

$$\left(\frac{\partial S}{\partial P}\right)_T = -\left(\frac{\partial V}{\partial T}\right)_P.$$

9.   The Helmholtz free energy, A, is a thermodynamic property. If $(\partial A/\partial V)_T = -P$ and $(\partial A/\partial T)_V = -S$, prove the following relation:

$$\left(\frac{\partial S}{\partial V}\right)_T = \left(\frac{\partial P}{\partial T}\right)_V.$$

10.   Examine the following functions for homogeneity and degree of homogeneity:

(a) $u = x^2 y + xy^2 + 3xyz.$

(d) $u = e^{y/x}.$

(b) $u = \dfrac{x^3 + x^2 y + y^3}{x^2 + xy + y^2}.$

(e) $u = \dfrac{x^2 + 3xy + 2y^3}{y^2}.$

(c) $u = \sqrt{x + y}.$

11.   Complete the calculations in Table 3 for the graphical differentiation of the data listed in Table 2.   Draw a graph corresponding to that of Fig. 2, but on a larger scale for more precise readings.

12.   Complete the calculations in Table 5 for the graphical integration of the data listed in Table 4.   Draw a graph of $\int C_p \, dT$ vs. temperature.

CHAPTER 3

The First Law of Thermodynamics

HAVING reviewed the mathematical background, we must now develop the basic concepts and postulates of chemical thermodynamics upon which we shall build the theoretical framework. In discussing these fundamental postulates, which are essentially concise statements of much experience, we shall try to emphasize at all times their applications to chemical objectives, rather than to problems in engineering. Since every postulate is expressed in terms of certain accepted concepts, however, it will be necessary first to define a few of the basic concepts of thermodynamics.

I. DEFINITIONS

Critical studies of the logical foundations of physical theory[1] of the twentieth century have emphasized the change in approach which is necessary in the definition of fundamental concepts if contradictions between theory and observation are to be avoided. The classical procedure has been to define a concept in terms of its properties. For example, Newton defined absolute time as follows: "Absolute, True, and Mathematical Time, of itself, and from its own nature flows equably without regard to anything external." The difficulty with a definition of this type, based on properties or attributes, is that we have no assurance that anything of the given description actually exists in nature. In fact, experimental observations frequently indicate that there definitely is nothing in nature corresponding to a concept that someone has defined. Such is the case with the concept of "absolute" or "true" time. Nevertheless, until this situation was pointed out by theoretical physicists during the first half of the twentieth century, experimentalists frequently found themselves led to contradictory conclusions from two apparently consistent sets of postulates and observations.

A new approach to the problem of definition of concepts has been emphasized by Mach, Poincaré, and Einstein,[2] and has been expressed in a

---

[1] P. W. Bridgman, *The Logic of Modern Physics*, Macmillan and Company, New York, 1927; *The Nature of Thermodynamics*, Harvard University Press, Cambridge, Mass., 1941.

[2] The operational concept had been used implicitly much earlier than the twentieth century. Thus Boyle defined a chemical element in terms of the experiments by which it might be recognized, in order to avoid the futile discussions of his predecessors, who identified elements with qualities or properties.

very clear form by Bridgman. In this new approach a concept is defined in terms of a set of experimental operations: "the concept is synonymous with the corresponding set of operations." When this notion of concept is applied to "absolute time," it soon becomes evident that there are no operations by which such a concept can be described. All operations by which time is measured are relative ones. The term "absolute time" thus becomes meaningless.

To avoid possible difficulties in thermodynamic applications, it is desirable that all thermal and energy concepts likewise be defined in operational terms.

A. Temperature. The earliest concept of temperature was undoubtedly physiological, that is, based on the sensations of hot and cold. Of necessity such an approach is very crude, both in sensitivity and accuracy. As men became more familiar with the properties of various substances, however, they realized that certain of these properties were not the same at all times but depended upon the temperature, as measured by crude physiological responses. Ultimately it became evident that one of these properties—for example, the volume of a gas—might be used as a more precise measure of the temperature. In this way it became possible to do away with any dependence upon physiological sensations.

The definition of temperature in terms of the volume of a gas is an operational one. The concept is defined in terms of an experimental procedure by which it may be measured. As the precision of physical measurements has increased, it has also become evident that the details of the experimental procedure are more involved than was originally realized. Thus it was soon recognized that not all gases indicate the same temperature, even if they have all been set originally at the same reference temperature. Consequently, it becomes necessary to take a series of volume readings at various pressures and to extrapolate a suitable function of the pressure-volume product to zero pressure. Similarly, difficulties arise if the thermometer is exposed to certain types of radiation. However, calculations indicate that under normal circumstances, these radiational fields raise the temperature by about 10^{-12} °C,[3] a quantity which is not detectable even with the most sensitive of present-day instruments. Similarly, we shall neglect relativistic corrections which arise at high velocities, for we shall not encounter such situations in ordinary thermodynamic problems.

So far we have considered the definition of the temperature concept in terms of a series of operations on gases. It is also possible to define the temperature concept in terms of certain heat quantities, as will be shown in Chapter 7, The Second Law of Thermodynamics. This too is an operational approach if the concept of heat is properly defined. For this pur-

---

[3] P. W. Bridgman, *The Nature of Thermodynamics*, Harvard University Press, Cambridge, Mass., 1941, page 16.

pose, therefore, as well as in preparation for the statement of the first law of thermodynamics, we shall consider the concept of heat in some detail.

B. Heat. Probably the best way of defining the concept of heat is in terms of measurements with ice calorimeters. In these calorimeters the absorption of heat is accompanied by a phase change, which in turn produces an alteration in the volume of the system. In this manner a quantity of heat is related to the measurement of a change in volume. In principle it would be necessary to use various substances in the ice calorimeters in order to cover a wide temperature range, and if one insists on a continuous range, it would be necessary to operate a given calorimeter under variable pressures.

The definition of heat in terms of volume changes in ice calorimeters has the special virtue of avoiding any dependence upon temperature measurements. Since the so-called "thermodynamic temperature scale" relates the temperature concept to measurements of heat quantities, it is necessary for this purpose to have a definition of heat which is independent of temperature measurements if we are to avoid the criticism of traveling in a logical circle.

For completeness it is also desirable to consider briefly the operational meaning of heat as it is more commonly defined in terms of temperature changes. This method of defining heat depends upon the use of the common type of calorimeter—a large body of material, usually water, which on absorbing heat suffers a small change in temperature. We say that the heat absorbed is proportional to the change in temperature,

$$Q = C(T_2 - T_1), \tag{3.1}$$

where Q is the heat absorbed, C is the proportionality constant, T_2 is the final temperature, and T_1 is the initial temperature. Experience teaches us that if the quantity of material (for example, water) within the calorimeter is doubled, the change in temperature is halved (if the heat capacity of the calorimeter itself can be ignored). Thus the constant, C, is proportional to the mass of the substance within the calorimeter. Since C is therefore a characteristic of the substance within the calorimeter, it is called the *heat capacity*. Also, since the magnitude of C is proportional to the quantity of matter within the calorimeter, we may define a heat capacity per unit mass of substance. The unit of mass may be one gram or one mole of material. In the former case we have defined the *specific heat;* in the latter case, the *molal heat capacity*.

As in the discussion of other concepts, our present description of the heat concept depends upon relatively crude experiments at first. Thus in the preceding paragraph we neglected the heat capacity of the calorimeter vessel. Similarly, we have assumed that the heat capacity, C, is independent of the temperature interval, whereas precise measurements indicate that C varies with temperature. These refinements introduce no fundamental difficulties, however, since, as in the case of the temperature concept, we may make use of the process of extrapolation-to-a-limit. Thus for heat capacity we can decrease the size of the temperature interval until the value of C per unit mass approaches a constant.

C. Work. For our purposes it will suffice to point out that work is defined as the product of a force by a displacement. Assuming, then, that force and displacement can be given suitable operational significance, the term "work" will also share this characteristic. The measurement of the displacement involves experimental determinations of a distance, which can be carried out, in principle, with a measuring rod. The concept of force is a little more complicated. It originated undoubtedly from the muscular sensation of resistance to external objects. A quantitative measure is readily obtained with an elastic body, such as a spring, whose deformation may be utilized as a measure of the force. This definition of force, however, is limited to static systems; for systems which are being accelerated, further refinements must be considered. Since these would take us too far from our main course, we shall merely make reference to Bridgman's critical analysis.[4] Nevertheless, for the definition of force in even the static situation, as well as in the definition of displacement, it should be emphasized that precision measurements require a number of precautions, particularly against changes in temperature. In dealing with these concepts, we generally assume implicitly that such sources of error have been recognized and accounted for.

II. ENERGY: THE FIRST LAW OF THERMODYNAMICS

The first law is generally stated as a relation between certain quantities, E (internal energy), Q (heat), and W (work), or actually their differentials, dE, DQ, and DW:

$$dE = DQ - DW. \tag{3.2}$$

Even in an elementary approach the differentials are distinguished in notation by capital and lower-case d's. This distinction is made in recognition of the fact that DQ and DW are inexact differentials; that is, the values of their line integrals depend upon the path as well as upon the initial and end points of a transformation.

The inexactness of these quantities is recognized quickly by consideration of one or two experimental examples. Thus in an electrical cell, the amount of heat evolved when useful work is done by the cell (for example, in the operation of a motor) differs substantially from that obtained when the electrodes are short-circuited, even though the initial and final chemical substances are the same in both cases. Similarly, the amount of work obtained from this system depends upon the rate at which the cell is dis-

---

[4] P. W. Bridgman, *The Logic of Modern Physics*, Macmillan and Company, New York, 1927, pages 102–108.

charged. If the counterpotential is near the potential of the cell, more work is obtained than when the electrodes are practically short-circuited. Again the nature of the chemical change is the same in both transformations.

Another example demonstrating the inexactness of DQ and DW is the isothermal expansion of a gas. When a confined gas is allowed to expand very slowly against an external pressure practically equal to that of the gas, the amount of work done and the quantity of heat absorbed, respectively, differ very greatly from the values obtained during the expansion of the gas into a vacuum, even though the initial pressure, temperature, and volume, and the final temperature, pressure, and volume are the same in both expansions. Obviously, then, the work and heat differentials are not exact.

Turning now to the internal energy quantity, dE, we might interpret equation (3.2) as follows. The change in internal energy, dE, within a bounded region of space is found as a matter of experiment to be equal to the quantity of heat absorbed, DQ, minus the amount of work done, DW, by the system. Upon careful consideration, however, such a statement appears too naïve, because it turns out that there are no independent operations that can be described for the measurement of the internal energy change, dE. The measurement of energy changes has meaning only in terms of equation (3.2), in other words, only in terms of measurements of work and heat. Internal energy changes have no independent operational significance. The significance of the first law cannot lie, therefore, in a statement of the equality of dE and $(DQ - DW)$; such a statement would be mere tautology in view of the fact that dE is measured in terms of DQ and DW. Hence the essence of the first law must lie in the difference in the character of the differential dE as contrasted with DQ and DW. The energy differential, dE, is exact. In other words, despite the fact that the heat absorbed, DQ, and the work done, DW, in going from one state to another depend upon the particular path used in the transformation, the difference in these two quantities, $DQ - DW$, defined as dE, is independent of the method by which the change is accomplished. Furthermore, making use of another property of an exact differential, we can say that the integral of dE around a closed path is zero. In other words, whenever the enclosed system is returned to its initial state, the difference between heat absorbed and work done, that is, the change in internal energy, is zero despite the fact that the heat and work quantities in themselves may differ very much from zero.

If we are dealing with large-scale changes for a given bounded region, we may state the first law in terms of the following equation in place of the differential form:

$$\Delta E = Q - W. \tag{3.3}$$

Since Q and W have operational meaning, ΔE is also operationally significant.

Thus the first law may be considered to consist of two parts. First, it defines the concept of energy in terms of previous concepts, heat and work. Secondly, it summarizes a wide variety of experience in the statement that the internal energy function, so defined, is a thermodynamic property; that is, it depends only on the state of a system and not on the previous history of the system.

It should also be pointed out that the very definition of the energy concept precludes the possibility of determining absolute values; that is, we have defined only a method of measuring *changes* in internal energy in terms of heat and work quantities. Classical thermodynamics, by itself, is incapable of determining the absolute zero of reference for any of its energy functions. In practice this is not a severe handicap, however, since interest is generally focused on chemical and physical *transformations*, and any convenient state may be chosen as the reference point.

III. SOME CONDITIONS UNDER WHICH W OR Q DEPENDS ONLY ON INITIAL AND FINAL STATES

There are a number of situations in which DW or DQ may become exact. In general we may say that whenever these differentials can be shown to be equal to an exact differential, for example, dE, then it follows that the work

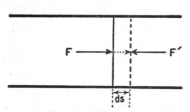

Fig. 1.   Element of work.

and heat differentials are exact. We shall consider two specific examples in which such exactness is obtained. For one of these, however, we shall need to consider the derivation of the expression for the reversible work of expansion.

A. Reversible work of expansion.   In Section I of this chapter it was pointed out that work is defined in terms of the product of a force, F, by a displacement, ds (Fig. 1):

$$DW = F\,ds. \tag{3.4}$$

The force is manifested by a pressure on the walls of the vessel and piston indicated schematically in Fig. 1. Since pressure, P, is force per unit area, A, of the surface of the wall, it is evident that the expression for the element of work becomes

$$DW = PA\,ds. \tag{3.5}$$

Since $A\,ds$ corresponds to the element of volume, dV, swept out by the force in moving the distance ds, equation (3.5) becomes

$$DW = P\,dV. \tag{3.6}$$

We must now turn our attention to the meaning of the term "reversible." For thermodynamic purposes one way of viewing a reversible process is as a succession of states each of which is in a condition of equilibrium.[5] For such a situation an infinitesimal change in external conditions will cause a reversal in the succession of states which make up the process. As an example let us consider the process illustrated in Fig. 1. If the work done is to differ only infinitesimally from that of a reversible process, there must be a force F' opposing the force F, and the magnitude of F' must be only infinitesimally smaller than F. If F' is made infinitesimally larger than F, then the direction of the displacement, ds, will be diametrically opposite to that of the forward change, and the succession of states in the expansion will be reversed. Similarly, in processes where an exchange of heat occurs, a reversible exchange implies that the reservoir giving up the heat is at a temperature T' which is only infinitesimally greater than T, the temperature of the body receiving the heat. Thus if dT is reversed, that is, if T' is made infinitesimally smaller than T, then the direction of heat flow is also reversed and heat goes from the body to the reservoir.

Strictly speaking, it would take infinite time to bring about a finite reversible change, since in any finite time only an infinitesimal change can occur. Reversible processes are thus only idealizations. All natural changes occur at finite speeds. Nevertheless, the concept of a reversible process is an exceptionally useful one because many a real change can be made to occur so slowly that it approximates the reversible one. For this reason we may consider the reversible process the limit which, in many cases, we can approach within any specified experimental error.

For a reversible process it is possible to draw a graph (or to derive a mathematical expression) to represent the change, since each state in the process is in a condition of equilibrium and the variables of state, such as pressure and temperature, are completely fixed. On the other hand, in a real process the system is not in equilibrium and hence we cannot assign unique values of the temperature and pressure to it. For example, in the

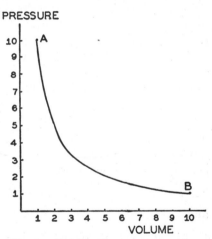

Fig. 2.   Graphical representation of a reversible process.

---

[5] Strictly speaking, a succession of states along a path of true equilibrium points is impossible. However, the real path can be made to approach the equilibrium one as closely as desired. The reversible path is thus the limit of the real path as the dynamic process is made slower and slower.

expansion process of Fig. 1, if the change occurs at a finite rate, the pressure immediately behind the piston will be somewhat less than that in the remainder of the medium (a gas, for example) because there is always a lag in the transmission of stresses. Hence no unique value can be assigned as the pressure within the medium. Thus if we use a graphical representation of the pressure and volume changes during the expansion of a substance, such as Fig. 2, we are dealing, in a strict sense, only with a truly reversible process. In the irreversible, real change, no graphical representation can be made, since no unique values can be assigned to all of the variables.

Returning to our consideration of the work of expansion, we see that only for the reversible process will P in equation (3.6) have a definite value. Thus, strictly speaking, equation (3.6) and its integral,

$$W = \int_{V_1}^{V_2} P \, dV, \tag{3.7}$$

apply only to the reversible work of expansion.

B. Constant-volume process, no nonmechanical work. Having derived an expression for the mechanical work of expansion, we can see quite readily from equation (3.7) that in a constant volume process ($dV = 0$), the work integral will be zero; that is, no net mechanical work is obtained. In equation (3.2), however, W represents all types of work, nonmechanical (electrical, for example) as well as mechanical. Nevertheless, in most chemical experiments the only significant work done is due to expansion or contraction, since the small contributions due to changes in surface area are generally negligible, and only in special cases are reactions used to obtain electrical energy. Consequently, in most situations of interest, if the reaction is carried out at constant volume, no work of any kind is obtained. It is evident, then, from equations (3.2) and (3.3), that

$$dE_V = DQ_V = dQ_V \tag{3.8}$$

and
$$\Delta E_V = Q_V; \tag{3.9}$$

that is, at constant volume dQ_V is exact, and the heat quantity accompanying the process depends only upon the initial and final states.

C. Adiabatic process. It is also evident from equation (3.2) that if DQ is zero, that is, if no heat is absorbed or evolved in a reaction, then

$$dE = -DW = -dW. \tag{3.10}$$

In other words, in an adiabatic process the differential dW is exact and the work done in the process depends only upon the initial and final states.

It should be pointed out that these two examples of situations where dW or dQ is exact are not isolated cases. Other processes will be mentioned later (for example, the isothermal expansion of an ideal gas) in which the work and heat differentials possess the property of exactness. It should be emphasized that in all these cases, by putting certain restrictions on the nature of the process which is to be treated, we are specifying, from the mathematical point of view, the path by which the change may take place. It is no surprise, then, that the heat and work quantities depend only upon the end points of these special processes.

Exercises

1. If the temperature of 1 cc of air at one atmosphere pressure and 0°C is raised to 100°C, the volume becomes 1.3671 cc. Calculate the value of absolute zero for a thermometer using air. Compare your result with that in Table 1 of Chapter 2.

2. Let Fig. 1 represent a plane surface which is being expanded in the direction indicated. Show that the work of reversible expansion is given by the expression

$$W = -\int \gamma \, dA,$$

where γ is the force per unit length.

3. From the first law of thermodynamics, show that DQ is exact for a process at constant pressure in which only mechanical work is done.

CHAPTER 4

Enthalpy and Heat Capacity

IN THE preceding chapter we defined a new function, the internal energy, and found that it is known from long experience (the first law of thermodynamics) to be a thermodynamic property; that is, dE is exact. Making use of this proposition, we were able to show that in constant-volume processes accompanied only by mechanical work, the heat absorbed is also a thermodynamic property. For example, in a given chemical reaction, carried out in a closed vessel of fixed volume, the heat absorbed (or evolved) depends only upon the nature and condition of the initial and of the final reactants; it does not depend upon the mechanism by which the reaction occurs. Therefore if a catalyst speeds up the reaction, by changing the mechanism, it does not affect the heat quantity accompanying the reaction.

Most chemical reactions are carried out at constant (atmospheric) pressure. It is of interest to know whether the heat absorbed in a constant-pressure reaction depends upon the path, that is, the method by which the reaction is carried out, or whether, on the contrary, it too is a function of the initial and final states only. If the latter were true, it would be possible to tabulate heat quantities for given chemical reactions and to use known values to calculate heats for new reactions which can be expressed as sums of known reactions.

Actually, of course, this question was answered on empirical grounds long before thermodynamics was established on a sound basis. In courses in elementary chemistry one becomes familiar with Hess's law of constant heat summation, enunciated in 1840. Hess pointed out that the heat absorbed (or evolved) in a given chemical reaction is the same whether the process occurs in one step or in several steps. Thus, to cite a familiar example, the heat of formation of CO_2 from its elements is the same if the process is the single step

$$C \text{ (graphite)} + O_2 \text{ (gas)} = CO_2 \text{ (gas)};$$
$$Q^1_{298°K} = -94.0518 \text{ kcal mole}^{-1}$$

or the series of steps

---

[1] *Tables of Selected Values of Chemical Thermodynamic Properties*, National Bureau of Standards, 1947. Series III, Table 23.

$$C \text{ (graphite)} + \tfrac{1}{2}O_2 \text{ (gas)} = CO \text{ (gas)};$$
$$Q_{298°K} = -26.4157 \text{ kcal mole}^{-1}$$

$$CO \text{ (gas)} + \tfrac{1}{2}O_2 \text{ (gas)} = CO_2 \text{ (gas)};$$
$$Q_{298°K} = -67.6361 \text{ kcal mole}^{-1}$$

$$C \text{ (graphite)} + O_2 \text{ (gas)} = CO_2 \text{ (gas)};$$
$$Q_{298°K} = -94.0518 \text{ kcal mole}^{-1}$$

Of course we could introduce Hess's generalization into thermodynamics as another empirical law, similar to the first law. A good theoretical framework, however, depends upon a minimum of empirical postulates. The power of thermodynamics lies in the fact that it leads to so many predictions, if one makes only two or three basic assumptions. Hess's law need not be among these postulates, since it can be derived directly from the first law of thermodynamics, perhaps most conveniently with the use of a new thermodynamic function.

I. ENTHALPY

A. Definition. This new thermodynamic quantity which we wish to introduce is known as the *enthalpy*, and sometimes also as *heat content*. It is defined in terms of thermodynamic variables which have been described already:

$$H = E + PV. \tag{4.1}$$

From the definition it is evident that H, the enthalpy, is a thermodynamic property, since it is defined by an explicit function. All of the quantities on the right-hand side of equation (4.1), E, P, and V, are properties of the state of a system, and consequently so is H.

Of course it is also evident from the definition, equation (4.1), that absolute values of H are unknown, since absolute values of E cannot be obtained from classical thermodynamics alone. From an operational point of view, therefore, it is more significant to consider changes in enthalpy, ΔH. Obviously, such changes can be defined readily by the expression

$$\Delta H = \Delta E + \Delta(PV). \tag{4.2}$$

B. Relation between Q and ΔH. Having defined this new thermodynamic property, the enthalpy, we may proceed to investigate the conditions under which it may become equal to the heat accompanying a process. Differentiating (4.1), we obtain

$$dH = dE + P \, dV + V \, dP. \tag{4.3}$$

From the first law of thermodynamics we may introduce

$$dE = DQ - DW \qquad (4.4)$$

and obtain $\qquad dH = DQ - DW + P\,dV + V\,dP. \qquad (4.5)$

Equation (4.5) is of general validity. Let us consider a set of restrictions, however, which are realized in many chemical reactions: (1) constant pressure and (2) no work other than mechanical (against the atmosphere). Under these conditions it is evident that equation (4.5) may be simplified considerably, since

$$DW = P\,dV \qquad (4.6)$$

and $\qquad\qquad\qquad dP = 0. \qquad (4.7)$

Hence $\qquad\qquad dH_P = DQ_P = dQ_P \qquad (4.8)$

and $\qquad\qquad\qquad \Delta H_P = Q_P, \qquad (4.9)$

where the subscript emphasizes the constancy of the pressure during the process. It should be emphasized, in addition, that equation (4.8) is valid only if no nonmechanical work is being done. Under these conditions dQ is evidently an exact differential. In other words, for chemical reactions carried out at constant pressure (at atmospheric pressure, for example) in the usual laboratory or large-scale vessels, the heat absorbed depends only on the nature and conditions of the initial reactants and of the final products. Evidently, then, it does not matter if a given substance is formed in one step or in many steps. So long as the starting and final materials are the same and so long as the processes are carried out at constant pressure and with no nonmechanical work, the net ΔH's will be the same. Thus Hess's law is a consequence of the first law of thermodynamics.

C. Relation between Q_V and Q_P. We have just proved that ΔH equals Q_P for a reaction at constant pressure. Although most calorimetric work is carried out at a constant pressure, some reactions must be observed in a closed vessel, that is, at constant volume. In such a closed system the heat quantity that is measured is Q_V. Yet for further chemical calculations it is frequently necessary to know Q_P. It is highly desirable, therefore, to derive some expression which relates these two heat quantities.

We shall make use of the following relation, which was proved in the preceding chapter,

$$\Delta E_V = Q_V, \qquad (4.10)$$

and equation (4.9), $\qquad \Delta H_P = Q_P, \qquad (4.9)$

together with equation (4.2) restricted to a constant-pressure process:

$$\Delta H_P = \Delta E_P + \Delta(PV). \qquad (4.11)$$

It has been found that ΔE_P is generally not significantly different from

ΔE_V. In fact, for ideal gases, as we shall see in a subsequent chapter, E is independent of the volume or pressure, at a fixed temperature. Hence, as a rule,

$$\Delta E_P \cong \Delta E_V = Q_V. \tag{4.12}$$

Substituting equations (4.9) and (4.12) into (4.11), we obtain

$$Q_P = Q_V + \Delta(PV). \tag{4.13}$$

In reactions involving only liquids and solids, the $\Delta(PV)$ term is usually negligible in comparison with Q, and hence the difference between Q_P and Q_V is slight. However, in reactions involving gases, $\Delta(PV)$ may be significant, since the changes in volume may be large. Generally this term may be estimated with sufficient accuracy by the use of the equation of state for ideal gases,

$$PV = nRT, \tag{4.14}$$

where n represents the number of moles of a particular gas. If the chemical reaction is represented by the expression

$$aA \text{ (g)} + bB \text{ (g)} + \ldots = lL \text{ (g)} + mM \text{ (g)} + \ldots, \tag{4.15}$$

where a, b, l, and m indicate the number of moles of each gas, then, since an isothermal change is being considered,

$$Q_P = Q_V + \Delta(PV) = Q_V + (n_L RT + n_M RT + \ldots - n_A RT - n_B RT - \ldots)$$

or

$$Q_P = Q_V + (\Delta n)RT. \tag{4.16}$$

It should be emphasized that Δn refers to the increase in number of moles *of gases only*.

II. HEAT CAPACITY

We have introduced the *enthalpy* function particularly because of its usefulness as a measure of the heat accompanying chemical reactions at constant pressure. We shall find it convenient also to have a function to relate heat quantities to temperature changes, at constant pressure or at constant volume. For this purpose we shall consider a new quantity, the *heat capacity*.

A. Definition.

1. *Fundamental statement.* It was pointed out in Chapter 3 that the heat absorbed by a body (not at a transition temperature) is proportional to the change in temperature:

$$Q = C(T_2 - T_1). \tag{4.17}$$

The proportionality constant, C, is called the *heat capacity* and is in itself proportional to the mass of the substance undergoing the temperature change. Hence the heat capacity per gram may be called the *specific heat*, and that for one mole of material the *molal heat capacity*.

It has also been pointed out previously that the value of C,

$$C = \frac{Q}{T_2 - T_1} = \frac{Q}{\Delta T}, \tag{4.18}$$

may itself depend on the temperature. For a rigorous definition of heat capacity, therefore, we must consider an infinitesimally small temperature interval. Consequently, we define the heat capacity by the expression

$$C = \frac{DQ}{dT}, \tag{4.19}$$

where the capital D in DQ emphasizes the inexactness of the heat quantity. It is evident, of course, that we may lay certain restrictions upon equation (4.19), for example, constancy of pressure or constancy of volume. For these situations we may modify equation (4.19) to the following expressions:

$$C_p = \left(\frac{DQ}{\partial T}\right)_P \tag{4.20}$$

and

$$C_v = \left(\frac{DQ}{\partial T}\right)_V. \tag{4.21}$$

2. *Derived relations.* Equations (4.19) to (4.21) are fundamental definitions. From these and our previous thermodynamic principles, new relations may be derived which are very useful in further work.

If we have a substance which is merely absorbing heat at a constant pressure, it is evident that the restrictions laid upon equation (4.8) are being fulfilled, and hence that

$$DQ_P = dH_P. \tag{4.8}$$

Simple substitution into equation (4.20) leads to the important expression

$$C_p = \left(\frac{\partial H}{\partial T}\right)_P. \tag{4.22}$$

Similarly, if we have a substance which is merely absorbing heat at constant volume, it is evident that the restrictions placed upon equation (3.8) are being fulfilled and hence that

$$DQ_V = dE_V. \tag{4.23}$$

Simple substitution into equation (4.21) leads to an additional basic relationship,

$$C_v = \left(\frac{\partial E}{\partial T}\right)_V. \tag{4.24}$$

B. Some relations between C_p and C_v. From the considerations of the preceding section there is no immediately apparent connection between the two heat capacities, C_p and C_v. Once again, however, we may illustrate the power of thermodynamic methods by developing several such relationships without any assumptions beyond the first law of thermodynamics and the definitions which have been made already.

1. Starting with the derived relation for C_p,

$$C_p = \left(\frac{\partial H}{\partial T}\right)_P,$$ (4.22)

we may introduce the definition of H:

$$C_p = \left[\frac{\partial(E + PV)}{\partial T}\right]_P = \left(\frac{\partial E}{\partial T}\right)_P + P\left(\frac{\partial V}{\partial T}\right)_P.$$ (4.25)

The partial derivative $(\partial E/\partial T)_P$ is not C_v, but if it could be expanded into some relation with $(\partial E/\partial T)_V$, we should have succeeded in introducing C_v into equation (4.25). The necessary relation may be derived readily by considering the internal energy, E, as a function of T and V and setting up the total differential:

$$dE = \left(\frac{\partial E}{\partial T}\right)_V dT + \left(\frac{\partial E}{dV}\right)_T dV.$$ (4.26)

By the methods described in Chapter 2, we may obtain the following:

$$\left(\frac{\partial E}{\partial T}\right)_P = \left(\frac{\partial E}{\partial T}\right)_V + \left(\frac{\partial E}{\partial V}\right)_T\left(\frac{\partial V}{\partial T}\right)_P.$$ (4.27)

Substituting equation (4.27) into (4.25) and factoring out the partial derivative $(\partial V/\partial T)_P$, we obtain the desired expression:

$$C_p = \left(\frac{\partial E}{\partial T}\right)_V + \left[P + \left(\frac{\partial E}{\partial V}\right)_T\right]\left(\frac{\partial V}{\partial T}\right)_P$$

$$= C_v + \left[P + \left(\frac{\partial E}{\partial V}\right)_T\right]\left(\frac{\partial V}{\partial T}\right)_P.$$ (4.28)

This expression will be of considerable value when we consider special cases for which values or equations for the partial derivatives $(\partial E/\partial V)_T$ and $(\partial V/\partial T)_P$ are available.

2. A second relationship may readily be derived by transformation of the second term in equation (4.28) so that the partial derivative $(\partial E/\partial V)_T$ is replaced by one containing H. Making use of the fundamental relationship between E and H, we obtain

$$\left(\frac{\partial E}{\partial V}\right)_T = \left[\frac{\partial(H - PV)}{\partial V}\right]_T = \left(\frac{\partial H}{\partial V}\right)_T - P - V\left(\frac{\partial P}{\partial V}\right)_T.$$ (4.29)

This expression may be placed in the bracketed factor in equation (4.28) to give

$$C_p = C_v + \left[P + \left(\frac{\partial H}{\partial V} \right)_T - P - V \left(\frac{\partial P}{\partial V} \right)_T \right] \left(\frac{\partial V}{\partial T} \right)_P, \qquad (4.30)$$

which may be reduced to

$$C_p = C_v + \left[\left(\frac{\partial H}{\partial P} \right)_T \left(\frac{\partial P}{\partial V} \right)_T - V \left(\frac{\partial P}{\partial V} \right)_T \right] \left(\frac{\partial V}{\partial T} \right)_P \qquad (4.31)$$

$$= C_v + \left[\left(\frac{\partial H}{\partial P} \right)_T - V \right] \left(\frac{\partial P}{\partial V} \right)_T \left(\frac{\partial V}{\partial T} \right)_P. \qquad (4.32)$$

Reference to equation (2.8) will show that

$$\left(\frac{\partial P}{\partial V} \right)_T \left(\frac{\partial V}{\partial T} \right)_P = - \left(\frac{\partial P}{\partial T} \right)_V. \qquad (4.33)$$

The insertion of this expression into equation (4.32) leads to the relation

$$C_p = C_v + \left[V - \left(\frac{\partial H}{\partial P} \right)_T \right] \left(\frac{\partial P}{\partial T} \right)_V, \qquad (4.34)$$

another very useful equation.

3. A third relation between C_p and C_v may be obtained by several operations on the second term within the bracket in equation (4.34). If we consider H as a function of T and P, then

$$dH = \left(\frac{\partial H}{\partial T} \right)_P dT + \left(\frac{\partial H}{\partial P} \right)_T dP. \qquad (4.35)$$

If H is held constant, $dH = 0$, and

$$0 = \left(\frac{\partial H}{\partial T} \right)_P \left(\frac{\partial T}{\partial P} \right)_H + \left(\frac{\partial H}{\partial P} \right)_T. \qquad (4.36)$$

Using equation (4.36) to substitute for $(\partial H / \partial P)_T$ in equation (4.34), we obtain

$$C_p = C_v + \left[V + \left(\frac{\partial H}{\partial T} \right)_P \left(\frac{\partial T}{\partial P} \right)_H \right] \left(\frac{\partial P}{\partial T} \right)_V. \qquad (4.37)$$

Several other general relations between C_p and C_v are obtainable by procedures similar to those just outlined.

C. Heat capacities of gases. From classical thermodynamics alone it is impossible to predict numerical values for heat capacities, and hence these quantities must be determined calorimetrically. With the aid of statistical mechanics, it is possible to determine heat capacities from spectroscopic data, instead of from direct calorimetric measurements.

Even with spectroscopic information, however, it is convenient to correlate data over a range of temperature in terms of empirical equations.[2] The following expressions have been used most generally:

$$C_p = a + bT + cT^2 + dT^3 \tag{4.38}$$

and

$$C_p = a + bT + \frac{c'}{T^2}. \tag{4.39}$$

The results of several critical surveys by H. M. Spencer and his collaborators are summarized in Table 1 and are illustrated for comparative purposes in Fig. 1.

TABLE 1

HEAT CAPACITIES AT CONSTANT PRESSURE [*],[†]

Substance	Temperature Range, °K	a	$b \times 10^3$	$c \times 10^7$	$c' \times 10^{-5}$	$d \times 10^9$
H_2 (g)	300–1500	6.9469	−0.1999	4.808	.....	...
O_2 (g)	300–1500	6.148	3.102	−9.23	.....	...
N_2 (g)	300–1500	6.524	1.250	−0.01	.....	...
Cl_2 (g)	300–1500	7.5755	2.4244	−9.650	.....	...
Br_2 (g)	300–1500	8.4228	0.9739	−3.555	.....	...
H_2O (g)	300–1500	7.256	2.298	2.83	.....	...
CO_2 (g)	300–1500	6.214	10.396	−35.45	.....	...
CO (g)	300–1500	6.420	1.665	−1.96	.....	...
$CNCl$ (g)	250–1000	11.304	2.441	.....	−1.159	...
HCl (g)	300–1500	6.7319	0.4325	3.697	.....	...
SO_2 (g)	300–1800	11.895	1.089	.....	−2.642	...
SO_3 (g)	300–1200	6.077	23.537	−96.87	.....	...
SO_3 (g)	300–1200	3.603	36.310	−288.28	.....	8.649
CH_4 (g)	300–1500	3.381	18.044	−43.00	.....	...
C_2H_6 (g)	300–1500	2.247	38.201	−110.49	.....	...
C_3H_8 (g)	300–1500	2.410	57.195	−175.33	.....	...
n-C_4H_{10} (g)	300–1500	4.453	72.270	−222.14	.....	...
n-C_5H_{12} (g)	300–1500	5.910	88.449	−273.88	.....	...
Benzene (g)	300–1500	−0.409	77.621	−264.29	.....	...
Pyridine (g)	290–1000	−3.016	88.083	−386.65	.....	...
Carbon (graphite)	300–1500	−1.265	14.008	−103.31	.....	2.751

* H. M. Spencer and J. L. Justice, *J. Am. Chem. Soc.*, 56, 2311 (1934); H. M. Spencer and G. N. Flannagan, *ibid.*, 64, 2511 (1942); H. M. Spencer, *ibid.*, 67, 1859 (1945); H. M. Spencer, *Ind. Eng. Chem.*, 40, 2152 (1948). Data on numerous other substances are given in these original papers.

† Constants are defined by equations (4.38) and (4.39) and apply when the unit is cal mole^{-1} deg^{-1}.

[2] A method has been described [B. L. Crawford, Jr., and R. G. Parr, *J. Chem. Phys.*, 16, 233 (1948)] by means of which one may proceed directly from spectroscopic data to an empirical equation of the form of (4.38).

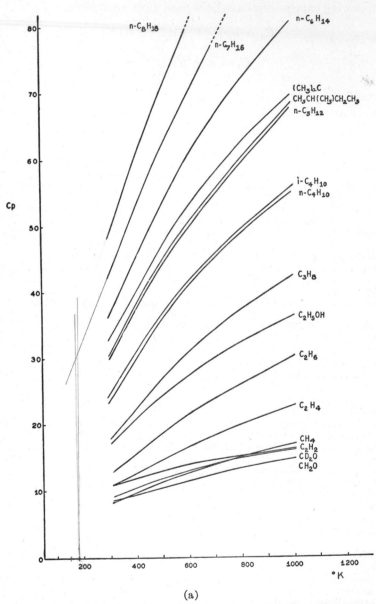

(a)

Fig. 1.   Parts (a) and (b): Variation of heat capacity, C_p, with temperature for some organic compounds.

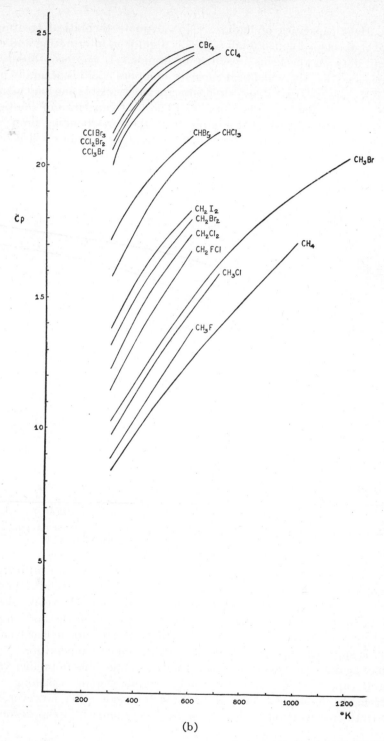

(b)

D. Heat capacities of liquids. No adequate theoretical treatment has been developed which might serve as a guide in interpreting and correlating data on the heat capacities of liquids. It has been observed, nevertheless, that the molal heat capacity of a pure liquid is generally near that of the solid, so that if measurements are not available one may assume that C_v is 6 cal deg^{-1} (gram-atom)$^{-1}$. The heat capacities of solutions, however, cannot be predicted reliably from the corresponding properties of the components. Empirical methods of treating solutions will be considered in later chapters.

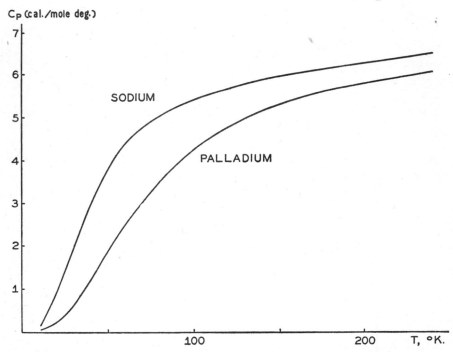

Fig. 2. Molal heat capacities of some solid elements. [Taken from data of G. L. Pickard and F. E. Simon, *Proc. Phys. Soc.*, **61**, 1 (1948).]

E. Heat capacities of solids. Early in the nineteenth century, Dulong and Petit observed that the molal heat capacity of a solid element is generally near 6 cal deg^{-1}. Subsequent investigations, however, showed that C_v (or C_p) varies strongly with the temperature, in the fashion indicated by Fig. 2, though the upper limiting value of about 6 cal mole^{-1} deg^{-1} is approached by the heavier elements at room temperature.

Once again it must be pointed out that it is impossible to predict values of heat capacities for solids by purely thermodynamic reasoning. The problem of the solid state has received much consideration, however, from an extrathermodynamic view, and several very important expressions for the heat capacity have been derived. For our purposes it will be sufficient to consider only the Debye equation, in particular its limiting form at very low temperatures:

$$C_v = \frac{12\pi^4}{5} R \frac{T^3}{\theta^3} = 464.5 \frac{T^3}{\theta^3} \text{ cal mole}^{-1} \text{ deg}^{-1}. \qquad (4.40)$$

θ is called the *characteristic temperature* and may be calculated from an experimental determination of the heat capacity at a low temperature. This equation has been very useful in the extrapolation of measured heat capacities down to the neighborhood of $0°K$, particularly in connection with calculations of entropies from the third law of thermodynamics. Strictly speaking, the Debye equation was derived only for an isotropic elementary substance. Nevertheless it has been found to be applicable to most compounds also, particularly in the region close to absolute zero.

F. Integration of heat capacity equations.

1. *Analytical method.* Later, in connection with calculations of enthalpies and entropies of substances, it will become necessary to integrate the values of the heat capacity over a given temperature range. If the heat capacity can be expressed as an analytic function of the temperature, the integration can be carried out in a simple, straight forward manner. For example, if C_p can be expressed as a power series,

$$C_p = a + bT + cT^2 + \ldots, \qquad (4.41)$$

where a, b, and c are constants, it is evident from equation (4.22) that

$$\int dH = \int C_p \, dT = \int (a + bT + cT^2) dT, \qquad (4.42)$$

if we neglect terms higher than second degree in equation (4.41), and that

$$H - H_0 = aT + \frac{b}{2} T^2 + \frac{c}{3} T^3, \qquad (4.43)$$

where H_0 is the constant of integration.

2. *Graphical method.* If the dependence of C_p on temperature is too complicated to be expressed by a simple function, as for example in solids at low temperatures, then a graphical method of integration can be used. A typical example was worked out partially in Chapter 2 (Fig. 3). As was pointed out there, if we take a small enough temperature interval, ΔT, the product of $\overline{C_p}$, the average value of the heat capacity in that region, by ΔT, gives the area under the C_p curve in that temperature interval. Consequently, the integral may be approximated to any desired degree by summation of the areas:

$$\sum \overline{C_p} \, \Delta T = \int C_p \, dT. \qquad (4.44)$$

The graphical method is applied very extensively in integrations of heat-capacity data for use in the prediction of the feasibility of chemical reactions on the basis of the third law of thermodynamics.

Exercises

1. Molal heat capacities of solid n-heptane are listed in Table 4 of Chapter 2.
   (a) Calculate $H_{182.5°K} - H_{0°K}$ by graphical integration.
   (b) If the heat of fusion at 182.5°K is 3356 cal mole^{-1}, calculate ΔH for the transformation

$$n\text{-heptane (solid, 0°K)} = n\text{-heptane (liquid, 182.5°K)}.$$

2. (a) Find the Debye equation for C_v for n-heptane. Use the given value of C_p at 15.14°K (see Problem 1) as a sufficiently close approximation to C_v.
   (b) Find $H_{15.14°K} - H_{0°K}$ by integrating the Debye equation for n-heptane. Using this analytical value in place of the graphical value for the same temperature range, and the graphical results above 15.14°K, determine $H_{182.5°K} - H_{0°K}$. Compare the answer with the result obtained entirely by graphical methods in Problem 1.

3. Calculate the differences between Q_P and Q_V in each of the following reactions:
   (a) $H_2 + \frac{1}{2}O_2 = H_2O$, at 25°C.
   (b) Ethyl acetate + water = ethyl alcohol + acetic acid, at 25°C.
   (c) Haber synthesis of ammonia, at 400°C.

4. Prove the following relations, using only definitions and mathematical principles:

$$\text{(a)} \quad \left(\frac{\partial E}{\partial V}\right)_P = C_p\left(\frac{\partial T}{\partial V}\right)_P - P; \qquad \text{(b)} \quad \left(\frac{\partial E}{\partial P}\right)_V = C_v\left(\frac{\partial T}{\partial P}\right)_V.$$

5. Solve the equations in Problem 4 for the special case of one mole of an ideal gas, for which $Pv = RT$.

CHAPTER 5

Heat of Reaction

IN CHAPTER 4 we were introduced to a new function, the enthalpy, and found among its properties a correspondence with the heat of reaction at constant pressure (when the only work is due to a volume change against that pressure). Most chemical reactions are carried out in some vessel exposed to the atmosphere, and under conditions such that no work other than that against the atmosphere is produced. For this reason, and because ΔH is independent of the path of a reaction and gives the heat absorbed in the reaction, the enthalpy change is a useful quantity.

Enthalpy changes also give pertinent information in several other problems in chemical thermodynamics. For a long time it was thought that the sign of ΔH, that is, whether heat is absorbed or evolved, is a criterion of the spontaneity of a reaction. When this misconception was cleared up, it still was evident that this criterion is useful within certain limitations. If the ΔH values are large, their signs may still be used as the basis of a first guess with respect to the feasibility of a reaction.

In more rigorous applications, such as in the use of the third law of thermodynamics for evaluating the feasibility of a chemical transformation, enthalpy changes must be known in addition to other quantities. ΔH values are also required to establish the magnitude of the temperature dependence of equilibrium constants. For all these reasons it is desirable to have tables of ΔH available, so that the heats of various transformations may be calculated readily. In many of these calculations we shall make use of the generalization of Hess, now firmly established on the basis of the first law of thermodynamics, and we shall use known values of ΔH to calculate heats for new reactions which can be expressed as sums of known reactions.

I. DEFINITIONS AND CONVENTIONS

It is evident that the heat of a reaction should depend upon the states of the substances involved. Thus in the formation of water,

$$H_2 \text{ (g)} + \tfrac{1}{2}O_2 \text{ (g)} = H_2O, \tag{5.1}$$

if the H_2O is in the liquid state, ΔH will differ from that observed if the H_2O is a vapor. Similarly, the heat of the reaction will depend upon the pressure of any gases involved. Furthermore, for reactions involving solids (for example, sulfur), the ΔH depends upon which crystalline form (for example, rhombic or monoclinic) participates in the reaction. For these reasons, if we are to tabulate values of ΔH, it is necessary to agree that the heats should refer to reactions with the compounds in certain standard states.

A. Some standard states. The states which by convention have been agreed upon as reference states in tabulating heats of reaction are summarized in Table 1. Other standard states may be adopted in special problems. When no state is specified, it may be assumed to be that listed in Table 1.

TABLE 1

STANDARDS AND CONVENTIONS FOR HEATS OF REACTION

Standard state of solid . . . . The most stable form at 1 atm pressure and the specified temperature (unless otherwise specified)*

Standard state of liquid . . . . The most stable form at 1 atm pressure and the specified temperature

Standard state of gas . . . . Zero pressure† and the specified temperature

Standard state of carbon . . . . Graphite‡

Standard temperature . . . . 25 °C‡

Sign of ΔH . . . . + if heat is absorbed‡

---

* Thus for some problems it may be convenient to assign a standard state to both rhombic sulfur and monoclinic sulfur, for example.

† It will be shown in Chapter 19 that internal consistency in the definition of standard states requires that zero pressure be the standard state for the enthalpy of a gas. Unfortunately, however, most reference sources use the convention of 1 atm pressure. In most cases, the difference in enthalpy between these two pressures is very small.

‡ It should be emphasized that these conventions with respect to carbon, temperature, and the sign of ΔH are not the ones used by Bichowsky and Rossini in their critical compilation of thermodynamic data, *Thermochemistry of the Chemical Substances*. These authors have taken diamond as the standard state of carbon and 18°C as the standard temperature in their tables. The heats of formation are labeled Q_f and are positive if heat is evolved. It is essential to keep these differences in mind when these authoritative reference tables are used.

B. Heat of formation. The tables of heats of reaction are generally listed in terms of the heats of formation of various compounds from their elements in their standard states at the specified temperature. Thus if the standard heat of formation, $\Delta Hf°$, of CO_2 at 25° is given as -94.0518 kcal per mole, the following equation is implied:

$$
\begin{array}{ccccc}
\text{C} & + & \text{O}_2 & = & \text{CO}_2; \qquad (5.2) \\
\text{(graphite at} & & \text{(gas at 298.16°K,} & & \text{(gas at 298.16°K,} \\
\text{298.16°K, 1 atm)} & & \text{zero pressure)} & & \text{zero pressure)}
\end{array}
$$

$$\Delta Hf° = -94.0518 \, \text{kcal mole}^{-1}.$$

From this definition it is evident that the heat of formation of an element in its standard state is zero, *by convention*. In other words, elements in their standard states are taken as reference states in the tabulation of heats of reaction, just as sea level is the reference point in measuring geographic heights.

TABLE 2

HEATS OF FORMATION* AT 25 °C

Substance	$\Delta Hf°$ kcal mole^{-1}	Substance	$\Delta Hf°$ kcal mole^{-1}
H (g)........	52.089	Methane (g)....	−17.889
O (g)........	59.159	Ethane (g).....	−20.236
Cl (g).......	29.01	Propane (g).....	−24.820
Br (g)........	26.71	n-Butane (g)....	−29.812
Br₂ (g).......	7.34	Ethylene (g)....	12.496
I (g)........	25.470	Propylene (g)...	4.879
I₂ (g)........	14.876	1-Butene (g)....	0.280
H₂O (g)......	−57.7979	Acetylene (g)...	54.194
H₂O (l)......	−68.3174	Benzene (g).....	19.820
HF (g).......	−64.2	Toluene (g).....	11.950
(HF)₆ (g).....	−426.0	o-Xylene (g)....	4.540
HCl (g)......	−22.063	m-Xylene (g)....	4.120
HBr (g)......	−8.66	p-Xylene (g)....	4.290
HI (g).......	6.20	Methanol (l)....	−57.036
ICl (g).......	4.20	Ethanol (l).....	−66.356
NO (g).......	21.600	Glycine (s).....	−126.33
CO (g).......	−26.4157	Acetic acid (l)...	−116.4
CO₂ (g)......	−94.0518	Taurine (s).....	−187.8

* Selected from *Tables of Selected Values of Chemical Thermodynamic Properties*, National Bureau of Standards, 1947.

A few standard heats of formation have been assembled in Table 2. Data on other substances may be obtained from the following critical compilations:

1. F. R. Bichowsky and F. D. Rossini, *Thermochemistry of the Chemical Substances*, Reinhold Publishing Corporation, New York, 1936.
2. *International Critical Tables*, McGraw-Hill Book Company, New York, 1933.
3. Landolt-Börnstein, *Physikalisch-chemische Tabellen*, Fifth Edition, Julius Springer, Berlin, 1936.
4. W. M. Latimer, *Oxidation Potentials*, Prentice-Hall, Inc., New York, 1938.
5. National Bureau of Standards, *Tables of Selected Values of Chemical Thermodynamic Properties*, 1947, *et seq.* These tables are being continually revised and expanded.

II. ADDITIVITY OF HEATS OF REACTION

It has been emphasized repeatedly that since the enthalpy is a thermodynamic property, the value of ΔH depends only upon the nature and state of the initial reactants and final products, and not upon the reactions which have been used to carry out the transformation. Thus one may use known ΔH's to calculate the heat of a reaction for which no data are available. Several examples of additivity will be given.

A. Calculation of heat of formation from heat of reaction. As an example we may calculate the heat of formation of $Ca(OH)_2$ (solid) from data for other reactions such as the following:

$$CaO \text{ (s)} + H_2O \text{ (l)} = Ca(OH)_2 \text{ (s)}; \quad \Delta H_{18°C} = -15,260 \text{ cal mole}^{-1}, \quad (5.3)$$

$$H_2 \text{ (g)} + \tfrac{1}{2}O_2 \text{ (g)} = H_2O \text{ (l)} \quad ; \quad \Delta H_{18°C} = -68,370 \text{ cal mole}^{-1}, \quad (5.4)$$

$$Ca \text{ (s)} + \tfrac{1}{2}O_2 \text{ (g)} = CaO \text{ (s)} \quad ; \quad \Delta H_{18°C} = -151,800 \text{ cal mole}^{-1}. \quad (5.5)$$

It is evident that the addition of the preceding three chemical equations leads to the following desired equation; hence the addition of the corresponding ΔH's gives the desired heat of formation:

$$Ca \text{ (s)} + O_2 \text{ (g)} + H_2 \text{ (g)} = Ca(OH)_2 \text{ (s)};$$
$$\Delta H_{18°C} = -235,430 \text{ cal mole}^{-1}. \quad (5.6)$$

B. Calculation of heat of formation from heat of combustion. This type of calculation is very common, since with most organic compounds the experimental data must be obtained for the combustion process, yet the heat of formation is the more useful quantity for further thermodynamic calculations. A typical example of such a calculation is outlined by the following equations:

$$C_2H_5OH \text{ (l)} + 3O_2 \text{ (g)} = 2CO_2 \text{ (g)} + 3H_2O \text{ (l)};$$
$$\Delta H_{298°} = -326,700 \text{ cal mole}^{-1}, \quad (5.7)$$

$$3H_2O \text{ (l)} = 3H_2 \text{ (g)} + \tfrac{3}{2}O_2 \text{ (g)} ;$$
$$\Delta H_{298°} = 204,952 \text{ cal mole}^{-1}, \quad (5.8)$$

$$2CO_2 \text{ (g)} = 2C \text{ (graphite)} + 2O_2 \text{ (g)};$$
$$\Delta H_{298°} = 188,104 \text{ cal mole}^{-1}, \quad (5.9)$$

$$C_2H_5OH \text{ (l)} = 3H_2 \text{ (g)} + \tfrac{1}{2}O_2 \text{ (g)} + 2C \text{ (graphite)};$$
$$\Delta H = 66,356 \text{ cal mole}^{-1}. \quad (5.10)$$

Reversing equation (5.10), we obtain the heat of formation of ethyl alcohol:

$$3H_2 \text{ (g)} + \tfrac{1}{2}O_2 \text{ (g)} + 2C \text{ (graphite)} = C_2H_5OH \text{ (l)};$$
$$\Delta H_f° = -66,356 \text{ cal mole}^{-1}. \quad (5.11)$$

C. Calculation of heat of transition from heat of formation. Calculations of this type are particularly important in consideration of changes of reference state from one allotropic form to another. As an example the case of carbon will be illustrated:

$$C \text{ (graphite)} + O_2 \text{ (g)} = CO_2 \text{ (g)}; \quad \Delta H_{298^\circ} = -94{,}051.8 \text{ cal mole}^{-1}, \quad (5.12)$$

$$C \text{ (diamond)} + O_2 \text{ (g)} = CO_2 \text{ (g)}; \quad \Delta H_{298^\circ} = -94{,}505.0 \text{ cal mole}^{-1}. \quad (5.13)$$

Subtracting the first reaction from the second, we obtain

$$C \text{ (diamond)} = C \text{ (graphite)}; \quad \Delta H_{298^\circ} = -453.2 \text{ cal mole}^{-1}. \quad (5.14)$$

III. BOND ENERGIES

The calculation of the heat of formation of a given compound depends upon the determination of the heat of at least one reaction of this substance. Frequently, however, it is desirable to estimate the heat of a chemical reaction involving a hitherto unsynthesized compound, and hence a substance for which no enthalpy data are available. For the solution of problems of this type a system of bond energies has been established, so that if the atomic structure of the compound is known, it is possible to *approximate* the heat of formation by summation of the appropriate bond energies.

A. Definition of bond energies. It is essential at the very outset to make a distinction between "bond energy" and the "dissociation energy" of a given linkage. The latter is a definite quantity referring to the energy required to break a given bond of some specific compound. On the other hand, bond energy is an intermediate value of the dissociation energies of a given bond in a series of different dissociating species.

The distinction between these two terms may be more evident if described in terms of a simple example, the O—H bond. The heat of dissociation of the O—H bond depends upon the nature of the molecular species from which the H atom is being separated. For example, in the water molecule,

$$H_2O \text{ (g)} = H \text{ (g)} + OH \text{ (g)}; \quad \Delta H_{298^\circ} = 119.95 \text{ kcal mole}^{-1}. \quad (5.15)$$

On the other hand, to break the O—H bond in the hydroxyl radical requires a different heat quantity,

$$OH \text{ (g)} = O \text{ (g)} + H \text{ (g)}; \quad \Delta H_{298^\circ} = 101.19 \text{ kcal mole}^{-1}. \quad (5.16)$$

The bond energy ϵ_{O-H} is defined as the average of these two values; in other words,

$$\epsilon_{O-H} = \frac{119.95 + 101.19}{2} = 110.57 \text{ kcal mole}^{-1}. \quad (5.17)$$

Thus ϵ_{O-H} is half the value of the enthalpy change of the following reaction:

$$H_2O \text{ (g)} = O \text{ (g)} + 2H \text{ (g)}; \quad \epsilon_{O-H} = \frac{\Delta H}{2} = 110.57 \text{ kcal mole}^{-1}. \quad (5.18)$$

In the case of diatomic elements, such as H_2, the bond energy and dissociation energy are identical, for each refers to the following reaction:

$$H_2 \text{ (g)} = 2H \text{ (g)}; \quad \epsilon_{H-H} = \Delta H_{298^\circ} = 104.178 \text{ kcal mole}^{-1}. \quad (5.19)$$

It should be noted that in general the bond energy refers to an intermediate value of the dissociation energy of a given linkage in a molecule, at the start with the compound in the gaseous state and at the end with the dissociated atoms in the gaseous state.

B. Calculation of bond energies. The principle of the method has been illustrated with water and the O—H bond. However, this is a relatively simple case. Since bond energies are particularly of value in problems involving organic compounds, it is desirable to consider an example from this branch of chemistry, where the fundamental data are obtained from heats of combustion. We shall calculate the C—H bond energy from data on the heat of combustion of methane.

To find ϵ_{C-H} we need to know the heat of the following reaction:

$$CH_4 \text{ (g)} = C \text{ (g)} + 4H \text{ (g)}; \quad \epsilon_{C-H} = \frac{\Delta H}{4}. \quad (5.20)$$

The ΔH for the preceding reaction can be obtained from the summation of those for the following reactions at $298\,^\circ K$:

$$CH_4 \text{ (g)} + 2O_2 \text{ (g)} = CO_2 \text{ (g)} + 2H_2O \text{ (l)} \quad ; \quad \Delta H = -212.80 \text{ kcal mole}^{-1} \quad (5.21)$$

$$CO_2 \text{ (g)} = C \text{ (graphite)} + O_2 \text{ (g)}; \quad \Delta H = 94.05 \text{ kcal mole}^{-1} \quad (5.22)$$

$$2H_2O \text{ (l)} = 2H_2 \text{ (g)} + O_2 \text{ (g)} \quad ; \quad \Delta H = 136.64 \text{ kcal mole}^{-1} \quad (5.23)$$

$$2H_2 \text{ (g)} = 4H \text{ (g)} \quad ; \quad \Delta H = 208.36 \text{ kcal mole}^{-1} \quad (5.24)$$

$$C \text{ (graphite)} = C \text{ (g)} \quad ; \quad \Delta H^1 = 171.698 \text{ kcal mole}^{-1} \quad (5.25)$$

$$CH_4 \text{ (g)} = C \text{ (g)} + 4H \text{ (g)} \quad ; \quad \Delta H = 398.0 \text{ kcal mole}^{-1}. \quad (5.26)$$

It follows that at $298\,^\circ K$

$$\epsilon_{C-H} = \frac{398.0}{4} = 99.5 \text{ kcal mole}^{-1}. \quad (5.27)$$

It is pertinent to point out that this value of the C—H bond energy does

---

[1] The value which should be used for this enthalpy change has been the subject of much controversy in recent years. See L. H. Long and R. G. W. Norrish, *Proc. Royal Soc.*, **A187, 337** (1946); B. Edlen, *Nature*, **159**, 129 (1947); S. Goudsmit, *ibid.*, **159, 742** (1947); A. G. Shenstone, *Phys. Rev.*, **72**, 411 (1947).

In the present calculation the value chosen is that for the normal ^3P state of the carbon atom, as given in Series I, Table 23-1, of the *Selected Values of Chemical Thermodynamic Properties*, National Bureau of Standards, Washington, D. C., March 31, 1948.

not correspond to the dissociation energy of the carbon-hydrogen link in methane,[2] 102 kcal mole^{-1}. However, it should also be emphasized that the latter energy refers to the equation

$$CH_4 \text{ (g)} = CH_3 \text{ (g)} + H \text{ (g)} \qquad (5.28)$$

and not to one-quarter of the ΔH associated with equation (5.26).

In the calculation of ϵ_{C-H} outlined above, enthalpy values at 298 °K have been used, and hence the bond energy also refers to this temperature. In practice it is more common to calculate bond energies at 0 °K, rather than 298 °K. For ϵ_{C-H} one would obtain 98.2 kcal mole^{-1} at 0 °K, a value slightly lower than that at 298 °K. In general the differences between the bond energies at the two temperatures are small. To conform with general usage, nevertheless, we too shall refer to values at 0 °K. A list of such bond energies is given in Table 3.

In connection with estimates of the reliability of bond energies in the table, it should be emphasized that in some cases the value obtained depends upon that calculated previously for some other bond. For example, to obtain ϵ_{C-C}, one combines the heat of combustion of ethane with the proper multiples of the ΔH's in equations (5.22) to (5.25) to obtain the enthalpy change for the reaction

$$C_2H_6 \text{ (g)} = 2C \text{ (g)} + 6H \text{ (g)}. \qquad (5.29)$$

On the basis of our definition of bond energy it is apparent that

$$\epsilon_{C-C} = \Delta H_{\text{equation (5.29)}} - 6\epsilon_{C-H}. \qquad (5.30)$$

Thus, the value of 80 kcal mole^{-1} listed in Table 3 is based on an ϵ_{C-H} of 98.2 kcal mole^{-1} at 0 °K. Other estimates of ϵ_{C-H} would lead to different values for ϵ_{C-C}.

C. Estimation of the heat of reaction from bond energies. It has been pointed out that the primary value of bond energies lies in the calculation of the heat of a reaction involving a compound for which no enthalpy data are available. For example, the heat of formation of Se_2Cl_2 (g) may be calculated from bond energies by the following steps. Since the bond energy refers to the *dissociation* of Cl—Se—Se—Cl gas into gaseous atoms, the enthalpy change for the *formation* of this gaseous molecule should be given by

$$2Se \text{ (g)} + 2Cl \text{ (g)} = Se_2Cl_2 \text{ (g)}; \quad \Delta H = -[\epsilon_{Se-Se} + 2\epsilon_{Se-Cl}] \qquad (5.31)$$
$$= -168 \text{ kcal mole}^{-1}.$$

To estimate the heat of formation, however, it is necessary to add two reactions to equation (5.31), since by definition the heat of formation refers to the elements in their standard states. If we assume that the bond energies used are not too much different from those at 298 °K, we may intro-

[2] G. B. Kistiakowsky and E. R. Van Artsdalen, *J. Chem. Phys.*, 12, 469 (1944).

duce the following enthalpy changes for converting the elements to their standard states at 298°K:

$$Cl_2 \text{ (g)} = 2Cl \text{ (g)}; \quad \Delta H = 57.1 \text{ kcal mole}^{-1}, \quad (5.32)$$
$$2Se \text{ (hexagonal)} = 2Se \text{ (g)}; \quad \Delta H = 2 \times 48.4 \text{ kcal mole}^{-1}. \quad (5.33)$$

The addition of equations (5.31)–(5.33) leads to the expression

$$2Se \text{ (hexagonal)} + Cl_2 \text{ (g)} = Se_2Cl_2 \text{ (g)}; \quad \Delta H = -14 \text{ kcal mole}^{-1}. \quad (5.34)$$

If we wish to know the heat of formation of liquid Se_2Cl_2, we can estimate the heat of condensation (perhaps from Trouton's rule, or by comparison with related sulfur compounds) and add it to the value of ΔH obtained in equation (5.34).

TABLE 3

BOND ENERGIES* AT 0°K

Bond	Kcal mole^{-1}	Bond	Kcal mole^{-1}	Bond	Kcal mole^{-1}
H—H	103.2	Li—H	58	K—Cl	101.4
Li—Li	26	C—H	98.2	Cu—Cl	83
C—C	80	N—H	92.2	As—Cl	69
C=C	145	O—H	109.4	Se—Cl	59
C≡C	198	F—H	141 (?)	Br—Cl	52.1
N—N	37	Na—H	47	Rb—Cl	101.0
N≡N	225.1	Si—H	76 (?)	Ag—Cl	71
O—O	34	P—H	77	Sn—Cl	76
O=O	117.2	S—H	87 (?)	Sb—Cl	75
F—F	50 (?)	Cl—H	102.1	I—Cl	49.6
Na—Na	17.8	K—H	42.9	Cs—Cl	103
Si—Si	(45)†	Cu—H	62	C—N	66
P—P	(53)†	As—H	56	C≡N	209
S—S	63 (?)	Se—H	67	C—O	79
S=S	101 (?)	Br—H	86.7	C=O	173
Cl—Cl	57.1	Rb—H	39	P≡N	138 (?)
K—K	11.8	Ag—H	53	S=O	120 (?)
Ge—Ge	(39.2)†	Te—H	59	Te=O	62.8
As—As	(39)†	I—H	70.6	.....	....
As≡As	90.8	Cs—H	41	.....	....
Se—Se	(50)†	Li—Cl	118.5	.....	....
Se=Se	65	C—Cl	78	.....	....
Br—Br	45.4	N—Cl	46 (?)	.....	....
Rb—Rb	11.1	O—Cl	49	.....	....
Sn—Sn	(35)†	F—Cl	60.3	.....	....
Sb—Sb	(42)†	Na—Cl	97.7	.....	....
Sb≡Sb	69	Si—Cl	87	.....	....
Te—Te	(49)†	P—Cl	77	.....	....
Te=Te	53	S—Cl	65 (?)	.....	....
I—I	35.6	....	....	.....	....
Cs—Cs	10.4	....	....	.....	....

* K. S. Pitzer, *J. Am. Chem. Soc.*, **70**, 2140 (1948).

† Values in parentheses are less reliable because they depend on data for elements in the solid state where semimetallic or van der Waals contributions to the bonding may be present. These contributions may be of no importance in compounds of these elements.

By these methods it is possible to obtain fairly reliable estimates of heats of formation of many compounds. Difficulties are sometimes encountered, particularly in applications to organic compounds, where discrepancies arise which may be attributed to special factors such as steric effects and resonance. It is possible to set up secondary rules to take account of such special situations. However, since we shall consider much more refined methods for estimating heats of formation of organic compounds in connection with the application of the third law of thermodynamics, no further attention will be paid here to possible refinements in the use of bond energies.

IV. HEAT OF REACTION AS A FUNCTION OF TEMPERATURE

In the preceding sections, methods of determining heats of reaction at a fixed temperature (generally 298.16°K) were discussed. In particular it was pointed out that it is possible to tabulate heats of formation and bond energies and to use these for calculating heats of reaction. These tables, of course, are available for only a few standard temperatures. Frequently it is necessary to know the heat of a reaction at a temperature different from those available in a reference table. It is pertinent, therefore, to consider at this point the procedures which may be used to determine the heat of reaction (at constant pressure) at one temperature from data at another temperature.

A. Analytical method. Since we are interested in the variation of enthalpy with temperature, we might recall first that

$$\left(\frac{\partial H}{\partial T}\right)_P = C_p. \tag{5.35}$$

It is evident that such an equation can be integrated, at constant pressure:

$$\int dH = \int C_p \, dT, \tag{5.36}$$

$$H = \int C_p \, dT + H_0, \tag{5.37}$$

where H_0 represents an integration constant. If we are considering a chemical transformation, represented in general terms by

$$A + B + \ldots = M + N + \ldots, \tag{5.38}$$

it is evident that we can write a series of equations of the form

$$H_A = \int C_{pA} \, dT + H_{0A}, \tag{5.39}$$

$$H_B = \int C_{pB} \, dT + H_{0B}, \tag{5.40}$$

$$H_M = \int C_{pM} \, dT + H_{0M}, \tag{5.41}$$

$$H_N = \int C_{pN} \, dT + H_{0N}. \tag{5.42}$$

For the chemical reaction (5.38) the enthalpy change, ΔH, is given by

$$\begin{aligned}
\Delta H &= H_M + H_N + \ldots - H_A - H_B - \ldots \\
&= (H_{0M} + H_{0N} + \ldots - H_{0A} - H_{0B} - \ldots) + \int C_{pM}\,dT \\
&\quad + \int C_{pN}\,dT + \ldots - \int C_{pA}\,dT - \int C_{pB}\,dT - \ldots \quad (5.43)
\end{aligned}$$

If we define the quantities inside the parentheses as ΔH_0, and if we group the integrals together, we obtain

$$\Delta H = \Delta H_0 + \int (C_{pM} + C_{pN} + \ldots - C_{pA} - C_{pB} - \ldots)dT, \quad (5.44)$$

or
$$\Delta H = \Delta H_0 + \int \Delta C_p\,dT. \quad (5.45)$$

Thus in order to obtain ΔH as a function of the temperature, it is necessary to know the dependence of the heat capacities of the reactants and products on the temperature, as well as one value of ΔH so that ΔH_0 can be evaluated.

As an example, let us consider the heat of formation of CO_2 (g),

$$C \text{ (graphite)} + O_2 \text{ (g)} = CO_2 \text{ (g)}; \quad \Delta H_{298.16} = -94{,}051.8 \text{ cal mole}^{-1}. \quad (5.46)$$

The heat capacities of the substances involved may be expressed by the following equations:

$$\begin{aligned}
C_{p(C)} &= -1.265 + 14.008 \times 10^{-3}\,T - 103.31 \times 10^{-7}\,T^2, \quad (5.47) \\
C_{p(O_2)} &= 6.148 + 3.102 \times 10^{-3}\,T - 9.23 \times 10^{-7}\,T^2, \quad (5.48) \\
C_{p(CO_2)} &= 6.214 + 10.396 \times 10^{-3}\,T - 35.45 \times 10^{-7}\,T^2. \quad (5.49)
\end{aligned}$$

Hence the difference in heat capacities of products and reactants is given by the equation

$$\Delta C_p = 1.331 - 6.714 \times 10^{-3}\,T + 77.09 \times 10^{-7}\,T^2, \quad (5.50)$$

and consequently

$$\begin{aligned}
\Delta H &= \Delta H_0 + \int (1.331 - 6.714 \times 10^{-3}\,T + 77.09 \times 10^{-7}\,T^2)dT \\
&= \Delta H_0 + 1.331T - 3.357 \times 10^{-3}\,T^2 + 25.70 \times 10^{-7}\,T^3. \quad (5.51)
\end{aligned}$$

Since ΔH is known at $298.16\,^\circ$K, it is possible to substitute into the preceding equation and to calculate ΔH_0:

$$\Delta H_0 = -94{,}218.3. \quad (5.52)$$

Thus we may now write a completely explicit equation for the heat of formation of CO_2 as a function of the temperature:

$$\Delta H = -94{,}218.3 + 1.331T - 3.357 \times 10^{-3}\,T^2 + 25.70 \times 10^{-7}\,T^3. \quad (5.53)$$

B. Arithmetic method. A second procedure, fundamentally no different from the analytic method, involves the addition of suitable equations to give the desired equation. For example, if we consider the freezing of water, the heat of the reaction is known at $0\,°C$ (T_1), but might be required at $-10\,°C$ (T_2). We may obtain the desired ΔH by the addition of the following equations:

$$H_2O \ (l, \ 0\,°C) \ - \ H_2O \ (s, \ 0\,°C); \qquad \Delta H = -1436 \text{ cal mole}^{-1} \quad (5.54)$$

$$H_2O \ (s, \ 0\,°C) = H_2O \ (s, \ -10\,°C); \qquad \Delta H = \int_{0°}^{-10°} C_{p(s)} \, dT$$
$$= C_{p(s)}(T_2 - T_1)$$
$$= -87 \text{ cal mole}^{-1} \quad (5.55)$$

$$H_2O \ (l, \ -10\,°C) = H_2O \ (l, \ 0\,°C); \qquad \Delta H = \int_{-10°}^{0°} C_{p(l)} \, dT$$
$$= C_{p(l)} \ (T_1 - T_2)$$
$$= 180 \text{ cal mole}^{-1} \quad (5.56)$$

$$H_2O \ (l, \ -10\,°C) = H_2O \ (s, \ -10\,°C); \quad \Delta H = -1436 - 87 + 180$$
$$= -1343 \text{ cal mole}^{-1}. \quad (5.57)$$

C. Graphical method. It is obvious from the discussion of enthalpy change in the preceding chapter that if analytic equations for the heat capacities of reactants and products are unavailable, one may still carry out the integration required by equation (5.45) by graphical methods. In essence one replaces equation (5.45) by the expression

$$\Delta H = \Delta H_0 + \sum (\overline{\Delta C_p})(\Delta T). \qquad (5.58)$$

In practice, however, this method is seldom necessary.

Exercises

1. Find the heat of formation of ethyl alcohol in the *International Critical Tables*, Bichowsky and Rossini's *Thermochemistry*, and the National Bureau of Standards *Tables*. Compare the respective values.

2. Estimate $\Delta Hf°$ of the gaseous N—H radical by using appropriate values from the table of bond energies.

3. Find the bond energy of the I—Cl bond at $25\,°C$ from the data listed in Table 2. Compare with the bond energy given in Table 3.

4. Taking the heat of combustion of ethane as -372.8 kcal mole^{-1}, find the C—C bond energy.

5. Find data for the standard heats of formation of Cl (g), S (g), S_8 (g), and S_2Cl_2 (g) from appropriate sources.
   (a) Calculate the energy of the S—S bond. Assume that S_8 consists of eight such linkages.
   (b) Calculate the energy of the S—Cl bond.
   (c) Estimate the heat of formation of SCl_2 (g).

6. Derive an equation for the dependence of ΔH on temperature for the reaction

$$CO \ (g) + \tfrac{1}{2}O_2 \ (g) = CO_2 \ (g).$$

Appropriate data can be found in the tables of Chapters 4 and 5.

CHAPTER 6

Application of the First Law to Gases

IT BECOMES of interest, as a prelude to the development of the second law of thermodynamics, to consider the information which is obtainable on the behavior of gases by application of the single thermodynamic postulate and the associated definitions which have been developed so far. Since the behavior of many gases may be approximated by the simple equation of state attributed to the *ideal gas*, it is convenient to begin our discussion with a consideration of this ideal substance.

I. IDEAL GASES

A. Definition. An ideal gas is one (1) which obeys the equation of state

$$PV = nRT, \tag{6.1}$$

where n is the number of moles and R is a universal constant; and (2) for which the internal energy, E, is a function of the temperature only, that is,

$$\left(\frac{\partial E}{\partial V}\right)_T = \left(\frac{\partial E}{\partial P}\right)_T = 0. \tag{6.2}$$

Although we shall consider these two requirements as sufficient for the definition of an ideal gas, occasionally an additional specification is made that the heat capacity at constant volume must be a constant, or

$$C_v = \text{constant.} \tag{6.3}$$

It is pertinent to point out that equation (6.1) is a combination of Boyle's and Charles' laws, as can be seen readily by the following procedure.[1] Boyle's law may be expressed by the relation

$$V = \frac{nk_T}{P} \qquad (T \text{ constant}) \tag{6.4}$$

or

$$\left(\frac{\partial V}{\partial P}\right)_T = -\frac{nk_T}{P^2}, \tag{6.5}$$

---

[1] The procedure described is introduced as an example of the use of an equation for the total differential. Equation (6.1) can also be derived from Boyle's and Charles' laws by simple algebraic manipulations.

where k_T is a constant at a fixed temperature. Similarly, Charles' law may be expressed by either of the following two expressions:

$$V = nk_P T \qquad \text{(P constant)} \quad (6.6)$$

or

$$\left(\frac{\partial V}{\partial T}\right)_P = nk_P. \qquad (6.7)$$

If we consider the total differential of the volume, $V = f(T,P)$, we obtain

$$dV = \left(\frac{\partial V}{\partial T}\right)_P dT + \left(\frac{\partial V}{\partial P}\right)_T dP. \qquad (6.8)$$

Equations (6.5) and (6.7) may be used to substitute for the partial derivatives in (6.8), so that we may obtain

$$dV = nk_P\, dT - \frac{nk_T}{P^2}\, dP. \qquad (6.9)$$

The constants k_P and k_T may be replaced by the introduction of equations (6.4) and (6.6).

$$dV = \frac{V}{T}\, dT - \frac{V}{P}\, dP. \qquad (6.10)$$

Rearranging expression (6.10), we obtain

$$\frac{dV}{V} = \frac{dT}{T} - \frac{dP}{P}, \qquad (6.11)$$

which may be integrated to give

$$PV = k'T, \qquad (6.12)$$

where k' is the constant of integration. If we identify k' with nR, we have equation (6.1), the ideal gas law.

It follows from equation (6.2) that if an ideal gas undergoes any isothermal transformation, its energy remains fixed. This restriction in the definition of an ideal gas actually is not independent, since it can be shown to follow from equation (6.1). However, this proof depends upon the use of the second law of thermodynamics; and since this principle has not yet been considered, equation (6.2) will be considered tentatively as independent.

B. Some relations derived from the definition.

1. *Enthalpy a function of the temperature only.* It is simple to prove that the enthalpy, as well as the internal energy, is constant in any isothermal change of an ideal gas. Since

$$H = E + PV, \qquad (6.13)$$

$$\left(\frac{\partial H}{\partial V}\right)_T = \left(\frac{\partial E}{\partial V}\right)_T + \left(\frac{\partial [PV]}{\partial V}\right)_T. \tag{6.14}$$

But from equations (6.1) and (6.2) it is evident that each term on the right-hand side of equation (6.14) is zero. Consequently,

$$\left(\frac{\partial H}{\partial V}\right)_T = 0. \tag{6.15}$$

By an analogous procedure it can also be shown readily that

$$\left(\frac{\partial H}{\partial P}\right)_T = 0. \tag{6.16}$$

2. *Relation between C_p and C_v.* In Chapter 4 the following expression [equation (4.28)] was shown to be a general relation between the heat capacity at constant pressure and that at constant volume:

$$C_p = C_v + \left[P + \left(\frac{\partial E}{\partial V}\right)_T\right]\left(\frac{\partial V}{\partial T}\right)_P. \tag{6.17}$$

For one mole of an ideal gas,

$$\left(\frac{\partial E}{\partial V}\right)_T = 0 \tag{6.2}$$

and

$$\left(\frac{\partial V}{\partial T}\right)_P = \frac{R}{P}. \tag{6.18}$$

It is evident immediately that the substitution of equations (6.2) and (6.18) into (6.17) leads to the familiar expression

$$C_p = C_v + R. \tag{6.19}$$

C. **Calculation of thermodynamic changes in expansion processes.**
1. *Isothermal.* An expansion at a fixed temperature may be carried out reversibly. The reversible process may be visualized as shown in Fig. 1. The general expression for the work done in the expansion is

$$W = \int_{V_1}^{V_2} P\, dV. \tag{6.20}$$

Using equation (6.1) to eliminate P from (6.20), we obtain

$$W = \int_{V_1}^{V_2} \frac{nRT}{V}\, dV = nRT \ln \frac{V_2}{V_1}. \tag{6.21}$$

In considerations of the change in energy, ΔE, it is merely necessary to keep in mind that the process is isothermal and that E depends only upon temperature. Hence

$$\Delta E = 0. \tag{6.22}$$

With this information and the use of the first law of thermodynamics, we can readily calculate the heat absorbed in the process:

$$Q = \Delta E + W = nRT \ln \frac{V_2}{V_1}. \tag{6.23}$$

If we pass finally to the enthalpy change, ΔH, it is evident that this quantity, like ΔE, must be zero, since

$$\Delta H = \Delta E + \Delta(PV) = 0 + \Delta(nRT) = 0. \tag{6.24}$$

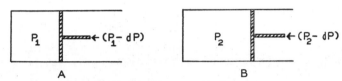

A   B

Fig. 1.   Schematic representation of an isothermal reversible expansion. The external pressure is maintained only infinitesimally below the internal pressure.

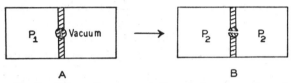

A   B

Fig. 2.   Schematic representation of a free expansion. A small valve separating the two chambers in A is opened so that the gas may rush in from left to right   The initial volume of the gas may be called V_1, and the final volume V_2.

We may also readily visualize a completely free isothermal expansion in which no work is done, such as is pictured in Fig. 2.   Under the conditions described, the process may be carried out isothermally.[2]   Hence

$$W = 0, \tag{6.25}$$

$$\Delta E = 0, \tag{6.26}$$

$$Q = \Delta E + W = 0, \tag{6.27}$$

and $$\Delta H = \Delta E + \Delta(PV) = 0. \tag{6.28}$$

In an actual case, some work would be obtained; but since the expansion would be carried out at finite speed, the pressure immediately adjacent to the piston wall would be slightly less than that in the interior at any given moment and hence the net work would be less than in the reversible expansion, though greater than zero.

---

[2] Although temperature differences may be set up temporarily, when equilibrium is reached the temperature is the same as the initial one.

$$0 < W < nRT \ln \frac{V_2}{V_1}. \tag{6.29}$$

On the other hand, if isothermal conditions are maintained, there is still no change in energy, since for an ideal gas

$$\Delta E = 0. \tag{6.30}$$

To determine Q, we again use the first law,

$$Q = \Delta E + W = W, \tag{6.31}$$

$$0 < Q < nRT \ln \frac{V_2}{V_1}. \tag{6.32}$$

The enthalpy change is found from the relation

$$\Delta H = \Delta E + \Delta(PV) = 0. \tag{6.33}$$

In summary, we may compare these three isothermal expansions of an ideal gas by tabulating the corresponding thermodynamic changes (Table 1). This table serves to emphasize the difference between an exact and an

TABLE 1
THERMODYNAMIC CHANGES IN ISOTHERMAL EXPANSIONS OF AN IDEAL GAS

Reversible	Free	Actual
$T_1 = T_2$	$T_1 = T_2$	$T_1 = T_2$
$W = nRT \ln \frac{V_2}{V_1}$	$W = 0$	$0 < W < nRT \ln \frac{V_2}{V_1}$
$\Delta E = 0$	$\Delta E = 0$	$\Delta E = 0$
$Q = nRT \ln \frac{V_2}{V_1}$	$Q = 0$	$0 < Q < nRT \ln \frac{V_2}{V_1}; \ Q = W$
$\Delta H = 0$	$\Delta H = 0$	$\Delta H = 0$

inexact thermodynamic function. Thus E and H, whose differentials are exact, undergo the same change in each of the three different paths used for the transformation. They are thermodynamic properties. On the other hand, the work and heat quantities depend upon the particular path chosen, even though the initial and final values of the temperature, pressure, and volume, respectively, are the same in all these cases. Thus the heat and work functions are not thermodynamic properties. It is desirable to speak not of the heat (or work) contained in a system, but rather of the energy (or enthalpy) of the system, since the system may evolve different quantities of heat in going by different paths from a given initial to a specified final state. The heat and work quantities are manifested only during a transformation and their magnitudes depend upon the manner in which the system goes from one state to another. Since the work obtained in going from State A to State B may not be equal to that required to return the system from State B to State A, it is misleading to speak of the work contained in the system.

2. *Adiabatic.* By definition, an adiabatic expansion is one accompanied by no transfer of heat. Evidently, then,

$$DQ = 0. \tag{6.34}$$

It follows immediately from the first law of thermodynamics that

$$dE = -DW = -dW. \tag{6.35}$$

This equality may be used, in turn, to specify the work done in a more explicit fashion, because if ΔE is known, W is obtained immediately. For an ideal gas, ΔE in the adiabatic expansion may be determined by a simple procedure.

Considering the energy as a function of temperature and volume, $E = f(T,V)$, we may write an equation for the total differential:

$$dE = \left(\frac{\partial E}{\partial T}\right)_V dT + \left(\frac{\partial E}{\partial V}\right)_T dV. \tag{6.36}$$

Since we are dealing with an ideal gas, it is evident that

$$\left(\frac{\partial E}{\partial V}\right)_T = 0, \tag{6.37}$$

and hence that equation (6.36) may be reduced to

$$dE = \left(\frac{\partial E}{\partial T}\right)_V dT = C_v\, dT = -dW. \tag{6.38}$$

Hence the work done, W, and the energy change, ΔE, may be obtained by integration of equation (6.38):

$$W = -\int_{T_1}^{T_2} C_v\, dT = -\Delta E. \tag{6.39}$$

If C_v is independent of temperature for this ideal gas, then

$$W = -\Delta E = -C_v(T_2 - T_1). \tag{6.40}$$

It is also simple to determine the enthalpy change in this adiabatic expansion, particularly if C_v is constant. The equation of state for one mole of an ideal gas affords a simple substitution for $\Delta(PV)$ in the equation

$$\Delta H = \Delta E + \Delta(PV) \tag{6.41}$$

to give

$$\Delta H = \Delta E + \Delta(RT)$$
$$= C_v(T_2 - T_1) + (RT_2 - RT_1),$$

or

$$\Delta H = (C_v + R)(T_2 - T_1) = C_p(T_2 - T_1). \tag{6.42}$$

So far we have not had to specify whether the adiabatic expansion under consideration is reversible or not. Equations (6.34), (6.40), and (6.42), for the calculation of the thermodynamic changes in this process, evidently apply, then, for the reversible expansion, free expansion, or actual expansion, so long as we are dealing with an ideal gas.

The numerical values of W, ΔE, and ΔH, however, will not be the same for each of the three types of adiabatic expansion, since, as we shall see shortly, T_2, the final temperature of the gas, will differ in each case, even though the initial temperature may be identical in all cases.

If we consider the free expansion first, it is apparent from equation (6.40) that since no work is done, there is no change in temperature; that is, $T_2 = T_1$. Thus ΔE and ΔH must also be zero for this process. Reference to Table 1 shows, then, that an adiabatic free expansion and an isothermal free expansion are two different names for the same process.

Turning to the reversible adiabatic expansion, we find that a definite expression can be derived to relate the initial and final temperatures to the respective volumes or pressures. Again we start with equation (6.35). Recognizing the restriction of reversibility, we obtain

$$-dE = dW = P\, dV. \tag{6.43}$$

Since we are dealing with an ideal gas, equation (6.38) is valid. Substitution from equation (6.43) into (6.38) leads to

$$C_v\, dT = -P\, dV. \tag{6.44}$$

We may now make use of the equation of state for an ideal gas. It is evident that whatever the value of n in equation (6.1), C_v in equation (6.44) will equal n times the molal heat capacity. Hence if we consider one mole of the substance,

$$C_v\, dT = -\frac{RT}{V}\, dV. \tag{6.45}$$

Rearranging, we obtain

$$\frac{C_v}{T}\, dT = -\frac{R}{V}\, dV, \tag{6.46}$$

which may be integrated within definite limits to give, when C_v is constant,

$$C_v \ln\left(\frac{T_2}{T_1}\right) = -R \ln\left(\frac{V_2}{V_1}\right). \tag{6.47}$$

This equation in turn may be converted to

$$\left(\frac{T_2}{T_1}\right)^{C_v} = \left(\frac{V_2}{V_1}\right)^{-R} = \left(\frac{V_1}{V_2}\right)^{R} \tag{6.48}$$

or
$$\frac{T_2}{T_1} = \left(\frac{V_1}{V_2}\right)^{R/C_v}.$$

(6.49)

Hence
$$T_2 V_2{}^{R/C_v} = T_1 V_1{}^{R/C_v}.$$

(6.50)

Equation (6.50) says in essence that the particular temperature-volume function shown is constant during a reversible adiabatic expansion. Hence we may write

$$TV^{R/C_v} = \text{constant}$$

(6.51)

or
$$T^{C_v/R} V = \text{constant}'.$$

Any one of the equations (6.47) to (6.51) may be used to calculate a final temperature from the initial temperature and the observed volumes.

If one measures pressures instead of volumes, it is possible to use the following equation instead:

$$PV^{C_p/C_v} = \text{constant}''.$$

(6.52)

Equation (6.52) may be derived readily from (6.51) by substitution from the equation of state of the ideal gas.

In any event it is evident, most directly perhaps from equation (6.49), that the final temperature, T_2, in the reversible adiabatic expansion (for the ideal gas) must be less than T_1, since V_1 is less than V_2, and R and C_v are both positive numbers. Thus the adiabatic reversible expansion is accompanied by a temperature drop which can be calculated readily from the measured volumes or pressures at the beginning and the end of the process. Knowing T_2 and T_1, we can then proceed immediately to the evaluation of W, ΔE, and ΔH by simple substitution into equations (6.40) and (6.42).

Considering, finally, an adiabatic expansion in any actual case, we can see again that W, the work done by the gas, would be less than in the reversible case, since in this actual process the lag in transmission of stresses would keep the pressure in the immediate vicinity of the moving piston somewhat lower than it would be in a corresponding reversible process. Thus if our final volume V_2 is the same as that in the reversible process (and if the initial state is the same in both processes), T_2 will not be as low in the actual expansion, since according to equation (6.38) the temperature drop, dT, depends directly upon the work done by the expanding (ideal) gas. Similarly, it is apparent from equations (6.40) and (6.42) that ΔE and ΔH, respectively, must also be numerically smaller in the actual expansion than in the reversible one.

Thus in the adiabatic expansion, from a common set of initial conditions to the same final volume, the values of the energy and enthalpy changes, as well as of the work done, appear to depend on the path. At first glance, such behavior may appear to be in contradiction to the assumption of exactness of dE and dH. However, careful consideration shows quickly that the basis of the difference lies in the different end point for each of the three paths; despite the fact that the final volume may be made the same in each case, the final temperature depends upon whether the expansion is free, reversible, or actual.

II. REAL GASES

A. Equations of state. We wish to turn our attention now to substances for which the ideal gas laws do not give an adequate description. Numerous equations of state have been used to describe real gases.[3] For our purposes, however, it will be sufficient to consider only three of the more common ones.

1. *van der Waals equation.* This equation was one of the first to be introduced to describe deviations from ideality. The argument behind it is discussed adequately in elementary textbooks. It is generally stated in the form

$$\left(P + \frac{a}{v^2}\right)(v - b) = RT, \tag{6.53}$$

where v is the volume per mole and a and b are constants (Table 2). Alternative methods of expression will be considered later.

2. *Virial function.* A very useful form of expression of deviations from the ideal gas law is the following equation,

$$Pv = A + BP + CP^2 + DP^3 + EP^4 + \ldots, \tag{6.54}$$

where A, B, C, D, and E are constants at a given temperature and are known as *virial coefficients* (Table 3). The term A must reduce, of course, to RT, since at very low pressures all gases approach ideal gas behavior.

3. *Berthelot equation.* This equation is too unwieldy to be used generally as an equation of state. However, it is very convenient in calculations of deviations from ideality near pressures of one atmosphere and hence has been used extensively in determinations of entropies from the third law of thermodynamics. This aspect of the equation will receive further attention in subsequent discussions.

The Berthelot equation may be expressed as

$$Pv = RT\left[1 + \frac{9}{128}\frac{P}{P_c}\frac{T_c}{T}\left(1 - 6\frac{T_c^2}{T^2}\right)\right], \tag{6.55}$$

[3] F. T. Gucker, Jr., and W. B. Meldrum, *Physical Chemistry*, American Book Company, New York, 1942, pages 103–105; C. F. Prutton and S. H. Maron, *Fundamental Principles of Physical Chemistry*, The Macmillan Company, New York, 1944, pages 29–36; F. Daniels, *Outlines of Physical Chemistry*, J. Wiley and Sons, Inc., New York, 1948, pages 15–21; E. D. Eastman and G. K. Rollefson, *Physical Chemistry*, McGraw-Hill Book Company, Inc., New York, 1947, pages 68–72.

where P_c and T_c are the critical pressure and critical temperature, respectively.

B. Joule-Thomson effect. One method of measurement of deviations from ideal behavior in a quantitative fashion is by determinations

TABLE 2

VAN DER WAALS CONSTANTS FOR SOME GASES*

Substance	a $\left(\dfrac{\text{atm liter}^2}{\text{mole}^2}\right)$	b $\left(\dfrac{\text{liter}}{\text{mole}}\right)$
Acetylene...............	4.39	0.0514
Ammonia...............	4.17	0.0371
Argon...................	1.35	0.0322
Carbon dioxide..........	3.59	0.0427
Carbon monoxide........	1.49	0.0399
Chlorine................	6.49	0.0562
Ethyl ether.............	17.4	0.1344
Helium.................	0.034	0.0237
Hydrogen...............	0.244	0.0266
Hydrogen chloride.......	3.67	0.0408
Methane................	2.25	0.0428
Nitric oxide............	1.34	0.0279
Nitrogen...............	1.39	0.0391
Nitrogen dioxide........	5.28	0.0442
Oxygen.................	1.36	0.0318
Sulfur dioxide..........	6.71	0.0564
Water.................	5.46	0.0305

* Values for other substances may be calculated from data in Landolt-Börnstein, *Physikalisch-chemische Tabellen*, Fifth Edition, Volume I, Julius Springer, Berlin, 1923, pages 253–263.

TABLE 3

VIRIAL COEFFICIENTS FOR SOME GASES*

Substance	t, °C	A	$B \times 10^2$	$C \times 10^5$	$D \times 10^8$	$E \times 10^{11}$
Hydrogen............	0	22.414	1.3638	0.7851	−1.206	0.7354
	500	63.447	1.7974	0.1003	−0.1619	0.1050
Nitrogen............	0	22.414	−1.0512	8.626	−6.910	1.704
	200	38.824	1.4763	2.775	−2.379	0.7600
Carbon monoxide......	0	22.414	−1.4825	9.823	−7.721	1.947
	200	38.824	1.3163	3.052	−2.449	0.7266

* Taken from C. F. Prutton and S. H. Maron, *Fundamental Principles of Physical Chemistry*, The Macmillan Company, New York, 1944, page 34. The values given for the constants apply to one mole of the gas, to the pressure expressed in atmospheres, and to the volume in liters.

of the change in temperature in the Joule-Thomson porous-plug experiment (Fig. 3). The enclosed gas, initially of volume V_1, flows very slowly from the left-hand chamber through a porous plug into the right-hand chamber. The pressure on the left side is maintained constant at a value of P_1, while that on the right side is also constant but at a lower value, P_2. The apparatus is jacketed with a good insulator so that no heat is exchanged with the surroundings. In general for real gases, it is observed that the final temperature, T_2, differs from the initial one, T_1.

1. *Isenthalpic nature.* Since the Joule-Thomson experiment is carried out adiabatically, we may write

$$Q = 0. \tag{6.56}$$

However, it does not follow that ΔH is also zero, since the process involves a change in pressure. Nevertheless, it can readily be shown that the process is an isenthalpic one, that is, that ΔH is zero.

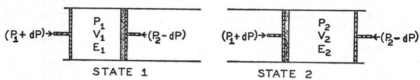

STATE 1   STATE 2

Fig. 3. Schematic representation of a Joule-Thomson porous-plug experiment. The entire apparatus is kept in a jacket and is well insulated from the surrounding environment.

The work done by the gas is that accomplished in the right-hand chamber,

$$W_2 = \int_0^{V_2} P_2 \, dV = P_2 V_2, \tag{6.57}$$

plus that done in the left-hand chamber,

$$W_1 = \int_{V_1}^0 P_1 \, dV = -P_1 V_1. \tag{6.58}$$

Hence the net work

$$W = W_1 + W_2 = P_2 V_2 - P_1 V_1. \tag{6.59}$$

Similarly, the net gain in energy, ΔE, is

$$\Delta E = E_2 - E_1. \tag{6.60}$$

Since $Q = 0$, it follows from the first law of thermodynamics that

$$E_2 - E_1 = -W = P_1 V_1 - P_2 V_2. \tag{6.61}$$

Therefore $\qquad E_2 + P_2 V_2 = E_1 + P_1 V_1, \tag{6.62}$

$$H_2 = H_1, \tag{6.63}$$

$$\Delta H = 0. \tag{6.64}$$

Thus we have proved that the Joule-Thomson experiment is isenthalpic as well as adiabatic.

2. *Joule-Thomson coefficient.* Knowing that the process is isenthalpic, we are in a position to formulate the Joule-Thomson effect in a quantitative manner.

a. *Definition.* Since it is the change in temperature which is observed as the gas flows from a higher to a lower pressure, the data are summarized in terms of a quantity, μ, which is defined by

$$\mu = \left(\frac{\partial T}{\partial P}\right)_H. \tag{6.65}$$

It is evident from this definition that the Joule-Thomson coefficient, μ, is positive when a cooling of the gas, that is, a temperature drop, is observed, because since dP is always negative, μ will be positive when dT is negative. Conversely, μ is a negative quantity when the gas warms on expansion, because dT is then a positive quantity.

b. *Derived relations.* For purposes of evaluation, it is frequently necessary to have the Joule-Thomson coefficient expressed in terms of other partial derivatives.

Considering the enthalpy as a function of temperature and pressure, $H(T,P)$, we may write the total differential

$$dH = \left(\frac{\partial H}{\partial P}\right)_T dP + \left(\frac{\partial H}{\partial T}\right)_P dT. \tag{6.66}$$

Placing a restriction of constant enthalpy on equation (6.66), we obtain

$$0 = \left(\frac{\partial H}{\partial P}\right)_T + \left(\frac{\partial H}{\partial T}\right)_P\left(\frac{\partial T}{\partial P}\right)_H, \tag{6.67}$$

which can be rearranged to give

$$\left(\frac{\partial T}{\partial P}\right)_H = -\frac{(\partial H/\partial P)_T}{(\partial H/\partial T)_P}, \tag{6.68}$$

or

$$\mu = -\frac{1}{C_p}\left(\frac{\partial H}{\partial P}\right)_T. \tag{6.69}$$

An additional relation of interest may be obtained by substitution of the fundamental definition of H into equation (6.69):

$$\mu = -\frac{1}{C_p}\left(\frac{\partial E}{\partial P}\right)_T - \frac{1}{C_p}\left(\frac{\partial [PV]}{\partial P}\right)_T. \tag{6.70}$$

From either of these last two expressions it is evident that $\mu = 0$ for an ideal gas, since each partial derivative is zero for such a substance.

c. *Inversion temperature.* For a gas which obeys the van der Waals equation, it can be shown readily that the sign of the Joule-Thomson coefficient depends upon the temperature and that at sufficiently high temperatures μ is negative for all gases, whereas at suitably low temperatures μ is positive. Thus each gas has at least one temperature at which μ changes sign, and hence is zero. The value of this inversion temperature,[4] T_i, may be predicted if the constants in the van der Waals equation are known, for it can be shown that the Joule-Thomson coefficient is related to these constants. It will be proved in Chapter 16 that

$$\left(\frac{\partial H}{\partial P}\right)_T = b - \frac{2a}{RT}.$$ (6.71)[5]

Hence it follows from equation (6.69) that

$$\mu = \frac{1}{C_p}\left(\frac{2a}{RT} - b\right).$$ (6.72)

From equation (6.72) it is evident that for a given gas:

(1)   At high temperatures, $2a/RT < b$; $\mu < 0$; gas warms on expansion.

(2)   At low temperatures, $2a/RT > b$; $\mu > 0$; gas cools on expansion.

(3)   At some intermediate temperature, T_i, $2a/RT_i = b$; $\mu = 0$; no temperature change on expansion in the porous-plug apparatus.

Obviously the value of T_i must be given by

$$T_i = \frac{2a}{Rb}.$$ (6.73)

Hence, if the van der Waals constants are known, T_i may be calculated. For all gases except hydrogen and helium, this inversion temperature is above common room temperatures.

C.   Calculation of thermodynamic quantities in reversible expansions.

1.   *Isothermal.* The procedure used in calculations of the work and energy quantities in an isothermal reversible expansion of a real gas is similar to that used for the ideal gas. Into the expression for the work done,

$$W = \int_{V_1}^{V_2} P \, dV,$$ (6.20)

---

[4] Actually the inversion temperature depends upon the pressure at which the Joule-Thomson experiment is carried out. In the discussion given at this point, the assumption is made implicitly that higher terms in the expansion of the van der Waals equation may be neglected. If these higher terms are not omitted (see Chapter 16), the inversion temperature obeys the following condition, in which the solutions clearly depend upon the pressure:

$$\frac{2a}{RT_i} - \frac{3abP}{R^2T_i{}^2} - b = 0.$$

It is apparent from this expression that there are *two* inversion temperatures for every pressure.

[5] This equation is printed in type of smaller size to emphasize that it has not yet been proved.

we may substitute for P (or for dV) from the equation of state of the gas and then carry out the required integration. For one mole of a van der Waals gas, for example,

$$W = \int_{v_1}^{v_2} \left(\frac{RT}{v - b} - \frac{a}{v^2}\right) dv = RT \ln \frac{v_2 - b}{v_1 - b} + \frac{a}{v_2} - \frac{a}{v_1}. \quad (6.74)$$

The change in energy in an isothermal expansion, however, cannot be expressed in a simple form without the introduction of the second law of thermodynamics. Nevertheless we shall anticipate this second basic postulate and use one of the deductions obtainable from it:

$$\left(\frac{\partial E}{\partial V}\right)_T = T\left(\frac{\partial P}{\partial T}\right)_V - P. \quad (6.75)$$

For a van der Waals gas, equation (6.75) reduces to

$$\left(\frac{\partial E}{\partial V}\right)_T = \frac{a}{v^2}. \quad (6.76)$$

ΔE may be obtained by integration of this equation:

$$\Delta E = \int_{v_1}^{v_2} \frac{a}{v^2} dv = -\frac{a}{v_2} + \frac{a}{v_1}. \quad (6.77)$$

With the aid of the first law of thermodynamics we may now calculate readily the heat absorbed in the isothermal reversible expansion:

$$Q = \Delta E + W = RT \ln \frac{v_2 - b}{v_1 - b}. \quad (6.78)$$

In turn, ΔH may be determined by integration of each of the terms in

$$dH = dE + d(PV). \quad (6.79)$$

The PV product may be obtained from the equation

$$P = \frac{RT}{v - b} - \frac{a}{v^2} \quad (6.80)$$

by multiplying each term by v:

$$Pv = RT \frac{v}{v - b} - \frac{a}{v}. \quad (6.81)$$

Differentiation gives

$$d(Pv) = -\frac{bRT}{(v - b)^2} dv + \frac{a}{v^2} dv, \quad (6.82)$$

which upon integration from v_1 to v_2 leads to

$$\Delta(Pv) = bRT \left[\frac{1}{v_2 - b} - \frac{1}{v_1 - b} \right] - \frac{a}{v_2} + \frac{a}{v_1}. \tag{6.83}$$

Adding equation (6.83) to (6.77), we obtain

$$\Delta H = bRT \left[\frac{1}{v_2 - b} - \frac{1}{v_1 - b} \right] - \frac{2a}{v_2} + \frac{2a}{v_1}. \tag{6.84}$$

2. *Adiabatic.* The restriction of no heat exchange may be expressed by the equation

$$Q = 0. \tag{6.85}$$

A calculation of the work and energy quantities, however, depends upon the solution of equation (6.36) for the change in energy. Furthermore, the specification of the equation of state for the gas does not automatically give an expression for the dependence of C_v on temperature. When adequate equations, empirical or theoretical, for the variation of E and of C_v with T and V are available, however, they may be used in an equation equivalent to (6.38), that for the ideal gas, and the resultant expression may be integrable. If they are, the energy and work quantities may be calculated by a procedure similar to that used for the ideal gas.

Exercises[6]

1. Derive an explicit equation for the reversible work of an isothermal expansion for each of the following cases:
   (a) P is a constant.
   (b) P is given by the equation of state of an ideal gas.
   (c) P is obtained from the van der Waals equation of state.
   (d) dV is obtained from the equation of state: $Pv = RT + aP + bP^2$.
   (e) dV is obtained from the Berthelot equation.

2. Rozen [*J. Phys. Chem. (USSR)*, **19**, 469 (1945), and *Chem. Abstracts*, **40**, 1712 (1946)] characterizes gases by "deviation coefficients" such as $T(\partial P/\partial T)_V/P$, $P(\partial V/\partial T)_P/R$, and $P^2(\partial V/\partial P)_T/RT$. Find the values of these coefficients for (a) an ideal gas, (b) a gas which obeys van der Waals'equation, and (c) a gas which obeys the Dieterici equation of state.

3. Find expressions for W, ΔE, Q, and ΔH in an isothermal reversible expansion of a gas which obeys each of the equations of state, respectively, (a) van der Waals and (b) $Pv = RT + aP$.
   In calculating ΔE, make use of the equation $(\partial E/\partial V)_T = T(\partial P/\partial T)_V - P$.

4. A gas obeys the equation of state

$$Pv = RT + aP.$$

   (a)   Find an expression relating T and V in an adiabatic reversible expansion.
   (b)   Find an equation for ΔH in an adiabatic reversible expansion.

---

[6] When derivations or proofs of equations are called for, start from fundamental definitions and principles.

(c) Find an equation for ΔH in an adiabatic free expansion.

5. (a) Given the equation

$$C_p = C_v + \left[V - \left(\frac{\partial H}{\partial P}\right)_T \right]\left(\frac{\partial P}{\partial T}\right)_V,$$

prove that for any substance

$$C_v = C_p \left[1 - \mu \left(\frac{\partial P}{\partial T}\right)_V \right] - V \left(\frac{\partial P}{\partial T}\right)_V.$$

(b) To what expression can this equation be reduced at the inversion temperature?

6. For one mole of an ideal monatomic gas, $C_v = \frac{3}{2}R$. Find the work done in an adiabatic reversible expansion of this gas by integration of equation (6.43), after appropriate substitution from equation (6.52).

7. An ideal gas absorbs 2250 cal of heat when it is expanded isothermally (at 25°C) and reversibly from 1.5 to 10 liters. How many moles of the gas are present?

8. Prove the following relation for an ideal gas:

$$\left(\frac{\partial E}{\partial V}\right)_P = \frac{C_v P}{R}.$$

(c)   Find an equation for ΔT in an adiabatic free expansion.

5.   (a)   Given the equation

CHAPTER 7

The Second Law of Thermodynamics

I. THE NEED FOR A SECOND LAW

For the chemist the primary interest in thermodynamics lies in its ability to establish a criterion of the feasibility of a given chemical or physical transformation under specified conditions. As yet, however, we have given little attention to this objective, primarily because the first law of thermodynamics, and its consequences, do not supply a basis upon which to establish the desired criterion. As we shall see shortly, the functions developed so far do not in themselves form an adequate foundation for chemical applications.

The first law of thermodynamics summarizes many experimental observations in terms of a statement about the function E, the internal energy of a system. Extensive experience has shown with no exceptions that E is a thermodynamic property; that is, its value depends only upon the state of the system. It follows, then, that any path used in going from an initial state A to a final state B will be accompanied by the same change in internal energy. It also follows that the change in internal energy in going from B to A must be equal, but opposite in sign, to that accompanying the forward change from A to B:

$$\Delta E_{AB} \text{ (forward)} = -\Delta E_{BA} \text{ (reverse)}.$$

But the first law makes no statement as to which reaction, the forward or the reverse, is the natural or spontaneous one. There are many spontaneous reactions for which ΔE is negative, that is, in which energy is evolved, for example the crystallization of water at $-10\,°C$:

$$H_2O \ (l, -10°) = H_2O \ (s, -10°); \qquad \Delta E = -1.3 \text{ kcal mole}^{-1}.$$

There are also some reactions for which ΔE is positive and which nevertheless proceed spontaneously, as in the melting of ice at $+10\,°C$:

$$H_2O \ (s, 10°) = H_2O \ (l, 10°); \qquad \Delta E = 1.5 \text{ kcal mole}^{-1}.$$

Clearly, then, ΔE cannot be used as a criterion of spontaneity.

Of the other thermodynamic quantities which have been introduced thus far, the enthalpy, H, is also a thermodynamic property. For a long time it was believed that ΔH could be used as a criterion of spontaneity, since most spontaneous reactions at constant pressure are accompanied by an evolution of heat. Thus both Thomsen and Berthelot believed that a reaction would proceed in a stated direction if $\Delta H < 0$. This rule is a useful approximation, within limits which will be discussed later. As more reactions were investigated, however, and as the precision of experimental technique increased, it became abundantly evident that there are many reactions which proceed spontaneously though accompanied by a positive ΔH. One example of these (in addition to the melting of ice at $10\,^{\circ}\text{C}$) is the following:

$$\text{Ag (s)} + \tfrac{1}{2}\text{Hg}_2\text{Cl}_2 \text{ (s)} \rightarrow \text{AgCl (s)} + \text{Hg (l)}; \quad \Delta H = 1.28 \text{ kcal mole}^{-1}.$$

Thus it is evident that both ΔE and ΔH fail as reliable criteria of the direction in which a reaction may proceed. Apparently the first law of thermodynamics does not contain within it the basis of any criterion of spontaneity. Some further principle is necessary which will summarize in a general form the observed tendency of systems of many different types to change in a given direction.

II. THE NATURE OF THE SECOND LAW

A. Natural tendencies toward equilibrium. Before attempting to express in some general statement the tendency of systems to proceed toward a condition of equilibrium, we shall find it desirable to recognize the many different forms in which this tendency exhibits itself. One of the most obvious examples is the flow of heat from a warm body to a colder body. Also, a clock always tends to run down; in fact, any spring tends to relax. Turning to phenomena in gases, we observe that effusion will always occur into a vacuum. And if a barrier separating two pure gases is removed, spontaneous mixing occurs. Similarly, if aqueous solutions of NaCl and AgNO$_3$ are mixed, AgCl is formed spontaneously. In chemical reactions a multitude of examples may be cited in which the mixing of two substances results in the formation of other substances.

It would be desirable to express this tendency toward a condition of equilibrium in all these systems in some common statement, which in turn should be convertible into a mathematical form capable of serving as a criterion of spontaneity. There are a number of possible statements which fulfill these requirements. They are all equivalent; that is, if one is taken as the basic expression, the others may be derived directly.

B. Statement of the second law as a principle of impotence. The statement which we shall choose as the fundamental expression of the second law of thermodynamics is in a form that resembles the statements of other fundamental principles of physical science. Our statement will be expressed in a form which has been called (Whittaker)[1] a "principle of

---

[1] M. Born, *Experiment and Theory in Physics*, Cambridge University Press, London, 1943.

impotence," that is, an assertion of the impossibility of carrying out a particular process. In physics such "principles of impotence" occur frequently. The impossibility of sending a signal with a speed greater than that of light forms the basis of the theory of relativity. Similarly, wave mechanics may be considered a consequence of the impossibility of measuring simultaneously the position and velocity of an elementary particle. In an analogous fashion, we may state the first law of thermodynamics in terms of man's impotence to construct a machine capable of producing energy, a so-called "perpetual-motion machine."

Therefore, for the *second law of thermodynamics* we shall use the following statement:

> *It is impossible to construct a machine which is able to convey heat by a cyclical process from one reservoir (at a lower temperature) to another at a higher temperature unless work is done on the machine by some outside agency.*   (I)

This expression of the second law is a condensed statement of much experience. The term "cyclical process" implies one in which the system carrying out the transfer of heat has returned to its initial state. Such a transfer may occur even in a cyclical process if the net result is an expenditure of work—for example, in a refrigerator heat from the inner compartment is transferred to the surroundings at a higher temperature, but this process is accompanied by a net input of work. The second law states that an equivalent process *without* an input of work is impossible.

It should be emphasized that the second law is not an a priori principle;[2] that is, it is not a statement which might be deduced from earlier principles, such as the first law. The second law is a statement summarizing the experience of many men over a long period of time. Except perhaps in submicroscopic phenomena, no exceptions to this law have ever been found.[3]

C. Mathematical counterpart of the verbal statement. The second law, in the form in which it has been expressed thus far, is not a statement which can be applied conveniently to chemical problems. The principal objective for most chemists is to use the second law of thermodynamics to establish a criterion by which we can determine whether a chemical reaction, or a phase change, will proceed spontaneously or not. Such a criterion would be available if we could obtain a function which, as a consequence of the second law of thermodynamics, possessed the following two characteristics:

(1) It should be a thermodynamic property; that is, its value should

---

[2] The other principles of impotence described above also are subject to this same statement.

[3] These submicroscopic phenomena must be treated by special methods which are outside the scope of classical thermodynamics.

depend only upon the state of the system and not upon the particular path by which the state has been reached.

(2)   It should change in a characteristic manner (for example, always increase) when a reaction proceeds spontaneously.

A function which satisfies these requirements has been devised and is known by the name "entropy."

III. DEFINITION OF S, THE ENTROPY OF A SYSTEM

The entropy of a system is defined in terms of a differential equation for dS, the infinitesimal change in entropy.   For any infinitesimal portion of a process which is carried out in a reversible manner,

$$dS = \frac{DQ_{rev}}{T},\qquad(7.1)$$

where DQ is the heat absorbed in this infinitesimal reversible portion and T is the absolute temperature at which the heat is absorbed.   It should be emphasized that the entropy has thus been defined in an "operational" manner, that is, in terms of the operations by which it is measured.   The definition specifies that the entropy change can be measured only when DQ is known for a *reversible* change and when the temperature has been specified.   No statement is made about entropy changes in irreversible processes.   However, as a consequence of the second law of thermodynamics it can be shown that dS is an exact differential,[4] or in other words that S is a thermodynamic property and depends only upon the state of a system. Hence the change in entropy in any system going from a specified initial state to a given final state is independent of the path by which the change is carried out.

The proof that the change in entropy is independent of the path will be carried out in four steps.

(1)   It will be proved that dS is an exact differential for an ideal gas carried through a Carnot cycle.

(2)   It will be proved that dS is an exact differential for any substance carried through a Carnot cycle.

(3)   It will be proved that dS is an exact differential for any substance carried through any reversible cycle.

(4)   It will be proved that the entropy is a function only of the state of a system.

IV. PROOF THAT S IS A THERMODYNAMIC PROPERTY

A.   Ideal gas in a Carnot cycle.

1.   *Net work and heat quantities.*   The Carnot cycle is the name for a

---

[4] Since DQ is an inexact differential and $(1/T) \times DQ$ makes the resultant function an exact differential, the factor $1/T$ is an "integration factor."

series of four changes which a substance may undergo. In the forward direction these are: isothermal expansion, adiabatic expansion, isothermal compression, and adiabatic compression, in that order. At the end of this series of changes, the substance must have been returned to its initial state. Such a cycle is represented most frequently by a pressure-volume diagram, as in Fig. 1. It is more convenient for the purposes of the following proof, however, to refer to a temperature-volume diagram, as in Fig. 2. The latter diagram emphasizes more strongly the isothermal nature of the first and third steps in the cycle.

When an ideal gas is carried through a Carnot cycle, the quantities of work done and heat absorbed in each of the steps may be calculated readily from relations previously derived. For an isothermal reversible expansion of one mole of an ideal gas the change in energy is zero, and hence

$$W = Q = RT \ln \frac{V_{\text{final}}}{V_{\text{initial}}} . \tag{7.2}$$

In the adiabatic reversible expansion the quantity of heat absorbed is zero and hence, if C_v is a constant,

$$W = -C_v(T_{\text{final}} - T_{\text{initial}}) = -\Delta E. \tag{7.3}$$

Using these relations, we may tabulate W and Q for the steps specified in the reversible cycle in Fig. 2:

Step	Work Done	Heat Absorbed	
I	$RT_2 \ln (V_2/V_1)$	$RT_2 \ln (V_2/V_1)$	(a)
II	$C_v(T_2 - T_1)$	0	
III	$RT_1 \ln (V_4/V_3)$	$RT_1 \ln (V_4/V_3)$	
IV	$C_v(T_1 - T_2)$	0	

We shall be interested in the net work obtained from the ideal gas in this complete cycle; this quantity may be obtained readily by summation of the works done in the individual steps. Thus

$$W = RT_2 \ln \frac{V_2}{V_1} + C_v(T_2 - T_1) + RT_1 \ln \frac{V_4}{V_3} + C_v(T_1 - T_2). \tag{7.4}$$

The second and fourth terms in equation (7.4) are equal but opposite in sign, so that we may reduce the expression to

$$W = RT_2 \ln \frac{V_2}{V_1} + RT_1 \ln \frac{V_4}{V_3}. \tag{7.5}$$

It can be shown readily that V_3 and V_4 are related in a very definite manner to V_2 and V_1, so that equation (7.5) may be simplified even further. A glance at Fig. 2 shows that V_2 and V_3 are end points of an adiabatic reversible process, Step II. Similarly, V_4 and V_1 are the initial and final volumes, respectively, of the adiabatic reversible compression, Step IV.

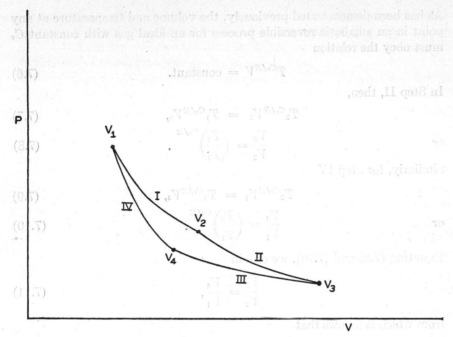

Fig. 1.   Carnot cycle; pressure-volume diagram.

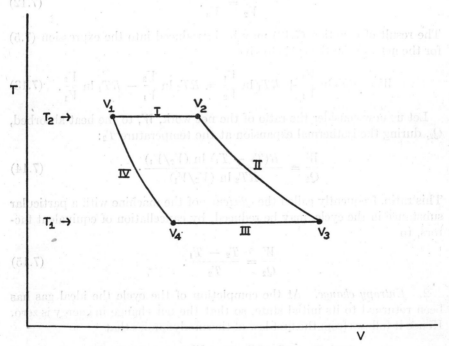

Fig. 2.   Carnot cycle; temperature-volume diagram.

As has been demonstrated previously, the volume and temperature at any point in an adiabatic reversible process for an ideal gas with constant C_v must obey the relation

$$T^{C_v/R}V = \text{constant.} \tag{7.6}$$

In Step II, then,

$$T_2{}^{C_v/R}V_2 = T_1{}^{C_v/R}V_3, \tag{7.7}$$

or

$$\frac{V_3}{V_2} = \left(\frac{T_2}{T_1}\right)^{C_v/R}. \tag{7.8}$$

Similarly, for Step IV

$$T_2{}^{C_v/R}V_1 = T_1{}^{C_v/R}V_4, \tag{7.9}$$

or

$$\frac{V_4}{V_1} = \left(\frac{T_2}{T_1}\right)^{C_v/R}. \tag{7.10}$$

Equating (7.8) and (7.10), we obtain

$$\frac{V_3}{V_2} = \frac{V_4}{V_1}, \tag{7.11}$$

from which it follows that

$$\frac{V_1}{V_2} = \frac{V_4}{V_3}. \tag{7.12}$$

The result of equation (7.12) may be introduced into the expression (7.5) for the net work in the cycle to give

$$W = RT_2 \ln \frac{V_2}{V_1} + RT_1 \ln \frac{V_1}{V_2} = RT_2 \ln \frac{V_2}{V_1} - RT_1 \ln \frac{V_2}{V_1}. \tag{7.13}$$

Let us now consider the ratio of the net work, W, to the heat absorbed, Q_2, during the isothermal expansion at the temperature T_2:

$$\frac{W}{Q_2} = \frac{R(T_2 - T_1) \ln (V_2/V_1)}{RT_2 \ln (V_2/V_1)}. \tag{7.14}$$

This ratio, frequently called the *efficiency* of the machine with a particular substance in the cycle, may be reduced, by cancellation of equivalent factors, to

$$\frac{W}{Q_2} = \frac{T_2 - T_1}{T_2}. \tag{7.15}$$

2. *Entropy change.* At the completion of the cycle the ideal gas has been returned to its initial state, so that the net change in energy is zero. Hence it follows from the first law of thermodynamics that

$$0 = Q_{\text{total}} - W_{\text{total}}. \tag{7.16}$$

W_{total} has been evaluated in terms of equation (7.13). Since Steps II and IV in the cycle are adiabatic, Q_{total} is given by the sum of heats Q_2 and Q_1, absorbed at the two temperatures T_2 and T_1, respectively. Therefore

$$W_{total} = W = Q_{total} = Q_2 + Q_1. \tag{7.17}$$

Substituting the result of (7.17) in the expression (7.15), we obtain

$$\frac{W}{Q_2} = \frac{Q_2 + Q_1}{Q_2} = \frac{T_2 - T_1}{T_2}. \tag{7.18}$$

Rearrangement leads to

$$Q_2 + Q_1 = Q_2 - \frac{T_1}{T_2} Q_2,$$

$$\frac{Q_1}{T_1} = -\frac{Q_2}{T_2},$$

$$\frac{Q_1}{T_1} + \frac{Q_2}{T_2} = 0, \tag{7.19}$$

or

$$\sum \frac{Q}{T} = 0. \tag{7.20}$$

If the ideal gas in the Carnot cycle is considered as being carried through a series of infinitesimally small steps, then the summation of equation (7.20) may be replaced by

$$\int \frac{DQ}{T} = 0. \tag{7.21}$$

Since this process has been carried out reversibly, we may make use of the definition of entropy in equation (7.1), and by indicating explicitly the cyclical nature of the process, we arrive at the following expression for an ideal gas being carried through the reversible Carnot cycle:

$$\oint dS = 0. \tag{7.22}$$

Thus for the conditions specified it has been proved that dS is an exact differential and, consequently, that the entropy, S, is a thermodynamic property.

B. Any substance in a Carnot cycle. At this stage we wish to eliminate any restrictions as to the nature of the substance being carried through the Carnot cycle and thus to show that dS is exact for any material in this reversible process. It is desirable to prepare for this proof by consideration of a few numerical examples of work and heat quantities involved in a Carnot cycle with an ideal gas.

1. *Some numerical examples with ideal gas in Carnot cycle.* Referring to Fig. 2, let us consider a situation in which the temperature of the upper heat reservoir is twice that of the lower, for example, 200°K and 100°K,

respectively. If V_2 is also taken as exactly twice V_1, then it can be shown from equation (7.13) that

$$W = 138 \text{ cal.}$$

It can also be shown readily [see Table (a) on page 80] that the heat absorbed at the upper temperature is

$$Q_2 = 276 \text{ cal.}$$

Since the energy change in a complete cycle is zero,

$$W = Q_2 + Q_1.$$

Therefore $\qquad\qquad Q_1 = -138 \text{ cal};$

that is, 138 cal of heat is *evolved* by the ideal gas during isothermal compression at the lower temperature, T_1. These results may be tabulated conveniently as follows:

$$\left.\begin{array}{rr} Q_2 = & 276 \text{ cal} \\ W = & 138 \text{ cal} \\ Q_1 = & -138 \text{ cal} \end{array}\right\} \qquad (b)$$

A Carnot cycle may be run in reverse also; that is, the series of stages may be in the following order: adiabatic expansion, isothermal expansion, adiabatic compression, isothermal compression. If Fig. 2 is used to represent this reverse cycle, the order of the steps would be IV, III, II, I. By methods analogous to those used in the description of the forward cycle, it can be shown readily that

$$\left.\begin{array}{rcl} W \text{ (reverse)} & = & -W \text{ (forward)} \\ Q_2 \text{ (reverse)} & = & -Q_2 \text{ (forward)} \\ Q_1 \text{ (reverse)} & = & -Q_1 \text{ (forward)} \end{array}\right\} \qquad (c)$$

For the numerical conditions specified above, the reverse Carnot cycle would be accompanied by an absorption of 138 cal of heat at the temperature T_1, 138 cal of work would be done *on* the ideal gas, and 276 cal of heat would be evolved by the gas at the temperature T_2. These changes may be tabulated as follows for comparison with the forward cycle:

$$\left.\begin{array}{rr} Q_2 = & -276 \text{ cal} \\ W = & -138 \text{ cal} \\ Q_1 = & 138 \text{ cal} \end{array}\right\} \qquad (d)$$

It is of interest to note in passing that the Carnot cycle in reverse corresponds to an ideal refrigerating engine, since the result of the cycle is the removal of a quantity of heat from a reservoir at low temperature and the deposition of a quantity of heat into a reservoir at a higher temperature. It should be emphasized, however, that this result does not contradict the second law of thermodynamics, because net work *was done* on the gas during the process.

2. *Principle of the proof.*   We may turn our attention now to the proof that dS is exact for any substance carried through the reversible Carnot

cycle. Reference to the proof for this cycle with an ideal gas shows that the basic relation which must be established is that

$$\text{Efficiency} \equiv \frac{W}{Q_2} = \frac{T_2 - T_1}{T_2}. \tag{7.15}$$

Once the applicability of this relation has been demonstrated, it is a simple matter to show, as has been done above, that $\oint dS = 0$.

The proof that the efficiency for any substance in a Carnot cycle is given by equation (7.15) may be carried out in two steps.

(1) We shall assume first that this other substance (not an ideal gas) gives an efficiency greater than $(T_2 - T_1)/T_2$. Using this assumption, we shall arrive at a contradiction of the second law. Consequently we shall have proved that the efficiency for this other substance cannot be greater than $(T_2 - T_1)/T_2$.

(2) We shall assume that this other substance (not an ideal gas) gives an efficiency less than $(T_2 - T_1)/T_2$. Using this assumption, we shall arrive at a contradiction of the second law. Consequently, we shall have proved that the efficiency for this other substance cannot be less than $(T_2 - T_1)/T_2$.

Since the efficiency for this other substance (not an ideal gas) can be neither greater than nor less than $(T_2 - T_1)/T_2$, it must be equal to $(T_2 - T_1)/T_2$. Equations (7.15) and (7.22) apply, therefore, for any substance in a Carnot cycle.

3. *Proof that efficiency for a real substance in the Carnot cycle cannot exceed that for the ideal gas.* The details of this proof may be understood more readily if a numerical special case is used first. Let us assume that a given Carnot cycle is being operated between reservoirs of which the upper is at a temperature, T_2, which is exactly twice that of the lower, T_1. If $T_2 = 2T_1$, then the efficiency for an ideal gas in this cycle is 0.5. If 100 cal of heat were absorbed at T_2, 50 cal would be converted into work at the completion of the cycle and 50 cal would be evolved to the reservoir at the lower temperature, T_1. If 120 cal were absorbed at T_2, 60 cal would be converted into work and 60 cal would be evolved at T_1. These changes may be tabulated conveniently as follows:

$$\left. \begin{array}{lll} Q_2 = & 100 \text{ cal} & Q_2 = & 120 \text{ cal} \\ W = & 50 \text{ cal} & W = & 60 \text{ cal} \\ Q_1 = & -50 \text{ cal} & Q_1 = & -60 \text{ cal} \end{array} \right\} \tag{e}$$

If the Carnot cycle were run in reverse and 50 cal of heat were absorbed at T_1, 50 cal of work would be done on the ideal gas and 100 cal of heat would be evolved at T_2. The results for the reverse cycles corresponding to the forward cycles tabulated in (e) may be summarized as follows:

$$\left. \begin{array}{lll} Q_2 = & -100 \text{ cal} & Q_2 = & -120 \text{ cal} \\ W = & -50 \text{ cal} & W = & -60 \text{ cal} \\ Q_1 = & 50 \text{ cal} & Q_1 = & 60 \text{ cal} \end{array} \right\} \tag{f}$$

Let us now take the other substance (not an ideal gas) assumed to give an efficiency greater than 0.5 (say 0.6), run it through a Carnot cycle in a forward direction, and use the work obtained from this forward cycle to carry an ideal gas through a reverse Carnot cycle. If 100 cal of heat is absorbed at T_2 in the forward cycle with the other substance, 60 cal of work will be obtained (since the efficiency is 0.6) and 40 cal of heat will be given up to the reservoir at the lower temperature. If the 60 cal of work is now used to operate the reverse Carnot cycle with the ideal gas, then since the efficiency with the ideal gas is 0.5, 60 cal of heat will be absorbed at T_1 and 120 evolved at T_2. These changes together with their net results may be tabulated as follows:

	Other Substance, Efficiency = 0.6 (Forward Cycle)	Ideal Gas, Efficiency = 0.5 (Reverse Cycle)	Net Change for Working Substances	Net Change in Reservoirs
Q at T_2	100 cal	−120 cal	−20 cal	20 cal
W for complete cycle	60 cal	− 60 cal	0 cal	0 cal
Q at T_1	−40 cal	60 cal	20 cal	−20 cal

(g)

At the conclusion of the two cycles both substances have been returned to their initial conditions. Yet the net result for both substances is an evolution of 20 cal of heat at T_2 and an absorption of 20 cal at T_1. The net result for the reservoirs is a loss of 20 cal by the low-temperature source and a gain of 20 cal by the source at high temperature. Since the net work done during these changes is zero, we have transferred 20 cal of heat from a low-temperature reservoir to one at a higher temperature without doing any net work. Since the second law of thermodynamics proclaims the impossibility of such an accomplishment, our initial assumption that the efficiency for the Carnot cycle containing the other substance is 0.6 must be wrong.

Strictly speaking, of course, we have succeeded in proving so far only that the efficiency for the other substance cannot be 0.6 when that for an ideal gas is 0.5. It is apparent, however, that if the efficiency for the other substance were any value greater than 0.5, the net result in (g) would be similar.

The proof may be given also in a generalized form in which the values of Q_2 are not specified numerically. Again we operate the Carnot cycle in a forward direction using the other substance (not an ideal gas). Q_2 cal of heat is absorbed at the higher temperature, T_2, and W cal of work is obtained from the complete cycle. Therefore Q_1 (or since ΔE is zero, $W - Q_2$) cal of heat is absorbed at T_1. (Actually, of course, heat is evolved at T_1, but this merely means that Q_1 is a negative number.) If an ideal gas were operated in a Carnot cycle in the forward direction, with the same two reservoirs at T_2 and T_1 and in a manner such that W is the same as for the other substance, then since the efficiency given by the latter is assumed greater than that given by the ideal gas, that is,

$$\frac{W}{Q_2} > \frac{W}{Q_2'}, \tag{7.23}$$

we may write $\qquad\qquad Q_2' > Q_2, \tag{7.24}$

where Q_2' is the heat absorbed at T_2 by the ideal gas in a forward cycle. If the work, W, obtained from the forward cycle with the other substance is used to operate the ideal gas in a reverse cycle, then Q_2' cal of heat is *evolved* at T_2.

Similarly, in the isothermal step at T_1 the other substance in the forward cycle evolves Q_1 cal; and if the ideal gas were operated in the forward direction to produce the same quantity of work, W, the heat evolved, Q_1', would be greater in absolute value than Q_1.[5] Since the Q's are negative numbers, Q_1' must be algebraically less than Q_1. Thus, remembering that $\Delta E = 0$ for a complete cycle, we can write

$$W = Q_2 + Q_1 = Q_2 + Q_1'; \tag{7.25}$$

and since $\qquad\qquad Q_2' > Q_2, \tag{7.26}$

$$Q_1' < Q_1. \tag{7.27}$$

Keeping these relative values of the heats and works in mind, let us tabulate the results of operating the other substance in a forward direction and using the work obtained to run the ideal gas in a reverse Carnot cycle:

	Other Substance, Efficiency $> \dfrac{T_2 - T_1}{T_2}$ (Forward Cycle)	Ideal Gas, Efficiency $= \dfrac{T_2 - T_1}{T_2}$ (Reverse Cycle)	Net Change for Working Substances	Net Change in Reservoir	
Q at T_2	Q_2	$-Q_2'$	$Q_2 - Q_2' < 0$	$-(Q_2 - Q_2') > 0$	
W for complete cycle	W	$-W$	0	0	(h)
Q at T_1	Q_1	$-Q_1'$	$Q_1 - Q_1' > 0$	$-(Q_1 - Q_1') < 0$	

[5] If this conclusion is not evident, it may be made clear by the following additional details. Since both substances are operating in closed cycles, $\Delta E = 0$. Therefore, $Q_1' = W - Q_2'$ and $Q_1 = W - Q_2$. But W is the same for both substances. On the other hand, $Q_2' > Q_2$. Hence Q_1' must be larger in absolute value than Q_1. However, since Q_1 and Q_1' are both negative in a forward cycle, Q_1' must be algebraically less than Q_1.

Again we observe in this general case, as in the specific numerical example outlined above, that the net result of our double cycle is the transfer of heat in a cyclical process from the low-temperature source at T_1 to the high-temperature reservoir without the expenditure of any work. Since the possibility of such a transfer is contrary to accumulated experience as expressed in the second law, the assumption upon which the transfer was predicated must be incorrect. It follows, then, that no substance carried through a Carnot cycle can give an efficiency greater than $(T_2 - T_1)/T_2$.

4. *Proof that efficiency for a real substance in a Carnot cycle cannot be less than that for the ideal gas.* The second step in our complete proof requires the establishment of the impossibility of the other substance giving an efficiency less than $(T_2 - T_1)/T_2$. Obviously, this step may be demonstrated in a manner analogous to that used in the proof of the first step. We assume that the other substance gives an efficiency less than that given by the ideal gas. Again we operate the more efficient substance (this time the ideal gas) in the forward direction and the less efficient substance (this time the other substance) in the reverse direction. Using a specific numerical example first, let us assume again that $T_2 = 2T_1$ (so that the efficiency for the ideal gas will remain 0.5) and that the other substance gives an efficiency of 0.40. The results of operating with the ideal gas in the forward direction and using the work obtained to carry the other substance through a reverse cycle may be tabulated as follows:

	Ideal Gas, Efficiency = 0.5 (Forward Cycle)	Other Substance, Efficiency = 0.4 (Reverse Cycle)	Net Change for Working Substances	Net Change in Reservoirs	
Q at T_2	80 cal	−100 cal	−20 cal	+20 cal	
W for complete cycle	40 cal	− 40 cal	0 cal	0 cal	(i)
Q at T_1	−40 cal	+ 60 cal	+20 cal	−20 cal	

Clearly the second law has been violated.

We may state the proof of the second step in general terms as follows. We assume that the ideal gas gives a greater efficiency than the other substance, so that if both are operated upon in a forward direction with the same reservoirs at T_2 and T_1 and give the same net work, W, the following relation must be valid (where the prime again refers to the ideal gas):

$$\frac{W}{Q_2'} > \frac{W}{Q_2}. \tag{7.28}$$

Therefore
$$Q_2 > Q_2'. \tag{7.29}$$

In a complete cycle $\Delta E = 0$, and hence, from the first law of thermodynamics,

$$W = Q_2 + Q_1 = Q_2' + Q_1'. \tag{7.30}$$

Since
$$Q_2 > Q_2', \tag{7.31}$$

$$Q_1 < Q_1'. \tag{7.32}$$

Thus Q_1 is algebraically less than Q_1'. (Since these heats are negative numbers in the forward cycles, Q_1 must be greater in absolute magnitude than Q_1'.) The results of operating with the ideal gas in a forward cycle and using the work, W, obtained to carry the other substance through in reverse may be summarized as follows:

	Ideal Gas, Efficiency $= \dfrac{T_2 - T_1}{T_2}$ (Forward Cycle)	Other Substance, Efficiency $< \dfrac{T_2 - T_1}{T_2}$ (Reverse Cycle)	Net Change for Working Substances	Net Change in Reservoirs	
Q at T_2	Q_2'	$-Q_2$	$Q_2' - Q_2 < 0$	$-(Q_2' - Q_2) > 0$	
W for complete cycle	W	$-W$	0	0	(j)
Q at T_1	Q_1'	$-Q_1$	$Q_1' - Q_1 > 0$	$-(Q_1' - Q_1) < 0$	

A glance at the last column in Table (j) will show that the second law of thermodynamics has been contradicted. Clearly our postulate that the other substance gives an efficiency less than that given by the ideal gas is incorrect.

Since the efficiency of the machine containing this other substance (not an ideal gas) can be neither greater than nor less than $(T_2 - T_1)/T_2$, it follows that for any substance carried through a Carnot cycle,

$$\frac{W}{Q_2} = \frac{T_2 - T_1}{T_2}. \tag{7.15}$$

Therefore
$$\oint dS = 0. \tag{7.22}$$

C. Any substance in any reversible cycle. Having removed the restrictions on the nature of the substance, we must proceed to remove any specifications as to the nature of the reversible cycle through which the substance is being carried. Let us represent such a cycle by the example illustrated in Fig. 3A, which may be approximated by Carnot cycles, as illustrated in Fig. 3B, if we proceed along the path $abcdefgha$. It can be shown readily, by the following procedure, that

$$\oint dS = 0 \tag{7.22}$$

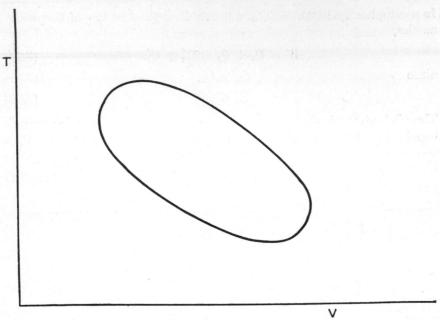

Fig. 3A.   A reversible cycle.

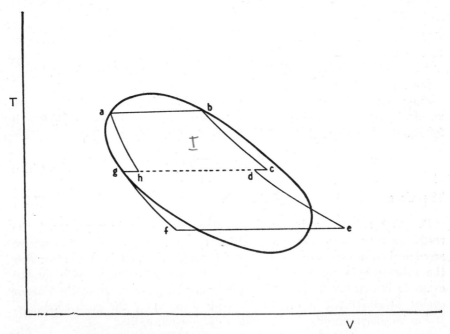

Fig. 3B.   A reversible cycle approximated by two Carnot cycles.

around this path, *abcdefgha*.

It has been proved already that equation (7.22) applies to any Carnot cycle. If we break up the integration over Carnot cycle 1 (upper) in Fig. 3B into two portions, one covering the solid line and the other the dotted section from d to h, we may write

$$\int_1 dS + \int_{\underset{\cdots}{1}} dS = 0, \tag{7.33}$$

where $\underline{1}$ represents paths *abcd* and *ha*, and $\underset{\cdots}{1}$ represents path *dh*. Similarly, for the second Carnot cycle in Fig. 3B,

$$\int_2 dS + \int_{\underset{\cdots}{2}} dS = 0, \tag{7.34}$$

where $\underline{2}$ represents the path *defgh* and $\underset{\cdots}{2}$ represents *hd*. From the principles of calculus it follows that

$$\int_{\underset{\cdots}{1}} dS = - \int_{\underset{\cdots}{2}} dS, \tag{7.35}$$

since the integrals are over the same path but in opposite directions. Using equations (7.35) and (7.34), respectively, to substitute for $\underset{\cdots}{1}$ in equation (7.33) and then for $\underset{\cdots}{2}$, we arrive at the expression

$$\int_{\underline{1}} dS + \int_{\underline{2}} dS = \oint_{\substack{\text{path} \\ abcdefgha}} dS = 0. \tag{7.36}$$

A better approximation to the actual path in Fig. 3A would be made by three Carnot cycles. Again it could be shown that

$$\int_{\underline{1}} dS + \int_{\underline{2}} dS + \int_{\underline{3}} dS = 0, \tag{7.37}$$

since the overlapping portions between each pair of cycles would cancel. At the limit of an infinite number of Carnot cycles, the approximation to the real cycle becomes perfect, and hence it follows that for any reversible path

$$\oint dS = 0. \tag{7.22}$$

D. The entropy, S, depends only upon the state of the system. In going from a state a to a state b, as is shown in Fig. 4, we may proceed by any one of an infinite number of paths. Two *reversible* paths, *acb* and *adb*, are illustrated in Fig. 4. Despite the great difference between the shapes of these two paths, the change in entropy must be the same for each, because in the complete cycle, *acbda*, the change in entropy is zero:

$$\int_{\substack{\text{path} \\ acb}} dS + \int_{\substack{\text{path} \\ bda}} dS = 0. \tag{7.38}$$

Hence

$$\int_{\substack{\text{path} \\ acb}} dS = - \int_{\substack{\text{path} \\ bda}} dS = \int_{\substack{\text{path} \\ adb}} dS. \tag{7.39}$$

The right-hand equality sign in equation (7.39) is valid because the path is the same but with the limits interchanged.   As a result of (7.39) it is obvious that all reversible paths from a to b are accompanied by the same change in entropy.   Since $S_b - S_a$ is independent of the reversible path used to calculate it, the entropies, S_b and S_a, must be functions only of the states of the systems b and a, respectively.

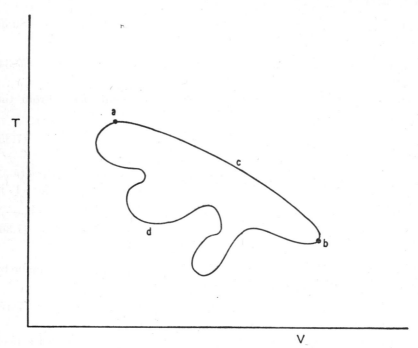

Fig. 4.   Two reversible paths from one point to another.

V. ALTERNATIVE STATEMENT OF THE SECOND LAW OF THERMODYNAMICS

There are a number of possible formulations of the second law in addition to the fundamental verbal statement which we have adopted.   In general, any one of these alternatives may be assumed as the basic expression of long experience and all of the others may be proved from it.   We shall continue to use the original statement we adopted [(I), page 78] and shall prove any others from it.

A very useful enunciation of the second law is the following:

> *It is impossible to take heat from a reservoir at constant temperature and convert it into work without accompanying changes*   (II)
> *in the reservoir or its surroundings.*

In other words, it is impossible to carry out a cyclical process in which heat from a reservoir at a fixed temperature has been converted into work.

The proof of this statement is straightforward. Once again we assume that it *is* possible to carry out the process which the new statement denies. As a result of this assumption we shall contradict our original statement of the second law. Since this contradiction violates all of our observations, the assumption must be incorrect.

If we could carry out a cyclical process in which heat from a reservoir at a constant temperature, T_1, would be converted into work, it still follows from the first law of thermodynamics that the net change in energy, ΔE, must be zero, since the system has returned to its initial state. Nevertheless, since a certain quantity of heat has been converted into work, W, we may use the work obtained to operate a reversible Carnot cycle in the reverse direction. The net result of this Carnot cycle in reverse (a refrigerating cycle) would be the transfer of a quantity of heat from a reservoir at the lower temperature, T_1, to one at a higher temperature, T_2, by a cyclical process. The net work done in the isothermal process and reverse Carnot cycle combined is zero. We have stipulated that the isothermal process is cyclical, and, of course, so is the Carnot cycle. Our net result, the transfer of heat from T_1 to T_2, therefore violates the original statement of the second law. Hence the new assumption that heat may be converted into work in an isothermal cyclical process is incorrect. Consequently statement (II) of the second law of thermodynamics stands proved.

VI. ENTROPY CHANGES IN REVERSIBLE PROCESSES

Having established the exactness of the entropy differential, let us turn our attention to some considerations of the value of this entropy function in specific situations.

A. Steps in the Carnot cycle. The four steps of the Carnot cycle fall into one or the other of two categories: (1) isothermal changes and (2) adiabatic changes. For each category the calculation of the entropy change is obvious from the definition of dS in equation (7.1).

For isothermal reversible changes it is clear that the entropy change for the working substance is given by

$$\Delta S_{\text{work subs}} = \int dS = \int \frac{DQ}{T} = \frac{1}{T} \int DQ = \frac{Q}{T}. \qquad (7.40)$$

For the specific case of the expansion of one mole of an ideal gas, since $\Delta E = 0$,

$$Q = W = RT \ln \frac{V'}{V}, \qquad (7.41)$$

where V' is the final volume and V the initial volume. Hence

$$\Delta S_{gas} = \frac{Q}{T} = R \ln \frac{V'}{V}. \tag{7.42}$$

It is pertinent to point out that in any isothermal stage of a Carnot cycle if Q is the heat absorbed by the working substance, then $-Q$ must be the heat released by the surroundings. Therefore

$$\Delta S_{surr} = -\frac{Q}{T}. \tag{7.43}$$

Hence for the system as a whole

$$\Delta S_{sys} = \Delta S_{work\ subs} + \Delta S_{surr} = 0. \tag{7.44}$$

In the adiabatic steps of the Carnot cycle there is no heat exchange. It is thus immediately obvious that

$$\Delta S_{work\ subs} = \Delta S_{surr} = \Delta S_{sys} = 0. \tag{7.45}$$

B. Phase transitions. A change from one phase to another, for example, from ice to water, may be carried out <u>reversibly</u> and at a <u>constant temperature</u>. Clearly, under these conditions equation (7.40) is applicable. Equilibrium phase transitions are also generally carried out at a <u>fixed pressure</u>. Since no work is expended in these transitions, except against the atmosphere, Q is given by the heat of transition, and hence

$$\Delta S_{subs} = \frac{\Delta H_{trans}}{T}. \tag{7.46}$$

In these isothermal, reversible phase transitions, as in the isothermal steps of a Carnot cycle, for every infinitesimal quantity of heat absorbed by the substance an equal quantity of heat must have been released by the surroundings. Consequently,

$$\Delta S_{surr} = -\Delta S_{subs}, \tag{7.47}$$

and hence again the entropy change for the system as a whole must be zero.

As a specific example of a calculation of ΔS for a phase transition we may consider the data on the fusion of ice at $0\,^{\circ}\text{C}$:

$$H_2O\ (s,\ 0\,^{\circ}\text{C}) = H_2O\ (l,\ 0\,^{\circ}\text{C}) \quad (\Delta H = 1436\ \text{cal mole}^{-1}),$$

$$\Delta S_{water} = \frac{1436}{273.16} = 5.257\ \text{cal mole}^{-1}\ \text{deg}^{-1},$$

$$\Delta S_{surr} = -5.257\ \text{cal mole}^{-1}\ \text{deg}^{-1},$$

$$\Delta S_{sys} = 0.$$

C. Isobaric temperature rise. In many cases it is necessary to calculate the change in entropy accompanying the rise in temperature of a substance. Such a temperature increase can be carried out in a reversible manner so that one may substitute an expression for the heat absorbed in the process into the equation for the calculation of the entropy change. Since the process is carried out at constant pressure,

$$\Delta S_{\text{subs}} = \int \frac{DQ}{T} = \int \frac{dH}{T} = \int \frac{C_p \, dT}{T} = \int C_p \, d \ln T. \qquad (7.48)$$

Should C_p be constant,

$$\Delta S = C_p \ln \frac{T_2}{T_1}, \qquad (7.49)$$

where T_2 is the final temperature and T_1 is the initial temperature.

Once again let us emphasize that the entropy change in the surroundings is equal but opposite in sign to that for the substance and consequently that ΔS for the system as a whole is zero.

D. Isochoric temperature rise. The entropy changes for a temperature rise at a constant volume are analogous to those at constant pressure, except that C_v replaces C_p. Thus since $P \, dV = 0$, $\quad E = Q = C_v dT$

$$\Delta S_{\text{subs}} = \int \frac{DQ}{T} = \int \frac{dE}{T} = \int \frac{C_v \, dT}{T} = \int C_v \, d \ln T. \quad (7.50)$$

Here, too, the entropy change for the system as a whole is zero.

E. General statement. From the examples cited, it is evident that whenever a substance absorbs a quantity of heat, DQ, the surroundings lose an equal quantity of heat. Thus

$$DQ_{\text{subs}} = -DQ_{\text{surr}}. \qquad (7.51)$$

It follows that in a reversible process

$$\frac{DQ_{\text{subs}}}{T} + \frac{DQ_{\text{surr}}}{T} = 0. \qquad (7.52)$$

Hence it is clear that

$$\int dS = 0$$

(1) for the *entire system* (substance + surroundings) undergoing *any reversible process*, and (2) for the *working substance* undergoing a *reversible cyclic process*.

VII. ENTROPY CHANGES IN IRREVERSIBLE PROCESSES

Thus far calculations of entropy changes have been described only for reversible paths. To determine the change of entropy in an irreversible process it is necessary to discover a reversible change between the same initial and final states. Since S is a property of a system, ΔS is the same for the irreversible as for the reversible process. Thus the entropy change must be calculated for the reversible path.

A. Isothermal expansion of an ideal gas. It has been shown already that in the reversible isothermal expansion of an ideal gas

$$\Delta S_{gas} = R \ln \frac{V'}{V}. \tag{7.42}$$

From the fact that S is a thermodynamic property, it follows that ΔS_{gas} is the same in an irreversible isothermal process from the same initial volume V to the same final volume, V'. The change in entropy of the surroundings, however, differs in the two types of processes. First let us consider an extreme case, an isothermal expansion into a vacuum with no work being done. Since the process is isothermal, ΔE for the perfect gas must be zero, and consequently the heat absorbed by the gas, Q, is also zero:

$$Q = \Delta E + W = 0. \tag{7.53}$$

Thus the surroundings have given up no heat and, in fact, have undergone no change in state. Consequently

$$\Delta S_{surr} = 0, \tag{7.54}$$

$$\Delta S_{sys} = R \ln \frac{V'}{V} + 0 > 0. \tag{7.55}$$

In other words, for the system as a whole this irreversible expansion has been accompanied by an increase in entropy.

In any actual isothermal expansion the work done by the gas is not zero but is less than that obtained by reversible means.[6] Since ΔE is still zero for an ideal gas and since

$$W_{irrev, \, gas} < RT \ln \frac{V'}{V}, \tag{7.56}$$

it follows that $\qquad Q_{irrev, \, gas} < RT \ln \frac{V'}{V}. \tag{7.57}$

Nevertheless *the entropy change for the gas is still given by equation (7.42), since it is equal to that for the reversible process between the same end points.*

Turning our attention to the surroundings, we can have the actual isothermal expansion occur with the gas in a vessel immersed in a large quan-

---

[6] It will be shown in Chapter 8 (pages 117–119) that in an isothermal process the reversible work is the maximum work.

tity of an ice-water mixture at equilibrium at constant pressure and temperature.[7] The heat lost by the surroundings must be numerically equal but opposite in sign to that gained by the gas:

$$Q_{surr} = - Q_{irrev, \, gas}. \tag{7.58}$$

For the ice-water mixture at constant pressure and temperature, however, the change in entropy depends only upon the quantity of heat evolved:

$$\Delta S_{surr} = \frac{Q_{surr}}{T}, \tag{7.59}$$

since the change in state of the ice-water mixture during this process is exactly the same as that which would occur if the same quantity of heat had been given up to the gas during a reversible isothermal expansion. In view of the conditions expressed by equations (7.57)–(7.59), it follows that Q_{surr}/T is smaller in absolute magnitude than is $R \ln (V'/V)$ and hence that the change in entropy for the system as a whole must be positive.

$$\Delta S_{sys} = \Delta S_{gas} + \Delta S_{surr} = R \ln \frac{V'}{V} + \frac{Q_{surr}}{T}$$

$$= R \ln \frac{V'}{V} - \frac{Q_{irrev, \, gas}}{T} > 0. \tag{7.60}$$

B. Irreversible adiabatic expansion of an ideal gas. Strictly speaking, an irreversible process cannot be represented by a path on a diagram such as Fig. 5, because in such a process not all parts of the substance have the same properties at any given instant of time. As has been pointed out previously, in an expansion, for example, there is a lag in the transmission of stresses, so that the pressure and temperature immediately adjacent to the piston may differ considerably from those in the body of the gas. Nevertheless, the initial state, a, and the final state, b, may be represented on the diagram.

To determine the entropy change in this irreversible adiabatic process, it is necessary to consider a reversible path from a to b. An infinite number of reversible paths are possible, and one is illustrated by the dotted lines in Fig. 5. This path consists of two steps: (1) an isothermal reversible expansion at the temperature T_a until the volume V' is reached; (2) an adiabatic reversible expansion from V' to V_b. The entropy change for the gas is given by the sum of the entropy changes for the two steps:

$$\Delta S_{gas} = R \ln \frac{V'}{V_a} + 0. \tag{7.61}$$

---

[7] This specification of an ice-water mixture does not in any way limit the isothermal temperature to 0°C. By a suitable change in pressure, the equilibrium temperature may be varied over a wide range. If this range is inadequate, other systems consisting of two phases in equilibrium may be used in place of ice and water.

Since $V' > V_a$, the entropy change for the gas is clearly positive for the reversible path and therefore also for the irreversible change.

The surroundings, on the other hand, undergo no change in state during the *irreversible* expansion, since the process is adiabatic. Hence for the surroundings

$$\Delta S_{surr} = 0. \tag{7.62}$$

Consequently for the system as a whole

$$\Delta S_{sys} = \Delta S_{gas} + \Delta S_{surr} = R \ln \frac{V'}{V} > 0. \tag{7.63}$$

Thus once again we have found that an irreversible process is accompanied by a net increase in entropy.

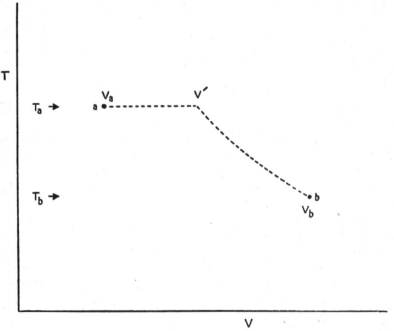

Fig. 5. Irreversible change from State a to State b. The dashed line represents one possible reversible path through the same end points.

C. Flow of heat from a higher to a lower temperature. For convenience in visualizing this process, imagine the flow of heat, by means of a conductor, from a very large reservoir at the higher temperature T_2 to a very large reservoir at a lower temperature T_1. By introducing large reservoirs into the mental picture, we may consider the heat sources to be at constant temperature, despite the gain or loss of a small quantity of heat, Q.

To calculate the change in entropy in this irreversible flow, it is necessary to consider a corresponding reversible process. There are an infinite number of ways in which this process may be carried out reversibly. One of them would be to allow an ideal gas to absorb reversibly the quantity of heat, Q, at the temperature T_2. The gas may then be expanded adiabatically and reversibly (therefore with no change in entropy) until it reaches the temperature T_1. At T_1 the gas is compressed reversibly and evolves the quantity of heat Q. During this reversible process the reservoir at T_2 loses heat and undergoes the entropy change

$$\Delta S_{\substack{\text{hot} \\ \text{reservoir}}} = - \frac{Q}{T_2}. \tag{7.64}$$

Since the same change in state occurs in the irreversible process, ΔS for the hot reservoir is still given by equation (7.64). During the reversible process the reservoir T_1 absorbs heat and undergoes the entropy change

$$\Delta S_{\substack{\text{cold} \\ \text{reservoir}}} = \frac{Q}{T_1}. \tag{7.65}$$

Since the same change in state occurs in the irreversible process, ΔS for the cold reservoir is still given by equation (7.65). In the irreversible process the two reservoirs are the only substances which undergo any changes. Since $T_2 > T_1$, the entropy change for the system as a whole is positive:

$$\Delta S_{\text{sys}} = \Delta S_{\substack{\text{hot} \\ \text{reservoir}}} + \Delta S_{\substack{\text{cold} \\ \text{reservoir}}} = - \frac{Q}{T_2} + \frac{Q}{T_1} > 0. \tag{7.66}$$

Again we see that an irreversible process is accompanied by an increase in entropy of the system as a whole.

D. Phase transition. A convenient illustration of an irreversible phase transition is the crystallization of water at $-10°C$ and constant pressure:

$$H_2O \ (l, -10°C) = H_2O \ (s, -10°C). \tag{7.67}$$

Here, too, to calculate the entropy changes it is necessary to consider a series of reversible steps leading from the liquid water at $-10°C$ to solid ice at $-10°C$. One such series might be the following: (1) heat supercooled water at $-10°C$ very slowly (reversibly) to $0°C$; (2) convert the water at $0°C$ very slowly (reversibly) to ice at $0°C$; (3) cool the ice very slowly (reversibly) from $0°C$ to $-10°C$. Since each of these steps is reversible, the entropy changes may be calculated readily by the methods discussed earlier. Since S is a thermodynamic property, the sum of these entropy changes is equal to ΔS for the process indicated by (7.67). The necessary calculations are summarized by the following table, where T_2 represents $0\,^{\circ}C$ and T_1 represents $-10\,^{\circ}C$.

H_2O (l, $-10\,^{\circ}C$) = H_2O (l, $0\,^{\circ}C$);

$$\Delta S_1 = \int \frac{DQ}{T} = \int \frac{C_p\,dT}{T} = C_p \ln \frac{T_2}{T_1} = 0.671 \text{ cal mole}^{-1} \text{ deg}^{-1}$$

H_2O (l, $0\,^{\circ}C$) = H_2O (s, $0\,^{\circ}C$);

$$\Delta S_2 = \int \frac{DQ}{T} = \frac{\Delta H}{T} = \frac{-1436}{273.16} = -5.257 \text{ cal mole}^{-1} \text{ deg}^{-1}$$

H_2O (s, $0\,^{\circ}C$) = H_2O (s, $-10\,^{\circ}C$);

$$\Delta S_3 = \int \frac{DQ}{T} = \int \frac{C_p'\,dT}{T} = C_p' \ln \frac{T_1}{T_2} = -0.324 \text{ cal mole}^{-1} \text{ deg}^{-1}.$$

Adding:
H_2O (l, $-10\,^{\circ}C$) = H_2O (s, $-10\,^{\circ}C$);

$$\Delta S_{H_2O} = \Delta S_1 + \Delta S_2 + \Delta S_3 = -4.910 \text{ cal mole}^{-1} \text{ deg}^{-1}.$$

It should be noted that there has been a decrease in the entropy of the water (that is, ΔS is negative) upon crystallization at $-10\,^{\circ}C$ despite the fact that the process is irreversible. This example emphasizes the fact that the sign of the entropy change *for the entire system*, and not merely for any component, is related to irreversibility. To obtain ΔS for the system, we must consider the entropy change in the surroundings, since the process described by equation (7.67) occurs irreversibly. If we consider the water as being in a large reservoir at a temperature of $-10\,^{\circ}C$, then the crystallization will evolve a certain quantity of heat, Q, which will be absorbed by the reservoir without a significant rise in temperature. The change in state of the reservoir thus occurs essentially reversibly, and hence ΔS is given by

$$\Delta S_{\text{reservoir}} = \int \frac{DQ}{T} = -\frac{\Delta H}{T} = -\frac{(-1343)}{263.16} = 5.103 \text{ cal deg}^{-1}, \qquad (7.68)$$

where ΔH represents the heat of crystallization of water at $-10\,^{\circ}C$. Clearly for the system as a whole the entropy has increased:

$$\Delta S_{\text{sys}} = \Delta S_{H_2O} + \Delta S_{\text{reservoir}} = -4.910 + 5.103 = 0.193 \text{ cal deg}^{-1}. \tag{7.69}$$

E. **General statement.** From the examples cited, it is evident that irreversible processes are accompanied by an increase in entropy for the system as a whole. This increase for isolated systems can be shown to occur generally by the following considerations.

Take any irreversible process in which an isolated system goes from State a to State b, as shown in Fig. 6. Since the process is irreversible, we can indicate only the end points (and not the path) on the diagram. Nevertheless, let us complete the cycle by going back from b to a by a series of reversible steps indicated by the dotted lines in Fig. 6. The adiabatic path bc is followed to some temperature T_c which may be higher or lower than

T_a. In Fig. 6, T_c is indicated as being above T_a, but in some specific case it may be lower. The only requirement in fixing T_c is that it be a temperature at which an isothermal reversible process can be carried out from State c to State d, where State d is chosen in such a way that a reversible adiabatic change will return the system to its initial state, a. By means of these three reversible steps, the system has been returned from State b to State a. The first and third steps in this reversible process being adiabatic, no changes in entropy are obtained. Consequently the entropy of State c must be the same as that of State b, namely, S_b. Similarly, the entropy

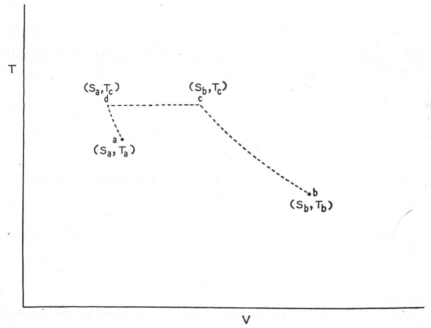

Fig. 6. Schematic diagram of general irreversible change. The dashed line represents one possible reversible path through the same end points.

of State d must be S_a. An entropy change does occur, however, along the path cd. Since this is an isothermal reversible process,

$$S_d - S_c = S_a - S_b = \frac{Q}{T_c}. \tag{7.70}$$

This entropy change, $S_a - S_b$, must be a negative number; that is, $Q < 0$, since if Q were a positive number ($Q > 0$), we should have violated the second law of thermodynamics as stated in the alternative form (II). Since in the complete cycle (irreversible process from a to b followed by the three reversible steps) ΔE must be zero, hence $Q = W$. If $Q > 0$, then $W > 0$; that is, work would have been done by the system. In other words, we should have succeeded in carrying out a cyclical process in which heat at a constant temperature, T_c, has been converted into work.   Since such a process cannot be carried out, Q cannot be a positive number.   Since Q must be a negative number,[8] $S_a - S_b$ must also be negative;   that is,

$$S_a - S_b < 0. \tag{7.71}$$

$S_a - S_b$ is the change in entropy in going from b to a.   Since entropy is a thermodynamic property, the change in entropy in going from a to b must fulfill the following restriction:

$$(S_b - S_a) = -(S_a - S_b). \tag{7.72}$$

Therefore $\qquad\qquad\qquad S_b - S_a > 0. \tag{7.73}$

Since in the irreversible process from a to b no heat was exchanged with the surroundings, the latter undergo no change in state and hence no change in entropy.   Therefore

$$\Delta S_{\text{sys}} = (S_b - S_a) > 0. \tag{7.74}$$

Thus we see that there must be an increase in entropy in an isolated system undergoing an irreversible transformation.

VIII.   GENERAL EQUATIONS FOR THE ENTROPY OF GASES

A.   Entropy of the ideal gas.   Using the mathematical statements of the two laws of thermodynamics, we can obtain an explicit equation for the entropy of an ideal gas.   In carrying out its derivation, we shall assume that the transformations which the gas undergoes are of a reversible nature.   Once we have obtained our final result, however, the equation will be applicable to irreversible processes also, since the entropy is a function only of the state of the system.

In a system where only pressure-volume work is possible, the first law may be stated as

$$dE = DQ - P\, dV. \tag{7.75}$$

Since we are considering a reversible transformation,

$$dS = \frac{DQ}{T}. \tag{7.76}$$

But $\qquad\qquad\qquad DQ = P\, dV + dE, \tag{7.77}$

and thus $\qquad\qquad\qquad dS = \frac{dE}{T} + \frac{P\, dV}{T}. \tag{7.78}$

---

[8] It may be observed that if Q is not greater than zero, then Q may still equal zero. However, if $Q = 0$, then it is clear that all the reversible steps in Fig. 6 must be adiabatic, and hence ΔS for going from b to a must be zero.   But if $S_a = S_b$, then the initial step from a to b must also be reversible adiabatic and hence cannot be irreversible.

For an ideal gas the internal energy, E, is a function of the temperature only. Consequently

$$dE = \left(\frac{\partial E}{\partial T}\right)_V dT = C_v\, dT. \tag{7.79}$$

Using (7.79) as a replacement for dE and substituting for P/T in (7.78) by means of the equation of state of an ideal gas, we obtain

$$dS = \frac{C_v\, dT}{T} + \frac{R\, dV}{V}. \tag{7.80}$$

If C_v is constant this latter expression may be integrated readily to give

$$S = C_v \ln T + R \ln V + S_0, \tag{7.81}$$

where S_0 is an integration constant characteristic of the gas. This integration constant cannot be evaluated explicitly by classical thermodynamic methods. It can be evaluated, however, with the aid of kinetic-molecular theory and statistical methods. For a monatomic gas, S_0 was formulated explicitly originally by Tetrode[9] and by Sackur.[10]

B. Entropy of a real gas. The procedure used to derive a general equation for the entropy of a real gas is analogous to that for an ideal gas. In the preceding section the discussion up to equation (7.78) is general and hence not restricted to ideal gases. This equation, therefore, may be used as the starting point for the consideration of a real gas. For nonideal gases we may no longer make the substitution of equation (7.79) or of the ideal equation of state. However, a suitable substitution may be made for the total differential, dE, in (7.78) if we use the partial derivatives obtained on considering the internal energy, E, as a function of V and T:

$$dE = \left(\frac{\partial E}{\partial T}\right)_V dT + \left(\frac{\partial E}{\partial V}\right)_T dV. \tag{7.82}$$

Thus
$$dS = \frac{1}{T}\left(\frac{\partial E}{\partial T}\right)_V dT + \frac{1}{T}\left(\frac{\partial E}{\partial V}\right)_T dV + \frac{P}{T}\, dV. \tag{7.83}$$

If we recognize that the entropy, S, may also be considered as a function of V and T, we may obtain a second equation for the total differential, dS:

$$dS = \left(\frac{\partial S}{\partial T}\right)_V dT + \left(\frac{\partial S}{\partial V}\right)_T dV. \tag{7.84}$$

---

[9] H. Tetrode, *Ann. Physik*, IV, **38**, 434; **39**, 255 (1912).
[10] O. Sackur, *Ann. Physik*, IV, **40**, 67 (1913).

A comparison of the coefficients of the dT terms in equations (7.83) and (7.84) leads to the following equality:

$$\left(\frac{\partial S}{\partial T}\right)_V = \frac{1}{T}\left(\frac{\partial E}{\partial T}\right)_V = \frac{1}{T}C_v. \tag{7.85}$$

It can be shown also, most readily by the procedure to be outlined in Chapter 8 (page 115), that the following relation is true:

$$\left(\frac{\partial S}{\partial V}\right)_T = \left(\frac{\partial P}{\partial T}\right)_V. \tag{7.86[11]}$$

Thus it follows that

$$dS = \frac{C_v}{T}\,dT + \left(\frac{\partial P}{\partial T}\right)_V dV. \tag{7.87}$$

Therefore $\quad S = \int C_v\,d\ln T + \int \left(\frac{\partial P}{\partial T}\right)_V dV + \text{constant}. \tag{7.88}$

To integrate this expression, it is necessary to know the equation of state of the gas and the dependence of C_v on temperature.

If a gas obeys the van der Waals equation of state, it can be shown readily that

$$S = \int C_v\,d\ln T + R\ln(v - b) + \text{constant}. \tag{7.89}$$

IX. TEMPERATURE SCALES

In addition to establishing the entropy function and its properties, the second law of thermodynamics permits us to set up a temperature scale without reference to the properties of an ideal gas, but rather to the properties of an ideal engine.   Nevertheless, the two scales turn out to be identical.   To show this identity we shall consider the gas scale first in some detail.

A. Ideal gas scale.   The ideal gas scale defines temperature in terms of measurements of the volume of an ideal gas at a fixed pressure;   that is, the ratio of two temperatures is determined by the ratio of the volumes of the gas at a fixed pressure:

$$\frac{T_2}{T_1} = \frac{V_2}{V_1}. \tag{7.90}$$

To complete the definition, it is necessary to choose the size of the unit on

---

[11] This equation is printed in type of smaller size to emphasize that it has not yet been proved.

the temperature scale. Conventionally this size is established by the equation[12]

$$T_{\text{b.p.}} - T_{\text{f.p.}} = 100, \tag{7.91}$$

where b.p. refers to the boiling point and f.p. to the freezing point of pure water at one atmosphere pressure and in the presence of air.

B. Thermodynamic temperature scale. Tentatively we shall allow thermodynamic temperature to be represented by the symbol θ. The ratio of two temperatures, θ_2 and θ_1, is given by the ratio of the absolute values of the heats absorbed in a Carnot cycle working between two reservoirs at the temperatures θ_2 and θ_1, respectively; that is,

$$\frac{\theta_2}{\theta_1} = \frac{|Q_2|}{|Q_1|}. \tag{7.92}$$

Here too we may choose any size for a unit difference on the scale, but conventionally the unit is established in terms of the freezing and boiling points of water:

$$\theta_{\text{b.p.}} - \theta_{\text{f.p.}} = 100. \tag{7.93}$$

The thermodynamic scale thus eliminates reference to the properties of a nonexistent ideal substance. On the other hand, it is defined in terms of the characteristics of a nonexistent perfect engine.

C. Equivalence of the thermodynamic and ideal gas scales. We have shown previously that when an ideal gas is carried through a Carnot cycle, the heats absorbed at the hot and cold reservoirs are related to each other by the expression

$$\frac{Q_2}{T_2} + \frac{Q_1}{T_1} = 0. \tag{7.94}$$

This relation may be rearranged readily to give

$$\frac{T_2}{T_1} = \frac{|Q_2|}{|Q_1|}. \tag{7.95}$$

Clearly, then,

$$\frac{\theta_2}{\theta_1} = \frac{T_2}{T_1}; \tag{7.96}$$

or, in other words, θ is proportional to T:

$$\theta = kT. \tag{7.97}$$

---

[12] At recent meetings of the International Unions of Chemistry and of Physics, the principle has been adopted that the absolute scale of temperature should have only one fixed point besides 0°K. This point should be the triple point of water. There is still some disagreement, however, about the second decimal place for this temperature. It will be some time, therefore, before the necessary modifications can take effect. See *Nature*, **163**, 427 (1949); C. Darwin, *Nature*, **164**, 262 (1949).

If we allow θ_2 in equation (7.96) to be the temperature of the boiling point of water, $\theta_{b.p.}$, and θ_1 to be $\theta_{f.p.}$, then it can be shown readily, by substitution of (7.96) into the expression

$$\theta_{b.p.} - \theta_{f.p.} = 100 = T_{b.p.} - T_{f.p.}, \qquad (7.98)$$

that

$$\theta_{f.p.} = T_{f.p.}. \qquad (7.99)$$

This equality establishes the value of the proportionality constant, k, in equation (7.97) as 1, so that

$$\theta = T. \qquad (7.100)$$

In other words, the thermodynamic temperature scale is identical with that defined by means of a thermometer containing an ideal gas.

D. Other temperature scales. On the thermodynamic temperature scale it is clear that there is no finite upper temperature limit. It should be emphasized also that although a low-temperature limit of $0\,°K$ is specified on the thermodynamic scale, this limit is just as unattainable as is the hot one.

The apparent closeness of approach to absolute zero on the thermodynamic scale is a property of this scale. This characteristic may be brought out more strikingly by consideration of a third temperature scale which is defined in the following manner. The temperature difference, $\alpha_2 - \alpha_1$, shall be one degree if

$$\frac{Q_2 + Q_1}{Q_2} = \frac{1}{2}, \qquad (7.101)$$

where Q_2 and Q_1 are the heats absorbed in a Carnot cycle operating between a hot reservoir at α_2 and a cold one at α_1. Since equation (7.101) defines only differences in temperature, we may establish the melting point of ice as a fixed point on the scale by setting it equal to the zero point. Substitution of a few numerical values will show that this new temperature scale is related to the ideal gas scale by the equation

$$T = (273.16) \times 2^\alpha. \qquad (7.102)$$

We may now compare a few representative temperatures on the two scales, which are listed in Table 1. On the new logarithmic scale both the

TABLE 1

LOGARITHMIC AND IDEAL GAS TEMPERATURE SCALES

°K	°C	α
∞	∞	∞
819	546	1.59
373	100	0.45
273	0	0.00
136.5	-136.5	-1.00
73	-200	-1.91
0	-273.16	$-\infty$

hot and cold ends tend toward infinity. Although this new scale is inconvenient for most practical purposes, it does have the pedagogical advantage of illustrating the similarity of the problems of obtaining very high and very low temperatures.

E. Value of absolute zero. Since the thermodynamic temperature scale is identical with the ideal gas scale, absolute zero may be determined numerically by extrapolation at constant pressure of a volume-(centigrade) temperature curve for an ideal gas to zero volume. Real determinations, however, depend upon the use of the thermodynamic definition and the equations of the Joule-Thomson process.[13] On the basis of such measurements a best value has been set as follows:[14]

$$0.000°C = +273.16 ± 0.01°K. \tag{7.103}$$

Frequently, however, we shall use the old value of 273.1°K, since many critical tables are based on this standard.

X. TEMPERATURE-ENTROPY DIAGRAM

In making diagrams of various reversible cycles, it is a common practice to plot pressure as a function of volume, because the area under the curve, $\int P \, dV$, gives the work done in any step. We have used, instead, temperature and volume as coordinates, because a diagram on this basis emphasizes the constancy of temperature in an isothermal process. However, it suffers the disadvantage that the area is not related to the work. Gibbs[15] pointed out that a diagram using temperature and entropy as coordinates is a particularly useful one, since it illustrates graphically not only the work involved in a reversible cycle but also the heats. In addition, this type of diagram emphasizes the isentropic nature of an adiabatic reversible process as well as the constancy of temperature in isothermal stages. A typical diagram—for a simple Carnot cycle—is illustrated in Fig. 7. The four stages in a forward cycle are labeled by the Roman numerals. In Step I the temperature is constant, heat Q_2 is absorbed by the working substance, and the entropy increases from S_1 to S_2. Since this stage is reversible and isothermal,

$$\frac{Q_2}{T_2} = \Delta S = S_2 - S_1 \tag{7.104}$$

and $$Q_2 = T_2(S_2 - S_1) = \text{area under line I.} \tag{7.105}$$

---

[13] P. S. Epstein, *Textbook of Thermodynamics,* John Wiley and Sons, Inc., New York, 1937, pages 74–76.

[14] Chapter 2, Table 1.

[15] *The Collected Works of J. Willard Gibbs,* Vol. I, Longmans, Green and Company, New York, 1931, page 9.

In Step II there is a drop in temperature in the adiabatic reversible expansion, but no change in entropy. The isentropic nature of II is emphasized by the vertical line. In Step III we have an isothermal reversible compression with a heat numerically equal to Q_1 being evolved. Since this step is reversible and isothermal,

$$\frac{Q_1}{T_1} = \Delta S = S_1 - S_2 = -(S_2 - S_1) \qquad (7.106)$$

and $\qquad Q_1 = -T_1(S_2 - S_1) = $ negative of area under line III. (7.107)

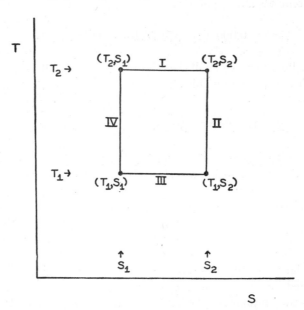

Fig. 7. Gibbs temperature-entropy diagram for a Carnot cycle.

In the fourth step, which is adiabatic and reversible, there is no entropy change, but the temperature rises to the initial value, T_2. Since the cycle has been completed,

$$\Delta E = 0 \qquad (7.108)$$

and $\qquad\qquad\qquad Q_2 + Q_1 = W. \qquad (7.109)$

Therefore $\quad W = T_2(S_2 - S_1) - T_1(S_2 - S_1)$

$$= (T_2 - T_1)(S_2 - S_1) = \text{area enclosed by cycle.} \qquad (7.110)$$

Thus the work and heats involved in the cycle are clearly illustrated by a

T-S diagram, and the nature of the isothermal and isentropic steps is strongly emphasized.

Exercises

1. A mole of an ideal monatomic gas ($C_v = \frac{3}{2}R$) at a pressure of 1 atm and a temperature of 273.1 °K is to be transformed to a pressure of 0.5 atm and a temperature of 546.2 °K. The change may be brought about in an infinite variety of ways. Consider the following four reversible paths, each consisting of two parts:

(1) Isothermal expansion and isobaric temperature rise.
(2) Isothermal expansion and isochoric temperature rise.
(3) Adiabatic expansion and isobaric temperature rise.
(4) Adiabatic expansion and isochoric temperature rise.

(a) Determine P, V, and T of the gas after the initial step of each of the four paths. Represent the paths on a T-V diagram. To facilitate plotting, some of the necessary data for the adiabatic expansions are presented in the following table. Supply the additional data required for the completion of the adiabatic curve.

V (liters)	P (atmosphere)	T (degrees K)
22.41	1.0000	273.1
44.80	0.5000	...
44.82	0.3150	172.0
67.23	0.1603	131.3
89.65	....	...

(b) Calculate for each portion of each path and for each complete path the following:

W, the work done by the gas
Q, the heat absorbed by the gas
ΔE of the gas
ΔH of the gas
ΔS of the gas.

Assemble your results in tabular form.

(c) Note which functions in (b) have values which are independent of the path used in the transformation.

2. An ideal gas is carried through a Carnot cycle. Draw diagrams of this cycle using each of the following sets of coordinates:

(a) P, V.   (d) E, S.
(b) T, P.   (e) S, V.
(c) T, S.   (f) T, H.

3. (a) By a procedure analogous to that used to obtain equation (7.85), show that

$$\left(\frac{\partial S}{\partial V}\right)_T = \frac{P + (\partial E/\partial V)_T}{T}. \tag{7.111}$$

(b) Knowing that dS is an exact differential, show that

$$\left(\frac{\partial E}{\partial V}\right)_T = T\left(\frac{\partial P}{\partial T}\right)_V - P. \tag{7.112}$$

(c) Derive the expression

$$\left(\frac{\partial E}{\partial V}\right)_P = C_v \left(\frac{\partial T}{\partial V}\right)_P + T\left(\frac{\partial P}{\partial T}\right)_V - P. \tag{7.113}$$

4. A gas obeys the equation of state

$$Pv = RT + AP,$$

where A is a constant at all temperatures.

(a) Show that the internal energy, E, is a function of the temperature only.

(b) Find $(\partial E/\partial V)_P$. Compare with the value obtained for this same partial derivative for an ideal gas.

(c) Derive an equation for the entropy of this gas which is analogous to the Sackur-Tetrode equation for the ideal gas.

5. Complete the steps missing between equations (7.98) and (7.99).

Optional 6. Show that the efficiency of a Carnot cycle in which each step is carried out irreversibly cannot be greater than that of a reversible Carnot cycle.

CHAPTER 8

The Free-Energy Functions

I. PURPOSE OF THE NEW FUNCTIONS

In Chapter 7 we formulated the second basic postulate of thermodynamics and derived from it a useful thermodynamic function, the entropy. In examining the properties of this new thermodynamic function we found that it has the very desirable characteristic of being a criterion of the spontaneity of a chemical or physical transformation. When the change in entropy of the substances involved in a transformation *and of their surroundings* is positive, the reaction is irreversible. A reaction with positive ΔS may (or may not) occur spontaneously; a reaction with negative ΔS (total) will never occur spontaneously.

Thus in principle we have attained our primary objective in chemical thermodynamics: the establishment of a criterion of spontaneity. Nevertheless, for practical purposes the criterion in its present form is generally an inconvenient one, since it requires a knowledge of the properties of the surroundings, as well as of the substances which are of primary interest. A modification of this criterion which does not deal directly with the surroundings would be much more useful. After several trials a suitable modification was proposed which introduced certain new thermodynamic quantities known as the free-energy functions. In addition to acting as criteria of spontaneity, these free-energy functions have been found to have the very useful properties of predicting the maximum yields obtainable in equilibrium reactions and the maximum work which may be obtained from a particular transformation.

II. DEFINITIONS

Two free-energy functions are in common use. The *Helmholtz free energy*, A, is defined by the relation

$$A = E - TS. \tag{8.1}$$

The *Gibbs free energy*, F, is defined by the expression

$$F = H - TS. \tag{8.2}$$

For most chemical problems the Gibbs free energy is the more useful, for reasons which will be evident shortly. Because of the much greater use of the Gibbs free-energy function, the term "free energy" without any name prefixed is frequently used to designate F, and will be so used in this text.

III. CONSEQUENCES OF DEFINITIONS AND DERIVED RELATIONS

A. Free energy a thermodynamic property. Since both F and A are defined by an explicit equation, in terms of variables which depend only upon the state of a system, it is evident that both of these new functions are thermodynamic properties. Hence their differentials are exact and we may write the expressions

$$\oint dF = 0, \tag{8.3}$$

$$\oint dA = 0. \tag{8.4}$$

B. Relation between F and A. From the definitions of these functions [equations (8.1) and (8.2)] it is clear that

$$F = H - TS = E + PV - TS = E - TS + PV, \tag{8.5}$$

or

$$F = A + PV. \tag{8.6}$$

C. Free-energy functions for isothermal conditions. Transformations at constant temperature are of very frequent interest in chemical problems. For finite changes at a fixed temperature $T \ (= T_2 = T_1)$, equation (8.2) becomes

$$
\begin{aligned}
\Delta F &= F_2 - F_1 = (H_2 - T_2 S_2) - (H_1 - T_1 S_1) \\
&= H_2 - H_1 - (T_2 S_2 - T_1 S_1) \\
&= H_2 - H_1 - T(S_2 - S_1) \\
&= \Delta H - T\,\Delta S.
\end{aligned} \tag{8.7}
$$

Similarly, for an infinitesimal change under isothermal conditions $(dT = 0)$

$$
\begin{aligned}
dF &= dH - T\,dS - S\,dT \\
&= dH - T\,dS.
\end{aligned} \tag{8.8}
$$

The equations for A can be derived in an analogous fashion:

$$\Delta A = \Delta E - T\,\Delta S, \tag{8.9}$$

$$dA = dE - T\,dS. \tag{8.10}$$

D. Equations for total differentials. Since the procedure is the same for both free-energy functions, we shall consider in detail the derivation only for the Gibbs free energy, F. Starting with the definition given by equation (8.2),

$$F = H - TS, \tag{8.2}$$

we may differentiate to obtain

$$dF = dH - T\,dS - S\,dT. \tag{8.11}$$

Since the enthalpy has been defined by the expression

$$H = E + PV, \tag{8.12}$$

it follows that dH is given by the relation

$$dH = dE + P\,dV + V\,dP. \tag{8.13}$$

Hence
$$dF = dE + P\,dV + V\,dP - T\,dS - S\,dT. \tag{8.14}$$

Considering a substance or a system which is doing no work except mechanical (that is, $P\,dV$ work), we may substitute for dE from the first law of thermodynamics,

$$dE = DQ - P\,dV, \tag{8.15}$$

and obtain
$$dF = DQ + V\,dP - T\,dS - S\,dT. \tag{8.16}$$

When the change in F is due to some reversible transformation, then the definition of entropy,

$$dS = \frac{DQ}{T}, \tag{8.17}$$

may be rearranged to

$$DQ = T\,dS, \tag{8.18}$$

which, upon substitution into equation (8.16), leads to the desired equation for dF:

$$dF = V\,dP - S\,dT. \tag{8.19}$$

By an entirely analogous procedure it is possible to show that the total differential of the Helmholtz free energy is given by the expression

Prove
$$dA = -P\,dV - S\,dT. \tag{8.20}$$

E. Pressure and temperature coefficients of the free energy. If we consider equation (8.19), it is obvious that if the temperature is maintained constant, the second term on the right-hand side of the expression disappears and the equation may be rearranged to

$$\left(\frac{\partial F}{\partial P}\right)_T = V. \tag{8.21}$$

Thus we find that the pressure coefficient of the free energy of a substance depends directly upon the volume of the substance.

More often we shall be interested in the change in ΔF of a chemical reaction, rather than merely in F, with variation in pressure. The required relation may be derived readily from equation (8.21). To be definite, let us represent the chemical transformation by the simple equation

$$A + B = C + D. \tag{8.22}$$

For each one of these substances we may write an equation corresponding to (8.21):

$$\left(\frac{\partial F_A}{\partial P}\right)_T = V_A, \tag{8.23}$$

$$\left(\frac{\partial F_B}{\partial P}\right)_T = V_B, \tag{8.24}$$

$$\left(\frac{\partial F_C}{\partial P}\right)_T = V_C, \tag{8.25}$$

$$\left(\frac{\partial F_D}{\partial P}\right)_T = V_D. \tag{8.26}$$

Subtracting the sum of the pressure coefficients for the reactants from that for the products, we obtain the desired relation:

$$\left(\frac{\partial F_C}{\partial P}\right)_T + \left(\frac{\partial F_D}{\partial P}\right)_T - \left(\frac{\partial F_A}{\partial P}\right)_T - \left(\frac{\partial F_B}{\partial P}\right)_T = V_C + V_D - V_A - V_B, \tag{8.27}$$

or

$$\left(\frac{\partial \Delta F}{\partial P}\right)_T = \Delta V. \tag{8.28}$$

Turning now to the temperature coefficient of the free energy, we may proceed in a manner analogous to that used to derive the equation for the derivative with respect to pressure. It is clear from equation (8.19) that if the pressure is maintained constant, the first term on the right-hand side drops out, and the expression may be rearranged to read

$$\left(\frac{\partial F}{\partial T}\right)_P = -S. \tag{8.29}$$

For a chemical transformation accompanied by a change in free energy, ΔF, it can be shown by a procedure similar to that outlined by equations (8.22) to (8.28) that

$$\left(\frac{\partial \Delta F}{\partial T}\right)_P = -\Delta S. \tag{8.30}$$

Since the derivations of the volume and temperature coefficients of the Helmholtz free energy are so similar to those just illustrated for the Gibbs free energy, only the final results will be recorded:

$$\left(\frac{\partial A}{\partial V}\right)_T = -P, \tag{8.31}$$

$$\left(\frac{\partial \Delta A}{\partial V}\right)_T = -\Delta P, \tag{8.32}$$

Prove $$\left(\frac{\partial A}{\partial T}\right)_V = -S,$$ (8.33)

Prove $$\left(\frac{\partial \Delta A}{\partial T}\right)_V = -\Delta S.$$ (8.34)

F. Equations derived from the reciprocity relation. Since the free energy, $F(T,P)$, is a thermodynamic property, that is, dF is an exact differential, the reciprocity relation (see Chapter 2) must hold. Thus we may write

$$dF = \left(\frac{\partial F}{\partial P}\right)_T dP + \left(\frac{\partial F}{\partial T}\right)_P dT$$ (8.35)

and

$$\frac{\partial}{\partial T}\frac{\partial F}{\partial P} = \frac{\partial}{\partial P}\frac{\partial F}{\partial T}.$$ (8.36)

But from previous considerations we have obtained

$$dF = V\,dP - S\,dT.$$ (8.19)

A comparison of equations (8.19) and (8.35) shows that

$$\left(\frac{\partial F}{\partial P}\right)_T = V$$ (8.21)

and

$$\left(\frac{\partial F}{\partial T}\right)_P = -S.$$ (8.29)

Hence

$$\frac{\partial}{\partial T}\frac{\partial F}{\partial P} = \left(\frac{\partial V}{\partial T}\right)_P = \frac{\partial}{\partial P}\frac{\partial F}{\partial T} = -\left(\frac{\partial S}{\partial P}\right)_T,$$ (8.37)

or

$$\left(\frac{\partial S}{\partial P}\right)_T = -\left(\frac{\partial V}{\partial T}\right)_P.$$ (8.38)

By a similar set of operations on the A function, we can show that

Prove. $$\left(\frac{\partial S}{\partial V}\right)_T = \left(\frac{\partial P}{\partial T}\right)_V.$$ Maxwellian Function (8.39)

IV. CRITERIA OF NATURE OF CHEMICAL CHANGE

Having considered the definitions of the free-energy functions and having derived some useful equations expressing their properties, we must proceed now on our primary course of relating the free-energy change to the nature of the physical or chemical transformation. We wish to find first a relation between ΔF and the spontaneity or lack of spontaneity of a transformation.

A. Criteria of equilibrium.

1. *At constant pressure and temperature.* Let us consider a system that is at equilibrium at a fixed temperature and is subject to no external

forces except the constant pressure of the environment. If the system is in equilibrium, it is reversible; that is, an infinitesimal stress can force it to go in one direction or the other. For a reversible change in a system in which only mechanical work may be done, we have shown previously that

$$dF = V\,dP - S\,dT. \tag{8.19}$$

If P and T are maintained constant, as in the system we are considering now, it is evident that

$$dF_{P,T} = 0. \tag{8.40}$$

We may derive an expression for large-scale changes in F, corresponding to equation (8.40) for infinitesimal changes. In a system at equilibrium (and therefore reversible) at fixed pressure and temperature and with only mechanical work possible,

$$\begin{aligned} \Delta F_{P,T} &= \Delta H - \Delta(TS) = \Delta E + P\,\Delta V - T\,\Delta S \\ &= Q - P\,\Delta V + P\,\Delta V - T\,\Delta S \\ &= T\,\Delta S - P\,\Delta V + P\,\Delta V - T\,\Delta S, \end{aligned}$$

or $\qquad \Delta F_{P,T} = 0. \tag{8.41}$

In the derivation of equation (8.41) we have assumed for simplicity that the pressure and temperature are maintained constant throughout the macroscopic change. Actually it is merely necessary that the pressure and temperature at the beginning of the change be the same as that at the end; for ΔF depends only upon the initial and final states of a system. Hence if ΔF is zero for the isothermal, isobaric transformation, it is also zero for any other path which has the same end points.

Thus we have arrived at a very simple criterion of physical or chemical equilibrium. For a system at equilibrium and subject to no external forces except the constant pressure of the environment, the change in free energy is zero for any transformation in which the initial temperature and pressure are the same as the respective final values.

Perhaps it should also be pointed out that in this new criterion expressed by equation (8.41), no mention is made of the surroundings. Thus we have been able to overcome the disadvantage of the entropy criterion. In reality, what we have done is to substitute for the entropy criterion, which is conveniently applicable to a system of fixed volume and internal energy, a new criterion which is applicable under conditions more commonly characteristic of chemical (or physical) transformations—fixed pressure and temperature.

It is of interest to consider also the change in the Helmholtz free energy, A, under the conditions just specified. Since the temperature is constant and the process is a reversible one,

$$dA_{P,T} = dE - T\,dS = DQ - P\,dV - T\,dS = T\,dS - P\,dV - T\,dS,$$

or
$$dA_{P,T} = -P \, dV \tag{8.42}$$

and
$$\Delta A_{P,T} = -P \, \Delta V. \tag{8.42a}$$

The quantity $P \, dV$ is the work necessary to push back the atmosphere during the infinitesimal transformation of the system at equilibrium. Clearly, then, $\Delta A_{P,T}$ is the negative of this work quantity for a macroscopic change, namely, $-P \, \Delta V$.

2. *At constant volume and temperature.* Another set of conditions frequently encountered in chemical transformations is that of fixed volume and constant temperature. For these conditions it turns out that the A function is the most suitable criterion of equilibrium. From previous considerations we have shown that in a reversible transformation with only mechanical work possible,

Prove. $\quad dA = -P \, dV - S \, dT. \tag{8.20}$

Obviously, if volume and temperature are to remain fixed,

$$dA_{V,T} = 0. \tag{8.43}$$

Similarly for a macroscopic transformation, by a procedure similar to that used in equation (8.41), we can show that

$$\Delta A_{V,T} = \Delta E - \Delta(TS) = Q - T \, \Delta S = T \, \Delta S - T \, \Delta S,$$

or
$$\Delta A_{V,T} = 0. \tag{8.44}$$

B. **Criteria of spontaneity.** Having established criteria of equilibrium, we may proceed to a consideration of the nature of ΔF and ΔA when transformations can occur spontaneously. Our considerations will depend upon the validity of the following statement: *In an isothermal process the maximum work is the reversible work.* We must proceed first, therefore, to prove this statement.

Our proof will depend upon a procedure used frequently in Chapter 7 for entropy. We shall consider two possible methods of going from State a to State b (Fig. 1) in an isothermal fashion: (1) a reversible transformation and (2) an irreversible transformation. For each path the first law of thermodynamics must be valid.

$$\Delta E_{rev} = Q_{rev} - W_{rev}, \tag{8.45}$$

$$\Delta E_{irrev} = Q_{irrev} - W_{irrev}. \tag{8.46}$$

Since E is a function of the state only and since both transformations have the same starting and end points,

$$\Delta E_{rev} = \Delta E_{irrev}, \tag{8.47}$$

or
$$Q_{rev} - W_{rev} = Q_{irrev} - W_{irrev}. \tag{8.48}$$

Let us *assume* now that the irreversible process gives more work than the reversible one. In that case

$$W_{irrev} > W_{rev}. \tag{8.49}$$

Hence[1]
$$Q_{irrev} > Q_{rev}. \tag{8.50}$$

Let us use the irreversible process to carry the system from State a to State b and the reversible process to return the system to its initial state. As in previous proofs, we may construct a table for the various steps:

	Irreversible Process (Forward)	Reversible Process (Backward)	Net for Both Processes
Heat absorbed.......	Q_{irrev}	$-Q_{rev}$	$Q_{irrev} - Q_{rev} > 0$
Work done..........	W_{irrev}	$-W_{rev}$	$W_{irrev} - W_{rev} > 0$

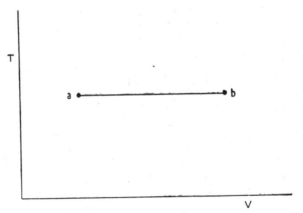

Fig. 1.   An isothermal process.

The net result, as we can see from the table, is that a positive amount of heat has been absorbed and a positive amount of work has been done in an isothermal cycle. However, such a consequence is in contradiction to our alternative statement of the second law of thermodynamics [Chapter 7 (page 92)], which denies the possibility of converting heat from a reservoir at constant temperature into work without some accompanying changes in the reservoir or its surroundings. In our postulated cyclical process no such accompanying changes have occurred. Hence our process cannot oc-

---

[1] If this consequence does not seem obvious when expressed in these general terms, it is suggested that a few numerical values be chosen for W_{irrev}, W_{rev}, and Q_{rev} [consistent with equation (8.49)] and that values of Q_{irrev} then be calculated from equation (8.48). The validity of equation (8.50) will soon become apparent.

cur, and consequently the isothermal irreversible work cannot be greater than the reversible work. We may conclude then that[2]

$$W_{\substack{irrev \\ isothermal}} < W_{\substack{rev. \\ isothermal}}$$ (8.51)

1. *Constant pressure and temperature.* Again we consider a chemical (or physical) transformation occurring at a fixed pressure and temperature, but this time a transformation which may occur spontaneously, though under conditions such that the only restraint is that of the constant pressure of the environment. In other words, no net useful work, such as electrical work, is being done. Such conditions are the ones of primary interest to the chemist, since generally he desires information on the feasibility of a given reaction if it is carried out in an apparatus open to the atmosphere (or at some fixed pressure).

A process which may occur spontaneously is clearly not reversible. However, if the same initial and final states are involved,

$$\Delta E_{irrev} = \Delta E_{rev},$$ (8.52)

where ΔE_{rev} refers to the change in internal energy which would be obtained if the transformation were carried out reversibly.

To be specific, let us consider the formation of one mole of aqueous HCl from the gaseous elements H_2 and Cl_2:

$$\tfrac{1}{2}H_2 \text{ (g)} + \tfrac{1}{2}Cl_2 \text{ (g)} = HCl \text{ (aq)}.$$ (8.53)

This reaction may be carried out in a spontaneous fashion if a container of H_2 and Cl_2 gas, at a fixed total pressure of, for example, one atmosphere, and a specified temperature, is exposed to the appropriate photochemical stimulus. The HCl gas formed could then be dissolved spontaneously in water. Such a process is not reversible. On the other hand, reaction (8.53) may be obtained also *reversibly* by the construction of a cell containing H_2 and Cl_2 electrodes, as illustrated in Fig. 2. This cell may then be connected to a potentiometer. If the electromotive force of this cell is opposed by the e.m.f. of the potentiometer-cell, maintained at an infinitesimally lower value than that of the H_2-Cl_2 cell, then the conversion to HCl can be carried out reversibly, though it would take an infinitely long time to obtain one mole. In either the reversible or the spontaneous method of carrying out reaction (8.53), the energy change is the same if the states of the initial and of the final substances are the same in both methods.

From the first law of thermodynamics, which is applicable to either reversible or irreversible changes, it follows that

$$DQ_{irrev} - DW_{irrev} = DQ_{rev} - DW_{rev}.$$ (8.54)

---

[2] The equality sign is ruled out, since in any isothermal process for which $W = W_{rev}$, it follows (since $\Delta E = \Delta E_{rev}$) that $Q = Q_{rev}$. Hence the entropy change for the substance must be (Q_{rev}/T), and that for the surroundings $(-Q/T)$. For the system as a whole, then, $\Delta S = 0$, which means that we must be dealing with a reversible change. Thus whenever $W = W_{rev}$, the process in which the work W was done must be reversible.

We have just shown that for an isothermal process, such as that which we are considering here, the reversible work is the maximum work.   Thus

$$DW_{irrev} < DW_{rev}. \qquad (8.55)$$

Since the equality expressed by (8.54) must still be valid, the following relation must also be true:[3]

$$DQ_{irrev} < DQ_{rev}. \qquad (8.56)$$

For the reversible change,

$$DQ_{rev} = T\,dS. \qquad (8.57)$$

For the irreversible change,

$$DQ_{irrev} = dH_{irrev}, \qquad (8.58)[4]$$

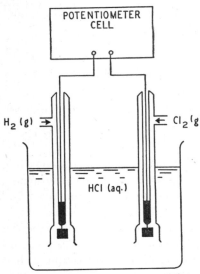

POTENTIOMETER
CELL

H_2 (g)

Cl_2 (g)

HCl (aq.)

Fig. 2.   Formation of aqueous HCl in a reversible manner.

since the pressure is constant.   Furthermore, since H is a thermodynamic property, depending only upon the initial and final states of a system,

$$dH_{irrev} = dH_{rev} = dH. \qquad (8.59)$$

Therefore $dQ_{irrev} = dH.$ (8.60)

Substituting the relations given by equations (8.57) and (8.60) into (8.56), we obtain

$$dH < T\,dS \qquad (8.61)$$

or $dH - T\,dS < 0,$ (8.62)

$$dF_{P,T} < 0. \qquad (8.63)$$

Thus for a spontaneous process occurring at constant pressure and temperature and subject only to the restraint of environmental pressure, there must be a decrease in free energy, that is, a negative free-energy change.   We thus have the complementary relation to equation (8.40), which is the criterion for a process at equilibrium.

---

[3] If this consequence does not seem obvious when expressed in these general terms, it is suggested that a few numerical values be chosen for DW_{irrev}, DW_{rev}, and DQ_{rev} [consistent with equation (8.55)] and that values of DQ_{irrev} then be calculated from equation (8.54).   The validity of equation (8.56) will soon become apparent.

[4] It is pertinent to point out that

$$DQ_{rev} \neq dH_{rev},$$

even though the pressure is constant, because if this transformation (such as the H_2-Cl_2 conversion to HCl) is carried out reversibly, electrical (or other) work is obtained in addition to the $P\,dV$ work.   The original proof of the equality of DQ and dH at constant pressure required that no work be done except that given by the $P\,dV$ term.   See Section **VI-D** of this chapter.

It can be shown also, by a process analogous to that used in equations (8.52) to (8.63), that for a macroscopic change in a spontaneous process subject only to environmental pressure,

$$\Delta F_{P,T} < 0. \tag{8.64}$$

2. *Constant volume and temperature.* The other set of conditions of particular interest to the chemist is that of fixed volume and constant temperature. Again we recognize the validity of equations (8.52) and (8.54) to (8.57) for the constant volume as well as for the constant pressure process. However, since we are now dealing with a process at constant volume, equation (8.58) is no longer applicable. On the other hand, at constant volume and where no nonmechanical work is produced,

$$DQ_{irrev} = dE_{irrev}. \tag{8.65}[5]$$

In view of equation (8.52), the following relation follows from (8.65):

$$DQ_{irrev} = dE_{rev} = dE. \tag{8.66}$$

Using equations (8.57) and (8.66) to substitute into (8.56), we obtain

$$dE < T\,dS, \tag{8.67}$$

or
$$dE - T\,dS < 0, \tag{8.68}$$

$$dA_{V,T} < 0. \tag{8.69}$$

By a similar procedure we obtain the following expression for a macroscopic spontaneous change at constant volume and temperature:

$$\Delta A_{V,T} < 0. \tag{8.70}$$

Thus we have succeeded in establishing criteria of spontaneity applicable to both sets of conditions of particular interest in chemical problems.

C. Heat of reaction as an approximate criterion of spontaneity. For many years it was thought, purely on an empirical basis, that if the enthalpy change for a given reaction was negative, that is, if heat were evolved at constant pressure, the transformation could occur spontaneously. This rule was verified for many reactions. Nevertheless, there are numerous exceptions, one of which was cited in the preceding chapter.

It is of interest to see under what conditions the enthalpy change might be a reliable criterion of spontaneity. Since we are generally interested in isothermal processes, we may refer to equation (8.7) for the free-energy change at constant temperature. It is evident from this expression that if $T\,\Delta S$ is small with respect to ΔH, there will be little difference between ΔH and ΔF. Usually $T\,\Delta S$ is of the order of magnitude of a few thousand

---

[5] The student is also reminded that

$$DQ_{rev} \neq dE_{rev}$$

because nonmechanical work is produced when this process is carried out reversibly.

calories. Evidently, then, if ΔH is sufficiently large, perhaps above 10 kcal, the sign of ΔH will be the same as that of ΔF. For such relatively large values of the heat of a reaction (at constant pressure), ΔH is a reliable criterion of spontaneity, since if ΔH were negative, ΔF, the fundamental criterion, would probably be negative also. It should be recognized, however, that ΔH is not the fundamental criterion, and judgments based on its sign may frequently be misleading, particularly when the magnitudes involved are small.

V. FREE ENERGY AND THE EQUILIBRIUM CONSTANT

In the preceding sections we have established the properties of the free-energy functions as criteria of equilibrium and spontaneity of transformations. From the sign of ΔF it is thus possible to predict whether a given chemical transformation can proceed spontaneously. Further considerations show that ΔF is capable of giving even more information. In particular, from the value of the free-energy change under certain standard conditions it is possible to calculate the equilibrium constant for a given reaction. First, however, we must define our standard conditions.

A. Definitions. It is evident that ΔF for a given reaction depends upon the state of the reactants and products. If we are to tabulate values of the change in free energy for chemical transformations, it will be necessary to agree on certain standard conditions to which this ΔF shall refer, since it would be of little use to have free-energy tables in which no two values referred to substances under the same conditions. The standard states which have been agreed upon by convention are assembled in Table 1.

TABLE 1

STANDARD STATES FOR FREE ENERGIES OF REACTION

State	Standard State
Solid.................	Pure solid in most stable form at one atmosphere pressure and the specified temperature
Liquid................	Pure liquid in most stable form at one atmosphere pressure and the specified temperature
Gas..................	Pure gas at unit fugacity;* for ideal gas fugacity is unity when the pressure is one atmosphere

* The term "fugacity" has yet to be defined. Nevertheless, it is used in this table because of reference to it in future problems. For the present the standard state of a gas may be considered to be that of an ideal gas—one atmosphere pressure.

In addition to standard states we shall speak of the *standard free energy of formation*, $\Delta Ff°$, of a substance. By this we shall mean *the change in free energy accompanying the formation of a substance in its standard state from its elements in their standard states, all of the substances being at the specified temperature.* For example, the standard free energy of formation of CO_2 refers to the reaction

$$C \text{ (graphite, 1 atm)} + O_2 \text{ (g, 1 atm)} = CO_2 \text{ (g, 1 atm)}; \quad \Delta F = \Delta Ff°.$$
$$(8.71)$$

It is a consequence of our definition that $\Delta Ff°$ for any element is zero.

We may also speak of the *standard free-energy change, $\Delta F°$,* for *any* reaction. Obviously we are referring to *the change in free energy accompanying the conversion of reactants in their standard states to products in their standard states.* It can be shown by simple additivity rules that since the free energy is a thermodynamic property and hence does not depend upon the path used to carry out a chemical transformation,

$$\Delta F° = \sum \Delta Ff° \text{ (products)} - \sum \Delta Ff° \text{ (reactants)}. \quad (8.72)$$

B. Relation between $\Delta F°$ and the equilibrium constant. Having established the requisite definition of $\Delta Ff°$, we may proceed to derive the important expression relating the standard free energy of a reaction to its equilibrium constant. For the present we shall restrict our discussion to pure ideal gases or to pure liquids and solids, because we are not yet in a position to consider the form of the free-energy functions for real gases.

For an ideal gas we may derive readily an equation for the change in free energy in an isothermal expansion by recognizing that equation (8.19) reduces to the following at fixed temperature ($dT = 0$):

$$dF = V \, dP. \quad (8.73)$$

For an ideal gas,
$$V = \frac{nRT}{P}, \quad (8.74)$$

and hence
$$dF = \frac{nRT}{P} \, dP, \quad (8.75)$$

$$\Delta F = nRT \ln \frac{P_2}{P_1}. \quad (8.76)$$

This expression will be essential to the following derivation.

Let us represent a chemical transformation by the equation (where lower-case letters represent the number of moles):

$$aA \text{ (gas, } P_A) + bB \text{ (gas, } P_B) = cC \text{ (gas, } P_C) + dD \text{ (gas, } P_D). \quad (8.77)$$

For the present, each substance is assumed to be a pure ideal gas at a given partial pressure. This reaction will be accompanied by a free-energy change, ΔF. Suppose we wish to calculate $\Delta F°$ for reaction (8.77). We may proceed to do so by adding suitable equations to equation (8.77), so that each substance is carried from its partial pressure to a pressure of one atmosphere, and then adding the free-energy changes accompanying these reactions to ΔF for equation (8.77). Thus we would add the following:

$$aA \ (P_A) + bB \ (P_B) = cC \ (P_C) + dD \ (P_D); \quad \Delta F \tag{8.77}$$

$$aA \ (P_A = 1) = aA \ (P_A); \quad \Delta F = aRT \ln \left(\frac{P_A}{1}\right) \tag{8.78}$$

$$bB \ (P_B = 1) = bB \ (P_B); \quad \Delta F = bRT \ln \left(\frac{P_B}{1}\right) \tag{8.79}$$

$$cC \ (P_C) = cC \ (P_C = 1); \quad \Delta F = cRT \ln \left(\frac{1}{P_C}\right) \tag{8.80}$$

$$dD \ (P_D) = dD \ (P_D = 1); \quad \Delta F = dRT \ln \left(\frac{1}{P_D}\right) \tag{8.81}$$

$$aA \ (P_A = 1) + bB \ (P_B = 1) = cC \ (P_C = 1) + dD \ (P_D = 1);$$

$$\Delta F^\circ = \Delta F + aRT \ln P_A + bRT \ln P_B + cRT \ln \left(\frac{1}{P_C}\right) + dRT \ln \left(\frac{1}{P_D}\right). \tag{8.82}$$

Equation (8.82) can be simplified by combining logarithmic terms to yield

$$\Delta F^\circ = \Delta F + RT \ln \frac{(P_A)^a (P_B)^b}{(P_C)^c (P_D)^d}. \tag{8.83}$$

Inversion of the ratio leads to

$$\Delta F^\circ = \Delta F - RT \ln \frac{(P_C)^c (P_D)^d}{(P_A)^a (P_B)^b}. \tag{8.84}$$

If the pressures in equation (8.77) correspond to equilibrium values, then since we are dealing with an isothermal, isobaric[6] transformation (with the only restraint being the pressure of the environment),

$$\Delta F = 0. \tag{8.85}$$

By definition, the ratio of the equilibrium pressures is symbolized by K:

$$\frac{(P_C)^c (P_D)^d}{(P_A)^a (P_B)^b} = K. \tag{8.86}$$

Equation (8.84) thus becomes

$$\Delta F^\circ = -RT \ln K. \tag{8.87}$$

---

[1] It should be emphasized that ΔF for equation (8.77) refers to the change in free energy when pure A at a *fixed* pressure of P_A and pure B at a *fixed* pressure of P_B react to give pure C at a *fixed* pressure of P_C and pure D at a *fixed* pressure of P_D, though the individual pressures are not necessarily the same for each gas.

Evidently, then, K must be a constant, because it is the antilogarithm of $\Delta F°$; and since $\Delta F°$ is the change in free energy under certain specified conditions (substances in their standard states), it must have a fixed value at a given temperature. K is commonly called the *thermodynamic equilibrium constant* and is truly a constant. Obviously, it can be evaluated numerically from equation (8.87), if the standard free energy for the reaction is known. Perhaps it should be emphasized that although K refers to equilibrium pressures, it is calculated by means of equation (8.87) from data for $\Delta F°$ referring to the reaction occurring between gases at individual pressures of one atmosphere.

C. Dependence of K upon the temperature. From the value of $\Delta F°$ at a single temperature it is possible to calculate the equilibrium constant, K. It is desirable, in addition, to be able to calculate K as a function of the temperature, so that it should not be necessary to have extensive tables of $\Delta F°$ at frequent temperature intervals. The derivation of the necessary functional relationship requires a direct relation between ΔF and ΔH.

1. *Direct relation between ΔF and ΔH.* Starting from the fundamental definition

$$F = H - TS, \tag{8.2}$$

we may obtain

$$\frac{F}{T} = \frac{H}{T} - S, \tag{8.88}$$

which may be differentiated at constant pressure to give

$$\left(\frac{\partial(F/T)}{\partial T}\right)_P = \left(\frac{\partial(H/T)}{\partial T}\right)_P - \left(\frac{\partial S}{\partial T}\right)_P$$

$$= H\left(\frac{\partial(1/T)}{\partial T}\right)_P + \frac{1}{T}\left(\frac{\partial H}{\partial T}\right)_P - \left(\frac{\partial S}{\partial T}\right)_P. \tag{8.89}$$

Each of these three terms may be reduced further to give the following expressions:

$$H\left(\frac{\partial(1/T)}{\partial T}\right)_P = -\frac{H}{T^2}, \tag{8.90}$$

$$\frac{1}{T}\left(\frac{\partial H}{\partial T}\right)_P = \frac{C_p}{T}, \tag{8.91}$$

$$dS_P = \frac{DQ_P}{T} = \frac{C_p\, dT}{T}, \tag{8.92}$$

or

$$\left(\frac{\partial S}{\partial T}\right)_P = \frac{C_p}{T}. \tag{8.93}$$

Substitution of equations (8.90), (8.91), and (8.93) into equation (8.89) leads to

$$\left(\frac{\partial(F/T)}{\partial T}\right)_P = -\frac{H}{T^2}. \tag{8.94}$$

It is obvious that by a procedure analogous to that used to derive equation (8.28), we may obtain the following from equation (8.94):

$$\left(\frac{\partial(\Delta F/T)}{\partial T}\right)_P = -\frac{\Delta H}{T^2}. \tag{8.95}[7]$$

2.   *ΔF as a function of temperature.*   We have shown in Chapter 5 that a general expression for ΔH as a function of temperature may be written in the form

$$\Delta H = \Delta H_0 + \int \Delta C_p \, dT. \tag{8.96}$$

If the heat capacities of the substances involved in the transformation can be expressed in the form of a simple power series,

$$C_p = a + bT + cT^2 + \ldots, \tag{8.97}$$

where a, b, and c are constants, then equation (8.96) takes the form

$$\Delta H = \Delta H_0 + \Delta aT + \frac{\Delta b}{2} T^2 + \frac{\Delta c}{3} T^3 + \ldots, \tag{8.98}$$

where the Δ's refer to the sums of the coefficients for the products minus the sums of the coefficients for the reactants.   Equation (8.98), in turn, may be inserted into (8.95), which may then be integrated, constant pressure being assumed [and terms higher than T^3 in equation (8.98) being dropped in this example]:

$$\int d\frac{\Delta F}{T} = -\int \frac{\Delta H}{T^2} \, dT = -\int \left(\frac{\Delta a}{T} + \frac{\Delta b}{2} + \frac{\Delta c}{3} T + \frac{\Delta H_0}{T^2}\right) dT. \tag{8.99}$$

If we represent the constant of integration by I, the preceding expression may be written

$$\frac{\Delta F}{T} = I - \Delta a \ln T - \frac{\Delta b}{2} T - \frac{\Delta c}{6} T^2 + \frac{\Delta H_0}{T}, \tag{8.100}$$

which leads finally to an equation explicit in ΔF:

$$\Delta F = \Delta H_0 - \Delta aT \ln T + IT - \frac{\Delta b}{2} T^2 - \frac{\Delta c}{6} T^3. \tag{8.101}[8]$$

---

[7] This equation is also valid if the free-energy change under discussion is $\Delta F°$.   In this case, however, $\Delta H°$ is the enthalpy change at zero pressure for gases and at infinite dilution for substances in solution.   For a rigorous discussion of these relationships between standard states, see Chapter 19.

[8] Strictly speaking, this equation is valid for $\Delta F°$ only if the heat capacities and the enthalpy change are those for the gases at zero pressure (see Chapter 19).

The constant ΔH_0 may be evaluated as described in Chapter 5 if one value of the heat of reaction is known. Similarly, the constant I may be determined if ΔH_0 and one value of ΔF are known.

The use of equation (8.101) may be illustrated best by a specific example. Consider the reaction

$$C \text{ (graphite)} + O_2 \text{ (g)} = CO_2 \text{ (g)}. \tag{8.102}$$

From the heat-capacity equations and $\Delta Hf°$, we have shown in Chapter 5, page 58, that

$$\Delta H = -94,218.3 + 1.331T - 3.357 \times 10^{-3} T^2 + 25.70 \times 10^{-7} T^3. \tag{8.103}$$

Substituting into equation (8.99) and integrating, we obtain the following expression for $\Delta F/T$:

$$\frac{\Delta F}{T} = I - 1.331 \ln T + 3.357 \times 10^{-3} T - 12.85 \times 10^{-7} T^2 - \frac{94,218.3}{T}. \tag{8.104}[9]$$

At $298.16°K$ the standard free energy of formation of CO_2, $\Delta F°$, is $-94,259.8$ cal mole^{-1}. Using this value for ΔF in equation (8.104), we can evaluate the integration constant, I:

$$I = 15.208. \tag{8.105}$$

Thus the explicit equation for the standard free-energy change in reaction (8.102) becomes

$$\Delta F° = 15.208T - 3.065T \log_{10} T + 3.357 \times 10^{-3} T^2$$
$$- 12.85 \times 10^{-7} T^3 - 94,218.3. \tag{8.106}$$

3. *K as a function of the temperature.* The equilibrium constant can be related to the temperature through either of two thermodynamic functions.

   a. *The differential relation.* It is evident that the substitution of equation (8.87) into (8.95) leads to

$$\left(\frac{\partial(\Delta F°/T)}{\partial T}\right)_P = -\frac{\Delta H°}{T^2} = \left(\frac{\partial(-RT \ln K/T)}{\partial T}\right)_P = -R\left(\frac{\partial \ln K}{\partial T}\right)_P. \tag{8.107}$$

From this equation we may obtain, upon rearrangement,

$$\left(\frac{\partial \ln K}{\partial T}\right)_P = \frac{\Delta H°}{RT^2}. \tag{8.108}$$

---

[9] Strictly speaking, this equation is valid for $\Delta F°$ only if the heat capacities and the enthalpy change are those for the gases at zero pressure (see Chapter 19). In the present example, the differences in C_p's between zero and one atmosphere pressure have been neglected.

b. *The integral relation.* If the heat capacities fit the power series specified in (8.97), ΔF may be expressed by equation (8.101). Obviously,
then, making use of equation (8.87), we obtain, in place of equation
(8.101),

$$\ln K = -\frac{\Delta H_0}{RT} + \frac{\Delta a}{R} \ln T - \frac{I}{R} + \frac{\Delta b}{2R} T + \frac{\Delta c}{6R} T^2. \qquad (8.109)$$

In the general case, when a function for the heat capacities is unavailable,
it is necessary to integrate equation (8.108):

$$\int_{T_1}^{T_2} d \ln K = \int_{T_1}^{T_2} \frac{\Delta H^\circ}{RT^2} dT. \qquad (8.110)$$

VI. USEFUL WORK AND FREE ENERGY

So far we have considered situations where the only restraint upon a
system is the constant pressure of the environment. Such situations are
of primary interest in connection with the general problem of spontaneity
or equilibrium in chemical reactions. Nevertheless, we must give some
attention also to systems which are so arranged that they can produce
work other than that against the atmosphere. Although in most cases
we shall consider only electrical manifestations of such additional work, it
should be pointed out that other restraints such as magnetic, centrifugal, or
surface forces may also assume significance in many problems.

**A.   Relation between ΔF and net useful work at constant pressure and
temperature.**   We shall direct our attention once again to the example
cited previously, the formation of a mole of aqueous HCl from the gases
H_2 and Cl_2:

$$\tfrac{1}{2}H_2 \text{ (g)} + \tfrac{1}{2}Cl_2 \text{ (g)} = HCl \text{ (aq)}. \qquad (8.53)$$

If the gaseous mixture is exposed to a photochemical stimulus, the reaction proceeds spontaneously. Since such a system is under no restraint
other than constant atmospheric pressure and since the temperature may
be maintained constant,

$$dF_{P,T} < 0. \qquad (8.111)$$

Furthermore, since no net *useful* work (that is, none other than that against
the atmosphere) can be obtained under the conditions described, we may
also write

$$DW_{\text{net}} = 0. \qquad (8.112)$$

On the other hand, the chemical transformation may be carried out reversibly, as in the cell illustrated by Fig. 2. The potential exerted by the

potentiometer-cell[10] is kept opposed to, but infinitesimally below, that of the H_2-Cl_2 cell, so that the chemical transformation represented by equation (8.53) is carried out very slowly. For the system as a whole, that is, cell plus potentiometer,

$$dF_{\text{cell + potentiometer}} = 0, \qquad (8.113)$$

since the pressure and temperature are fixed and the system as a whole is subject to no restraint other than atmospheric pressure. Nevertheless, the H_2-Cl_2 cell itself must undergo a decrease in free energy if the reaction indicated by equation (8.53) is allowed to proceed, even though very slowly and in a reversible fashion, because the free energy is a thermodynamic property, and hence the change in free energy must be the same as for an equivalent spontaneous process.

From the reversible discharge of the H_2-Cl_2 cell with the e.m.f. of the potentiometer-cell kept opposed to, but infinitesimally below, the potential of the cell, some useful (electrical) work is obtained other than that expended against the atmosphere. This electrical work is used to charge the potentiometer-cell. We shall call this work $W_{\text{net, rev}}$. Clearly

$$DW_{\text{total}} = DW_{\text{net, rev}} + P\, dV. \qquad (8.114)$$

Since the process is one at constant pressure and temperature, we may write

$$dF = dH - T\, dS = dE + P\, dV - T\, dS. \qquad (8.115)$$

According to the first law of thermodynamics,

$$dE = DQ - DW. \qquad (8.116)$$

We now have, in view of equation (8.114),

$$dE = DQ - DW_{\text{net, rev}} - P\, dV. \qquad (8.117)$$

The substitution of equation (8.117) into (8.115) leads to

$$dF = DQ - DW_{\text{net, rev}} - T\, dS. \qquad (8.118)$$

Since the process under discussion is being carried out in a reversible fashion,

$$DQ = T\, dS. \qquad (8.119)$$

Hence

$$dF_{P,T} = -DW_{\text{net, rev}}. \qquad (8.120)$$

---

[10] For the purposes of the present discussion, the conventional potentiometer cannot be used, for the rate of discharge of the operating battery is only slightly decreased, or increased, *but is not reversed* when the potential of the potentiometer is made slightly below, or above, that of the H_2-Cl_2 cell. Thus the common potentiometer is not operated in a reversible fashion. For the present discussion, therefore, we shall assume that the potential of the H_2-Cl_2 cell is opposed by a working battery, which we shall call a *potentiometer-cell*, whose voltage, near that of the H_2-Cl_2 cell, can be varied by changes in pressure. Such a potentiometer-cell can be operated in a reversible manner.

Thus we have derived the very significant consequence that the change in free energy is the negative of the net (that is, nonatmospheric) reversible work available in a transformation at constant pressure and temperature. Since in isothermal processes the reversible work is the maximum work, it is evident that the free-energy change gives the negative of the maximum possible net work available from a specified reaction. For any real process at constant pressure and temperature,

$$DW_{net, \, irrev} \; < \; dW_{net, \, rev}. \tag{8.55}$$

Therefore $\qquad\qquad DW_{net, \, irrev} \; < \; -dF_{P,T}. \tag{8.121}$

Thus the free-energy change gives the limit which may be approached in obtaining net useful work in any real process at constant temperature and pressure.

B. Relation between ΔA and total work at constant temperature. For the reversible transformation just considered in connection with the Gibbs free energy, it is evident that the Helmholtz free energy must obey the following relations under isothermal conditions:

$$\begin{aligned}
dA &= dE - T \, dS \\
&= DQ - DW_{total, \, rev} - T \, dS \\
&= T \, dS - DW_{total, \, rev} - T \, dS, \tag{8.122}
\end{aligned}$$

or $\qquad\qquad dA_T = -DW_{total, \, rev}. \tag{8.123}$

Thus the change in Helmholtz free energy in an isothermal process is the negative of the maximum total work (that is, net useful work as well as that against the atmosphere) available from a specified reaction.

C. Gibbs-Helmholtz equation. We have seen that, at constant pressure and a fixed temperature, $-\Delta F$ is a measure of the maximum net work available from a chemical reaction. The particular value obtained for ΔF, however, depends upon the specific temperature at which the isothermal transformation is carried out. Thus $\Delta F_{P,T}$ may differ at different temperatures, varying perhaps in a fashion such as is indicated in Fig. 3.

Knowing a value of $\Delta F_{P,T}$ at one temperature, we may wish to calculate it at another. The required relation may be derived readily from expressions available already.

For an isothermal reaction,

$$\Delta F = \Delta H - T \, \Delta S. \tag{8.7}$$

But we have also proved that at fixed pressure

$$\Delta S = - \left(\frac{\partial \Delta F}{\partial T} \right)_P. \tag{8.30}$$

Hence it follows that

$$\Delta F_{P,T} = \Delta H_{P,T} + T \left(\frac{\partial \Delta F_{P,T}}{\partial T} \right)_P. \tag{8.124}$$

Equation (8.124) frequently is called the *Gibbs-Helmholtz equation*. Clearly

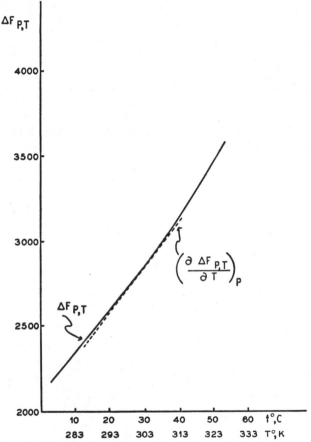

Fig. 3. Standard free-energy change for the isothermal, isobaric reaction
$$HSO_4^- \text{ (aq)} = H^+ \text{ (aq)} + SO_4^= \text{ (aq)}$$
as a function of the temperature. The broken line gives the slope, or derivative, at 25 °C.

the temperature coefficient of the free energy, $(\partial \Delta F_{P,T}/\partial T)_P$, can be obtained if ΔF and ΔH are known.

When electrical work is obtained from the reaction, the Gibbs-Helmholtz equation is generally used in a modified form. If the electrical work is obtained under reversible conditions, that is, against a counterpotential only infinitesimally smaller than that of the cell, then

$$W_{\text{elec}} = W_{\text{net, rev}} = (\text{potential}) \times (\text{charge})$$

$$= (\mathcal{E})(n\mathcal{F}), \tag{8.125}$$

where $\mathcal{E}$ is the potential obtained from the cell under reversible conditions, $\mathcal{F}$ is the number of charges in one faraday of electricity, and n is the number of faradays of charge which have passed from the cell. It follows from the integrated equivalent of equation (8.120) that

$$\Delta F = -n\mathcal{F}\mathcal{E}. \tag{8.126}$$

By differentiation we obtain

$$\left(\frac{\partial \Delta F}{\partial T}\right)_P = -n\mathcal{F}\left(\frac{\partial \mathcal{E}}{\partial T}\right)_P, \tag{8.127}$$

since n and $\mathcal{F}$ are temperature-independent quantities. Substitution of the preceding two expressions into equation (8.124) leads to

$$-n\mathcal{F}\mathcal{E} = \Delta H - n\mathcal{F}T\left(\frac{\partial \mathcal{E}}{\partial T}\right)_P \tag{8.128}$$

or, upon rearrangement, to a somewhat more familiar form of the Gibbs-Helmholtz equation,

$$\Delta H = n\mathcal{F}\left[T\left(\frac{\partial \mathcal{E}}{\partial T}\right)_P - \mathcal{E}\right]. \tag{8.129}$$

D. Relation between ΔH_P and Q_P when useful work is done. We have repeatedly made use of the relation

$$\Delta H_P = Q_P \tag{8.130}$$

with the very careful stipulation that not only must the pressure on the system be constant, but also there must be no nonmechanical work. Such are the conditions under which most chemical reactions are carried out, and hence the enthalpy change has been of much value as a measure of the heat of a reaction. It should be emphasized, however, that if non-atmospheric work is also being obtained, for example in the form of electrical work, equation (8.130) is no longer valid. That such must be the case is readily evident from the following considerations. According to the first law of thermodynamics,

$$dE = DQ - DW_{\text{total}}, \tag{8.116}$$

which we may now write as

$$dE = DQ - DW_{\text{net}} - P\,dV. \tag{8.131}$$

Since at constant pressure

$$dH_P = dE + P\,dV, \tag{8.132}$$

we may substitute equation (8.131) into (8.132) and obtain

$$dH_P = DQ_P - DW_{net} \qquad (8.133)$$

or
$$\Delta H_P = Q_P - W_{net}. \qquad (8.134)$$

Thus we see that in general the enthalpy change under isobaric conditions equals the heat absorbed *plus* the net useful (nonatmospheric) work. Only when W_{net} is zero do we get the common expression given by equation (8.130).

Exercises

1. Prove the validity of equations (8.20), (8.31) to (8.34), and (8.39).

2. Derive the following expressions:

(a)
$$\left(\frac{\partial(\Delta F/T)}{\partial(1/T)}\right)_P = \Delta H. \qquad (8.135)$$

(b)
$$dF = V\left(\frac{\partial P}{\partial V}\right)_T dV + \left[V\left(\frac{\partial P}{\partial T}\right)_V - S\right]dT. \qquad (8.136)$$

Optional 3. One mole of an ideal gas in an isolated system at 273.16°K is allowed to expand isothermally from a pressure of 100 atm to 10 atm.

(a) Calculate (and arrange in tabular form) the values of W, Q, ΔE, ΔH, ΔS, ΔF, and ΔA of the gas if the expansion is reversible.

(b) Calculate (and arrange in tabular form adjacent to the preceding table) the values of W, Q, ΔE, ΔH, ΔS, ΔF, and ΔA of the entire isolated system if the expansion is reversible.

(c) Calculate (and arrange in tabular form adjacent to the preceding tables) the values of the same thermodynamic quantities for the gas if it is allowed to expand freely so that no work whatever is done by it.

(d) Calculate (and arrange in tabular form adjacent to the preceding tables) the values of the same thermodynamic quantities for the entire isolated system if the expansion is free.

4. A mole of steam is condensed reversibly to liquid water at 100°C and 760 mm pressure. The heat of vaporization of water is 539.4 cal gram^{-1}. Assuming that steam behaves as an ideal gas, calculate W, Q, ΔE, ΔH, ΔS, ΔF, and ΔA for the condensation process.

5. Using thermal data available in preceding chapters, find an expression for $\Delta F°$ as a function of temperature for the reaction

$$CO \text{ (g)} + \tfrac{1}{2}O_2 \text{ (g)} = CO_2 \text{ (g)}.$$

6. If the heat capacities of reactants and products are expressed by equations of the form

$$C_p = a + bT - \frac{c'}{T^2},$$

where a, b, and c' are constants, what will be the form of the equation for ΔF as a function of temperature?

7. Consider a reaction such as (8.77) in which A is a pure solid at one atmosphere pressure, and the other substances are as indicated in the text. Derive an expression corresponding to equation (8.84).

CHAPTER 9

Application of the Free-Energy Function to Some Phase Changes

HAVING established convenient criteria for equilibrium and for spontaneity, we are now in a position to apply the fundamental laws of thermodynamics to problems which occur in chemistry. We shall direct our attention first to transformations involving merely phase changes.

I. TWO PHASES AT CONSTANT PRESSURE AND TEMPERATURE

The familiar equations describing equilibrium conditions between two phases of a specified substance are readily derivable from the two laws of thermodynamics, with the aid of the new functions which we have defined in the preceding chapter.

Let us represent the equilibrium at any given temperature and pressure by the equation

$$\text{State } A \rightleftharpoons \text{State } B. \tag{9.1}$$

Since the system is in equilibrium,

$$\Delta F_{P,T} = F_B - F_A = 0. \tag{9.2}$$

The temperature is now changed by an amount dT. If the system is to be maintained in equilibrium, the pressure must be altered by some quantity, dP. Nevertheless, if equilibrium is maintained at the new temperature, T',

$$T' = T + dT, \tag{9.3}$$

then
$$\Delta F'_{P,T'} = F'_B - F'_A = 0. \tag{9.4}$$

Obviously, then,
$$d(\Delta F) = d(F_B - F_A) = 0 \tag{9.5}$$

and
$$dF_B = dF_A. \tag{9.6}$$

A. Clapeyron equation. From the preceding chapter we may adopt the general equation for the total differential, dF, and write

$$dF_A = V_A \, dP - S_A \, dT, \tag{9.7}$$

$$dF_B = V_B \, dP - S_B \, dT. \tag{9.8}$$

Setting up the equality demanded by (9.6), we obtain

$$V_B \, dP - S_B \, dT = V_A \, dP - S_A \, dT, \tag{9.9}$$

which can be rearranged to give

$$(V_B - V_A) \, dP = (S_B - S_A) \, dT \tag{9.10}$$

and consequently
$$\frac{dP}{dT} = \frac{S_B - S_A}{V_B - V_A}. \tag{9.11}$$

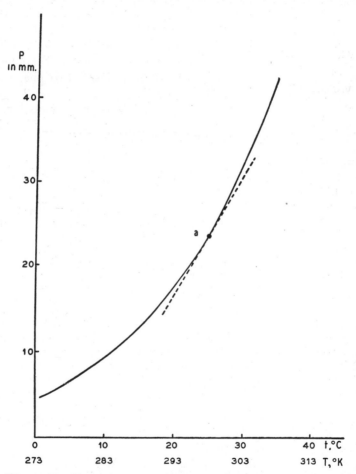

Fig. 1.   Equilibrium vapor-pressure curve for water.   The broken
line gives the slope at a specified pressure and temperature.

We are interested in the value of this derivative, dP/dT, at a specified
temperature and pressure such as is indicated by point a in Fig. 1.   For an
isothermal reversible (that is, equilibrium) condition at constant pressure,

$$S_B - S_A = \Delta S = \int \frac{DQ_P}{T} = \frac{1}{T} \int DQ_P = \frac{\Delta H}{T}. \tag{9.12}$$

Therefore equation (9.11) may be converted to

$$\frac{dP}{dT} = \frac{\Delta H}{T \, \Delta V}, \tag{9.13}$$

which is generally known as the *Clapeyron equation.*

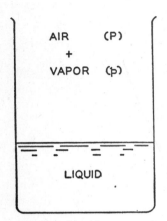

AIR (P)
+
VAPOR (p)

LIQUID

Fig. 2. Liquid-vapor equilibrium in the presence of an inert gas.

So far we have made no special assumptions as to the nature of the phases A and B in deriving equation (9.13). Evidently, then, the Clapeyron equation is applicable to equilibrium between any two phases of one component.

B. Clausius-Clapeyron equation. The Clapeyron equation can be reduced to a particularly convenient form if we restrict the equilibrium between A and B to that of a gas and condensed phase (liquid or solid). In this situation

$$V_B - V_A = V_{\text{gas}} - V_{\text{cond}}. \tag{9.14}$$

In general, the volume of a given weight of a gas, V_{gas}, is much larger than that of a corresponding weight of the condensed phase, V_{cond}:

$$V_{\text{gas}} \gg V_{\text{cond}}. \tag{9.15}$$

When equation (9.15) is valid, V_{cond} may be neglected[1] with respect to V_{gas} in equation (9.14). Furthermore, if we approximate the volume of the gas by the ideal gas law,

$$V_{\text{gas}} = \frac{RT}{P}, \tag{9.16}$$

then we may use equation (9.16) to substitute for ΔV in equation (9.13) and obtain

$$\frac{dP}{dT} = \frac{\Delta H}{T(RT/P)} = \frac{P(\Delta H)}{RT^2}. \tag{9.17}$$

It follows that
$$\frac{1}{P}\frac{dP}{dT} = \frac{d \ln P}{dT} = \frac{\Delta H}{RT^2}. \tag{9.18}$$

For many substances, over not too large a region of temperature, the heat of

---

[1] For example, the molal volume of liquid H_2O near the boiling point is about 18 cc, whereas that for water vapor is near 30,000 cc.

vaporization is substantially constant. Equation (9.18) may then be integrated readily as follows:

$$d \ln P = \frac{\Delta H}{R} \frac{dT}{T^2} = -\frac{\Delta H}{R} d\left(\frac{1}{T}\right), \qquad (9.19)$$

$$\ln \frac{P_2}{P_1} = -\frac{\Delta H}{R}\left[\frac{1}{T_2} - \frac{1}{T_1}\right], \; = \; -\frac{\Delta H}{R}\left(\frac{T_1 - T_2}{T_2 T_1}\right) \quad (9.20)$$

or
$$\log P = -\frac{\Delta H}{2.303 R} \frac{1}{T} + \text{constant}. \qquad (9.21)$$

Any one of the equations (9.18) to (9.21) is known as the *Clausius-Clapeyron equation* and may be used either to obtain ΔH from known values of the vapor pressure as a function of temperature or, vice versa, to predict vapor pressures of a liquid (or a solid) when the heat of vaporization (or sublimation) and one vapor pressure are known.

II. TWO PHASES AT SAME TEMPERATURE BUT DIFFERENT PRESSURES[2]

Another situation of interest occurs in liquid-vapor equilibria when the system is exposed to the atmosphere (Fig. 2). We shall assume that the air is essentially insoluble in the liquid phase and that the atmospheric pressure is constant at some value P. The partial pressure of the vapor, however, is some value, p, which we can show must differ from that of the saturation vapor pressure of the liquid in the absence of any foreign gas such as air.

If the liquid and vapor are in equilibrium at a fixed pressure and temperature, in the absence of any foreign gas,

$$\Delta F = F_{\text{gas}} - F_{\text{liquid}} = 0. \qquad (9.22)$$

We admit an infinitesimal amount of the foreign gas and thereby change the pressure on the liquid by an amount dP_{liq}. Since the temperature is fixed, this will result in a change in free energy given by the first term of equation (8.19).

$$dF_{\text{liq}} = V_{\text{liq}} \, dP_{\text{liq}}. \qquad (9.23)$$

If equilibrium is to be maintained, the vapor must undergo an equivalent change in free energy. Since the temperature is maintained constant, the free-energy change must be due to a change in partial pressure of the vapor, p_{gas}.

$$dF_{\text{gas}} = V_{\text{gas}} \, dp_{\text{gas}}. \qquad (9.24)$$

In view of the maintenance of equilibrium, the relations expressed by (9.23) and (9.24) may be equated:

[2] For a more precise treatment of this situation, see G. N. Lewis and M. Randall, *Thermodynamics*, McGraw-Hill Book Company, Inc., New York, 1923, pages 181–183.

$$V_{\text{gas}}\, dp_{\text{gas}} = V_{\text{liq}}\, dP_{\text{liq}}. \tag{9.25}$$

This expression can be rearranged to give

$$\left(\frac{\partial p_{\text{gas}}}{\partial P_{\text{liq}}}\right)_T = \frac{V_{\text{liq}}}{V_{\text{gas}}}. \tag{9.26}$$

Obviously, since the volume of a given weight of liquid, V_{liq}, is much smaller than that of a corresponding quantity of the vapor, V_{gas}, the change of vapor pressure with increasing pressure on the liquid is relatively small.

III. TWO PHASES AT DIFFERENT PRESSURES; VARIATION OF VAPOR PRESSURE WITH TEMPERATURE

Another situation of interest occurs when a liquid is exposed to the fixed pressure of the atmosphere and the temperature of the system is varied. The vapor pressure changes with the temperature, but the pressure on the liquid is fixed, since the atmospheric pressure is constant. It is of interest to derive the expression for the temperature dependence of the vapor pressure under these conditions.

For the vapor, the variation in free energy as we change the temperature of the system at equilibrium is given by

$$dF_{\text{gas}} = V_{\text{gas}}\, dp_{\text{gas}} - S_{\text{gas}}\, dT. \tag{9.27}$$

On the other hand, for the liquid, the dP_{liq} term vanishes, since the atmospheric pressure is constant. Hence

$$dF_{\text{liq}} = -S_{\text{liq}}\, dT. \tag{9.28}$$

Since equilibrium is maintained despite the change in temperature, we may equate the free-energy changes of equations (9.27) and (9.28):

$$V_{\text{gas}}\, dp_{\text{gas}} - S_{\text{gas}}\, dT = -S_{\text{liq}}\, dT. \tag{9.29}$$

Upon rearrangement, this expression leads to

$$\left(\frac{\partial p_{\text{gas}}}{\partial T}\right)_{P_{\text{liq}}} = \frac{S_{\text{gas}} - S_{\text{liq}}}{V_{\text{gas}}} = \frac{\Delta S}{V_{\text{gas}}}. \tag{9.30}$$

As in the case of the derivation of the Clapeyron equation, we are interested in the temperature coefficient of the vapor pressure at some specified temperature and fixed pressure.

$$\Delta S = \int \frac{DQ}{T} = \frac{\Delta H}{T} \tag{9.31}$$

and

$$\left(\frac{\partial p_{\text{gas}}}{\partial T}\right)_{P_{\text{liq}}} = \frac{\Delta H}{T V_{\text{gas}}}. \tag{9.32}$$

Equation (9.32) is very similar to the Clapeyron relation [equation (9.13)], and the results obtained by either differ very little.   Nevertheless, for systems which are exposed to the atmosphere, the variation of vapor pressure with temperature should be calculated, for greater accuracy, from equation (9.32).

IV. CALCULATION OF ΔF FOR SPONTANEOUS PHASE CHANGE

So far we have considered problems in phase changes in which equilibrium is maintained.   It is also pertinent to describe some procedures for the calculation of the change in free energy in transformations which are known to be spontaneous.   For example, we might consider the freezing of supercooled water at $-10\,°C$:

$$H_2O \ (l, \ -10\,°C) \ = \ H_2O \ (s, \ -10\,°C). \tag{9.33}$$

At $0\,°C$ and one atmosphere pressure of air, the process is at equilibrium. Hence

$$\Delta F_{0\,°C} \ = \ 0. \tag{9.34}$$

At $-10\,°C$, the supercooled water may freeze spontaneously.   Therefore we can say immediately that

$$\Delta F_{-10\,°C} \ < \ 0. \tag{9.35}$$

Nevertheless, we still wish to evaluate ΔF numerically.

A. Arithmetic method.   The simplest procedure to calculate the free-energy change at $-10\,°C$ makes use of the relation

$$\Delta F \ = \ \Delta H \ - \ T\,\Delta S, \tag{9.36}$$

for any isothermal process. ΔH and ΔS at $-10\,°C$ (T_2) are calculated from the known values at $0\,°C$ (T_1) and from the temperature coefficients of these thermodynamic quantities.   Since the procedure may be represented by the sum of a series of equations, the method may be called an arithmetic one.   The series of equations is the following:

$H_2O \ (l, 0\,°C) \ = \ H_2O \ (s, 0\,°C);$ $\Delta H \ = \ -1436 \ \text{cal mole}^{-1};$ $\Delta F = o$

$$\Delta S \ = \ \frac{-1436}{273.16} \ = \ -5.257 \ \text{cal mole}^{-1} \ \text{deg}^{-1}$$

$H_2O \ (s, 0\,°C) \ = \ H_2O \ (s, -10\,°C);$ $\Delta H \ = \ \int_{T_1}^{T_2} C_p \, dT \ = \ 8.7 \ (-10)$

$$= \ -87 \ \text{cal mole}^{-1};$$

$$\Delta S \ = \ \int_{T_1}^{T_2} \frac{C_p}{T} \, dT \ = \ C_p \ \ln \frac{T_2}{T_1}$$

$$= \ -0.324 \ \text{cal mole}^{-1} \ \text{deg}^{-1}$$

$H_2O \ (l, -10\,°C) \ = \ H_2O \ (l, 0\,°C);$ $\Delta H \ = \ \int_{T_2}^{T_1} C'_p \, dT \ = \ 18(10)$

$$= \ 180 \ \text{cal mole}^{-1};$$

$$\Delta S \ = \ \int_{T_2}^{T_1} \frac{C'_p}{T} \, dT \ = \ 0.671 \ \text{cal mole}^{-1} \ \text{deg}^{-1}$$

$H_2O \ (l, -10\,°C) \ = \ H_2O \ (s, -10\,°C);$ $\Delta H \ = \ -1343 \ \text{cal mole}^{-1};$

$$\Delta S \ = \ -4.910 \ \text{cal mole}^{-1} \ \text{deg}^{-1}$$

From the values calculated for ΔH and ΔS, it is evident that

$$\Delta F = -1343 + (263.16)(4.910) = -51 \text{ cal mole}^{-1}. \qquad (9.37)$$

B. Analytic method. The proposed problem could also be solved in a straightforward manner by integration of the equation derived in the preceding chapter,

$$\left(\frac{\partial(\Delta F/T)}{\partial T}\right)_P = -\frac{\Delta H}{T^2}. \qquad (9.38)$$

As in the arithmetic method, we may assume that the heat capacities of ice and water, respectively, are substantially constant over the small temperature range under consideration. Thus from the relation

$$\left(\frac{\partial \Delta H}{\partial T}\right)_P = \Delta C_p \qquad (9.39)$$

we obtain, upon integration,

$$\begin{aligned} \Delta H &= \Delta H_0 + \int (C_{p_\text{ice}} - C_{p_\text{water}})\, dT \\ &= \Delta H_0 - 9.3T. \end{aligned} \qquad (9.40)$$

Since at $0°C$ ΔH is -1436 cal mole^{-1}, we can determine ΔH_0:

$$\Delta H_0 = -1436 + 9.3(273.16) = 1104.39. \qquad (9.41)^3$$

Hence $\qquad\qquad \Delta H = 1104.39 - 9.3T \qquad\qquad\qquad\qquad (9.42)$

and

$$\frac{\Delta F}{T} = -\int \frac{(1104.39 - 9.3T)}{T^2}\, dT$$

$$= \frac{1104.39}{T} + 9.3 \ln T + I, \qquad (9.43)$$

where I is a constant. Rearrangement leads to

$$\Delta F = 1104.39 + 9.3T \ln T + IT. \qquad (9.44)$$

Since ΔF is known to be zero at $0°C$, the constant I may be evaluated:

$$I = \frac{-1104.39 - 9.3(273.16)(2.303) \log (273.16)}{273.16}$$

$$= -56.226. \qquad (9.45)^3$$

Thus we have an explicit equation for ΔF as a function of temperature:

$$\Delta F = 1104.39 + 9.3T \ln T - 56.226T. \qquad (9.46)$$

At $-10°C$ this equation leads to

---

[3] More significant figures are retained in these numbers than may be justified by the precision of the data upon which they are based. Such a procedure is necessary, however, in calculations involving small differences between large numbers.

$$\Delta F_{-10°C} = -51 \text{ cal mole}^{-1}. \qquad (9.47)$$

This is the same result as was obtained by the arithmetic method.

Other methods, depending upon other expressions for ΔF, may be used also. However, the preceding procedures are among the simplest and most direct.

Exercises

1. For each of the following seven transformations:
   (a) H_2O (s, $-10°C$, 1 atm) $\rightarrow H_2O$ (l, $-10°C$, 1 atm);
      (*Note:* No specification is made that this process is carried out isother-mally, isobarically, or reversibly.)
   (b) same as part (a) but restricted to a reversible change;
   (c) same as part (a) but restricted to isothermal and isobaric conditions;
   (d) the two-step, isobaric, reversible transformation
      H_2O (l, $25°C$, 1 atm) $\rightarrow H_2O$ (l, $100°C$, 1 atm) *isobaric* ⎱ *reversibl*
      H_2O (l, $100°C$, 1 atm) $\rightarrow H_2O$ (g, $100°C$, 1 atm); *isothermal,isobaric* ⎰
   (e) ideal gas ($25°C$, 100 atm) $\rightarrow$ ideal gas ($25°C$, 1 atm), reversible;
   (f) ideal gas ($25°C$, 100 atm) $\rightarrow$ ideal gas ($25°C$, 10 atm), no work done;
   (g) adiabatic reversible expansion of a perfect gas from 100 atm to 10 atm;
   consider each of the following equations:

   (1) $\displaystyle\int \frac{DQ}{T} = \Delta S;$

   (2) $Q = \Delta H;$

   (3) $\displaystyle\frac{\Delta H}{T} = \Delta S;$

   (4) $-\Delta F =$ actual net work;
   (5) $-\Delta F =$ maximum net work.

   For each transformation, list the equations which are valid. If your decision depends upon the existence of conditions which have not been specified, state what these conditions are.

2. Calculate $\Delta F°$ for each of the following transformations:
   (a) H_2O (l, $100°C$) $\rightarrow H_2O$ (g, $100°C$).
   (b) H_2O (l, $25°C$) $\rightarrow H_2O$ (g, $25°C$). The vapor pressure of H_2O at $25°C$ is 0.0313 atm.
   (c) Br_2 (l, $25°C$) $\rightarrow Br_2$ (0.01 molal aqueous solution, $25°C$). The vapor pres-sure of bromine in a dilute aqueous solution at $25°C$ obeys the equation $p = 1.45m$. The vapor pressure of pure bromine at $25°C$ is 0.280 atm.

3. An equation for ΔF for the freezing of supercooled water may be obtained also by integration of the equation

$$\left(\frac{\partial \Delta F}{\partial T}\right)_P = -\Delta S.$$

An expression for ΔS as a function of temperature, for substitution into the preced-ing equation, may be derived from

$$\left(\frac{\partial \Delta S}{\partial T}\right)_P = \frac{\Delta C_p}{T}.$$

(a)   Show that

$$\Delta F = I' - \Delta C_p(T \ln T) + (\Delta C_p - \Delta S_0)T,$$

where I' and ΔS_0 are constants.

(b)   Evaluate the constants from data for the freezing process at $0\,°C$.

(c)   Calculate ΔF at $-10\,°C$ and compare the result with the values calculated by the methods described in the text.

CHAPTER 10

Application of the Free-Energy Function to Chemical Reactions: Summary of Methods for Determining $\Delta F°$

WE MUST direct our attention next to the application of the free-energy criterion to chemical transformations. Since most chemical reactions are carried out at constant pressure and temperature, and with no restraints other than the pressure of the atmosphere, it is the Gibbs free energy, F, in which we shall be most interested. For application to chemical transformations, tables of free-energy data generally are assembled in terms of $\Delta F°$, so that the equilibrium constant of a reaction may be calculated.

The standard free-energy change for a specified reaction may be obtained by several procedures. Although most of these have become familiar from elementary work in physical chemistry, it will be convenient to discuss all of them briefly at one point, so that their advantages and limitations may be compared.

I. ADDITION OF KNOWN $\Delta F°$'s FOR SUITABLE CHEMICAL EQUATIONS LEADING TO THE DESIRED EQUATION

Since the free energy is a thermodynamic property, values of ΔF depend only upon the initial and final states and not upon the particular path used to go from one set of substances to another. Hence values of $\Delta F°$ do not depend upon the intermediate chemical reactions which have been used to transform a set of reactants, under specified conditions, to a series of products. Thus one may add known free energies to obtain values for reactions for which direct data are not available.

This method is illustrated best, perhaps, by a specific example. Let us consider as a problem[1] the determination of $\Delta F°_{298.16}$ for the reaction

$$CO_2 \text{ (g)} + H_2 \text{ (g)} = H_2O \text{ (l)} + CO \text{ (g)}. \tag{10.1}$$

---

[1] In problems involving gases, the student is expected at this point to think of ideal gases only, for which the standard state is the gas at a pressure of one atmosphere. For a real gas, the standard state is that of the gas at unit fugacity (see Chapter 15). All the statements of this chapter apply to a real gas at unit fugacity, as well as to an ideal gas in its standard state.

We shall assume that the standard free energies of formation of CO (g), CH_4 (g) and H_2O (l), respectively, are known and also that standard free energies are available for the condensation of water vapor at 25 °C and for the reaction

$$CO_2 \text{ (g)} + 4H_2 \text{ (g)} = CH_4 \text{ (g)} + 2H_2O \text{ (g)}. \tag{10.2}$$

The solution of our problem may then be obtained by the following summation process:

CO_2 (g) + $4H_2$ (g)	= CH_4 (g) + $2H_2O$ (g)	; ΔF°_{298} =	$-27{,}150$ cal mole^{-1}
CH_4 (g)	= C (graphite) + $2H_2$ (g);	ΔF°_{298} =	$12{,}140$ cal mole^{-1}
C (graphite) + $\frac{1}{2}O_2$ (g)	= CO (g)	; ΔF°_{298} =	$-32{,}807.9$ cal mole^{-1}
$2H_2O$ (g)	= $2H_2O$ (l)	; ΔF°_{298} =	-4109.6 cal mole^{-1}
H_2O (l)	= H_2 (g) + $\frac{1}{2}O_2$ (g)	; ΔF°_{298} =	$56{,}689.9$ cal mole^{-1}
CO_2 (g) + H_2 (g)	= H_2O (l) + CO (g)	; ΔF° =	4762 cal mole^{-1}

Once the standard free-energy change is known, it is of course possible to calculate the equilibrium constant for reaction (10.1):

$$\Delta F^\circ = -RT \ln K, \tag{10.3}$$
$$4762 = -(1.987)(298.16)(2.303) \log K,$$
$$K = 3.24 \times 10^{-4}.$$

II. DETERMINATION OF ΔF° FROM EQUILIBRIUM MEASUREMENTS

Frequently the standard free energies required to calculate ΔF° for a specified reaction are not available. It is necessary then to resort to more direct relations between ΔF° and experimental measurements.

One of these direct methods depends upon the determination of the equilibrium constant of a given reaction. As an example we shall consider the dissociation of isopropyl alcohol:

$$(CH_3)_2CHOH \text{ (g)} = (CH_3)_2CO \text{ (g)} + H_2 \text{ (g)}. \tag{10.4}$$

With a suitable catalyst, equilibrium pressures can be measured for this dissociation. At 452.2 °K and a total pressure, P, of 0.947 atm, the degree of dissociation, α, at equilibrium has been found[2] to be 0.564.

If we start with one mole of isopropyl alcohol, α moles each of acetone and hydrogen are formed. The quantity of alcohol remaining at equilibrium must be $1 - \alpha$. The total number of moles of all three gases is

$$\text{Total moles} = (1 - \alpha) + \alpha + \alpha = 1 + \alpha. \tag{10.5}$$

---

[2] H. J. Kolb and R. L. Burwell, Jr., *J. Am. Chem. Soc.*, **67,** 1084 (1945).

Hence the mole fraction, N, of each substance is

$$N_{(CH_3)_2CHOH} = \frac{1 - \alpha}{1 + \alpha}, \tag{10.6}$$

$$N_{(CH_3)_2CO} = \frac{\alpha}{1 + \alpha}, \tag{10.7}$$

$$N_{H_2} = \frac{\alpha}{1 + \alpha}. \tag{10.8}$$

The equilibrium constant, K, being a function of the equilibrium partial pressures,[3] is given by

$$K = \frac{P_{(CH_3)_2CO}P_{H_2}}{P_{(CH_3)_2CHOH}} = \frac{\dfrac{\alpha}{1 + \alpha}P \; \dfrac{\alpha}{1 + \alpha}P}{\dfrac{1 - \alpha}{1 + \alpha}P} = \frac{\alpha^2}{1 - \alpha^2}P, \tag{10.9}$$

$$K = 0.444. \tag{10.10}$$

The standard free-energy change may then be calculated from equation (10.3):

$$\Delta F^{\circ}_{452.2^{\circ}K} = -RT \ln (0.444) = 730 \text{ cal mole}^{-1}. \tag{10.11}$$

This is an appropriate occasion to emphasize that a positive value of ΔF° does not imply that the reaction under consideration may not proceed spontaneously under any conditions. ΔF° refers to the reaction

$$(CH_3)_2CHOH \text{ (g, } P = 1 \text{ atm)} = (CH_3)_2CO \text{ (g, } P = 1 \text{ atm)}$$
$$+ H_2 \text{ (g, } P = 1 \text{ atm)}, \tag{10.12}$$

where each substance is in its standard state, that is, at a partial pressure of one atmosphere.   Under these conditions the positive value of ΔF° allows us to state categorically that reaction (10.12) will not proceed spontaneously.   On the other hand, if we were to start with isopropyl alcohol at a partial pressure of one atmosphere and no acetone or hydrogen, the alcohol might decompose spontaneously at 452.2°K, because, as the value of the equilibrium constant and the experimental data upon which it is based indicate, over 50 per cent dissociation may occur in the presence of a suitable catalyst.   Yields can be made even greater if one of the products is removed continually.

---

[3] We are still using the assumption that the behavior of these gases may be described adequately by the ideal gas law.   For precise formulations one must substitute fugacity of the gas in place of pressure.   In problems involving solutions, it will be necessary to use a relative fugacity called the *activity* (see Chapter 19).

Thus, if one is considering a given reaction in connection with the preparation of some substance, it is important not to be misled by positive values of $\Delta F°$, for $\Delta F°$ refers to the reaction under standard conditions. It is quite possible that appreciable yields can be obtained, even though a reaction may not go to "completion." Such a case is illustrated well by the example of isopropyl alcohol just cited. Only if the values of $\Delta F°$ are very large positive ones, perhaps greater than 10 kcal, can one be assured without calculations of the equilibrium constant that no significant degree of transformation can be obtained.

In addition, it is worth while to emphasize again that although the standard free-energy change, $\Delta F°$, refers to reactions with every substance under standard conditions, for example reaction (10.12), the equilibrium constant, K, calculated from equation (10.3) refers to the equilibrium state:

$$(CH_3)_2CHOH \ (g, P_{equil}) = (CH_3)_2CO \ (g, P_{equil}) + H_2 \ (g, P_{equil}). \quad (10.13)$$

For reaction (10.13), $\Delta F = 0$, according to the criterion developed in Chapter 8. Nevertheless, the equilibrium constant for (10.13) can be calculated from $\Delta F°$ for reaction (10.12).

III. DETERMINATION FROM MEASUREMENTS OF ELECTROMOTIVE FORCE

This method, like the one just described, depends upon the ability of the system to undergo a transformation *reversibly* in an electrical cell. In this case the system will be opposed by a counter electromotive force just sufficient to balance the electromotive force (e.m.f.) obtained in the electrical cell. The e.m.f. observed under such circumstances is related to the free-energy change for the reaction by the expression derived in Chapter 8:

$$\Delta F = -n\mathcal{F}\mathcal{E}. \quad (10.14)$$

If the substances involved in the transformation are in their standard states,[4] then the measured e.m.f., called $\mathcal{E}°$, may be used to calculate the standard free-energy change, $\Delta F°$:

$$\Delta F° = -n\mathcal{F}\mathcal{E}°. \quad (10.15)$$

A very simple example of a reaction to which this method is applicable is

$$\tfrac{1}{2}H_2 \ (g) + AgCl \ (s) = HCl \ (aq) + Ag \ (s). \quad (10.16)$$

---

[4] Actually, the measurements need not be made with every substance in its standard state. Methods of calculating $\mathcal{E}°$ from e.m.f. measurements are described in Chapters 20 and 21.

From measurements of the e.m.f. of a cell containing hydrogen and silver-silver chloride electrodes, respectively, dipping into a solution of hydrochloric acid, it is possible to calculate that $\mathcal{E}°$ is 0.22239 volt.   Hence

$$\Delta F° = \quad (1)\ (96{,}485)\ (0.22239)\ \text{volt-coulomb}$$

or $\Delta F° = -5160$ cal mole^{-1}.   (10.17)

IV. CALCULATION FROM THERMAL DATA AND THE THIRD LAW OF THERMODYNAMICS

The methods described so far depend directly or indirectly upon the reversible character of at least one reaction for every substance of interest. For some time it was a challenge to theoretical chemists to devise some method of calculation of free energies from thermal data alone (that is, enthalpies and heat capacities) so that the need for experiments under equilibrium conditions might be avoided.   One of the equations for ΔF derived in Chapter 8 is

$$\Delta F = \Delta H - T\Delta S, \qquad (10.18)$$

applicable to any isothermal reaction.   Clearly, if it were possible to obtain ΔS from thermal data alone, it would be a simple matter to calculate ΔF.   The calculation of ΔS from thermal data alone, however, cannot be made without the introduction of a new assumption beyond the first two laws of thermodynamics.   We shall discuss the nature of this new assumption and the consequences of it in the following chapter.

V. CALCULATION FROM SPECTROSCOPIC DATA AND STATISTICAL MECHANICS

Many free-energy changes, particularly for gaseous reactions, may be calculated from theoretical analyses of vibrational and rotational energies of molecules.   The parameters used in these calculations are obtained from spectroscopic data.   This method, however, depends upon assumptions beyond those of classical chemical thermodynamics and hence will not be described in this textbook.   The results of these calculations can be used, nevertheless, even prior to an understanding of the methods by which they have been obtained.

Exercises

1.   According to D. P. Stevenson and J. H. Morgan [*J. Am. Chem. Soc.*, **70**, 2773 (1948)] the equilibrium constant, K, for the isomerization reaction

cyclohexane (l) = methylcyclopentane (l)

may be expressed by the equation

$$\ln K = 4.814 - \frac{2059}{T}.$$

(a) Derive an equation for $\Delta F°$ as a function of T.

(b) Find $\Delta H°$ and $\Delta S°$ at 25°C.

(c) Find $\Delta H°$ and $\Delta S°$ at 0°C.

2. The standard potentials for a galvanic cell in which the reaction

$$\tfrac{1}{2}H_2 \text{ (g)} + AgCl \text{ (s)} = Ag \text{ (s)} + HCl \text{ (aq)}$$

is being carried on are given in Table 2 of Chapter 2.

(a) By graphical differentiation, find the curve for $(\partial \mathcal{E}°/\partial T)_P$ as a function of temperature.

(b) Calculate $\Delta F°$, $\Delta H°$, and $\Delta S°$ at 0, 25, and 60°C, respectively. Tabulate the values you obtain.

(c) The empirical equation for $\mathcal{E}°$ as a function of temperature is

$$\mathcal{E}° = 0.22239 - 645.52 \times 10^{-6} \, (t - 25)$$
$$- 3.284 \times 10^{-6} \, (t - 25)^2 + 9.948 \times 10^{-9}(t - 25)^3,$$

where t is centigrade temperature. Using this equation, determine $\Delta F°$, $\Delta H°$, and $\Delta S°$ at 60°C. Compare with the values obtained by the graphical method.

CHAPTER 11

The Third Law of Thermodynamics

I. PURPOSE OF THE THIRD LAW OF THERMODYNAMICS

As was pointed out in the preceding chapter, if it were possible to calculate $\Delta S°$ for a reaction from thermal data alone, it would be a simple matter to evaluate $\Delta F°$ from the relation

$$\Delta F° = \Delta H° - T \, \Delta S°. \tag{11.1}$$

For example, we might wish to determine $\Delta F°_{298.16}$ for the reaction

$$C \text{ (graphite)} + 2H_2 \text{ (g)} = CH_4 \text{ (g)}. \tag{11.2}$$

It is obvious that $\Delta H°$ can be determined from the following heats of combustion:

$$CH_4 \text{ (g)} + 2O_2 \text{ (g)} = CO_2 \text{ (g)} + 2H_2O \text{ (l)}; \ \Delta H°_{\text{combustion of } CH_4} \tag{11.3}$$

$$CO_2 \text{ (g)} = C \text{ (graphite)} + O_2 \text{ (g)}; \ -\Delta H°_{\text{combustion of } C} \tag{11.4}$$

$$2H_2O \text{ (l)} = 2H_2 \text{ (g)} + O_2 \text{ (g)}; \ -2\Delta H°_{\text{combustion of } H_2} \tag{11.5}$$

$$CH_4 \text{ (g)} = C \text{ (graphite)} + 2H_2 \text{ (g)};$$
$$\Delta H° = \Delta H°_{CH_4} - \Delta H°_C - 2\Delta H°_{H_2}. \tag{11.6}$$

Clearly, $\Delta H°$ for reaction (11.2) is merely the negative of that for reaction (11.6). It remains, however, to calculate $\Delta S°_{298.16}$.

The problem of calculating $\Delta S°_{298.16}$ can be reduced to an alternative problem by the following considerations. Let us assume for the moment that $\Delta S°_{0°K}$ is known. It is evident that we could add the following equations, for each of which ΔS at a pressure of one atmosphere can be determined from thermal data alone, to lead to $\Delta S°_{298.16}$:[1]

$$C \text{ (graphite, 0°K)} + 2H_2 \text{ (s, 0°K)} = CH_4 \text{ (s, 0°K)} ; \ \Delta S_{0°K} \tag{11.7}[2]$$

---

[1] In this introductory discussion, it is assumed that CH_4 and H_2 behave as ideal gases at 298°K, and hence the calculation has been outlined for a pressure of one atmosphere, that of the standard state of an ideal gas. Methods of making corrections for deviations from ideal behavior are described later in this chapter.

[2] Note that at 0°K hydrogen and methane would be in the solid state.

$$\text{CH}_4 \text{ (s, } 0°\text{K)} \qquad = \text{CH}_4 \text{ (s, } T_{\text{m.p.}}) \quad ; \quad \Delta S_1 = \int_0^{T_{\text{m.p.}}} \frac{(C_p)_1 \, dT}{T}$$

$$\text{(11.8)}$$

$$\text{CH}_4 \text{ (s, } T_{\text{m.p.}}) \qquad = \text{CH}_4 \text{ (l, } T_{\text{m.p.}}) \quad ; \quad \Delta S_2 = \frac{\Delta H_{\text{fusion}}}{T_{\text{m.p.}}} \quad \text{(11.9)}$$

$$\text{CH}_4 \text{ (l, } T_{\text{m.p.}}) \qquad = \text{CH}_4 \text{ (l, } T_{\text{b.p.}}) \quad ; \quad \Delta S_3 = \int_{T_{\text{m.p.}}}^{T_{\text{b.p.}}} \frac{(C_p)_2 \, dT}{T}$$

$$\text{(11.10)}$$

$$\text{CH}_4 \text{ (l, } T_{\text{b.p.}}) \qquad = \text{CH}_4 \text{ (g, } T_{\text{b.p.}}) \quad ; \quad \Delta S_4 = \frac{\Delta H_{\text{vaporization}}}{T_{\text{b.p.}}}$$

$$\text{(11.11)}$$

$$\text{CH}_4 \text{ (g, } T_{\text{b.p.}}) \qquad = \text{CH}_4 \text{ (g, } 298.16°) \quad ; \quad \Delta S_5 = \int_{T_{\text{b.p.}}}^{298.16°} \frac{(C_p)_3 \, dT}{T}$$

$$\text{(11.12)}$$

$$2\text{H}_2 \text{ (s, } T'_{\text{m.p.}}) \qquad = 2\text{H}_2 \text{ (s, } 0°\text{K)} \quad ; \quad \Delta S_6 = \int_{T'_{\text{m.p.}}}^{0°} \frac{2(C_p)_4 \, dT}{T}$$

$$\text{(11.13)}$$

$$2\text{H}_2 \text{ (l, } T'_{\text{m.p.}}) \qquad = 2\text{H}_2 \text{ (s, } T'_{\text{m.p.}}) \quad ; \quad \Delta S_7 = \frac{-2\,\Delta H'_{\text{fusion}}}{T'_{\text{m.p.}}}$$

$$\text{(11.14)}$$

$$2\text{H}_2 \text{ (l, } T'_{\text{b.p.}}) \qquad = 2\text{H}_2 \text{ (l, } T'_{\text{m.p.}}) \quad ; \quad \Delta S_8 = \int_{T'_{\text{b.p.}}}^{T'_{\text{m.p.}}} \frac{2(C_p)_5 \, dT}{T}$$

$$\text{(11.15)}$$

$$2\text{H}_2 \text{ (g, } T'_{\text{b.p.}}) \qquad = 2\text{H}_2 \text{ (l, } T'_{\text{b.p.}}) \quad ; \quad \Delta S_9 = \frac{-2\Delta H'_{\text{vaporization}}}{T'_{\text{b.p.}}}$$

$$\text{(11.16)}$$

$$2\text{H}_2 \text{ (g, } 298.16°) \qquad = 2\text{H}_2 \text{ (g, } T'_{\text{b.p.}}) \quad ; \quad \Delta S_{10} = \int_{298.16°}^{T'_{\text{b.p.}}} \frac{2(C_p)_6 \, dT}{T}$$

$$\text{(11.17)}$$

$$\text{C (graphite, } 298.16°) = \text{C (graphite, } 0°\text{K)}; \quad \Delta S_{11} = \int_{298.16°}^{0°} \frac{(C_p)_7 \, dT}{T}$$

$$\text{(11.18)}$$

$$\text{C (graphite, } 298.16°) \quad +2\text{H}_2 \text{ (g, } 298.16°) = \text{CH}_4 \text{ (g, } 298.16°) \quad ;$$

$$\Delta S^{\circ}_{298.16} = \Delta S^{\circ}_{0°\text{K}} + \sum_1^{11} \Delta S_i \quad \text{(11.19)}$$

The determination of the sum, $\sum_{1}^{11} \Delta S_i$, requires only a knowledge of heat capacities, heats of fusion and vaporization, and the respective transition temperatures. If the solids existed in more than one crystalline form, it would be necessary to know also the heats and temperatures of such transitions. Nevertheless, all these data are merely thermal data. If, then, we can obtain some information on $\Delta S_{0°K}$ without introducing nonthermal data, we shall have fulfilled our present objective.

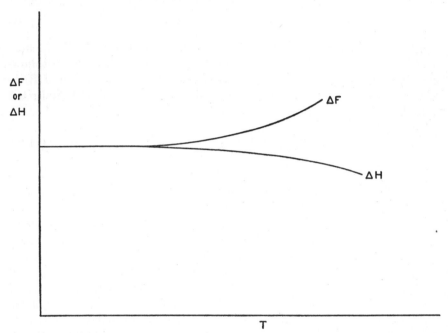

Fig. 1.   Limiting approach of ΔF and of ΔH as the temperature approaches absolute zero.

The first and second laws of thermodynamics give no indication of the value of $\Delta S_{0°K}$. Apparently it will be necessary to obtain a new principle, formulated from experimental observations, in order to solve this problem.

II. FORMULATION OF THE THIRD LAW

We have pointed out previously that for many reactions the contribution of the $T\ \Delta S$ term in equation (11.1) is relatively small, even at room temperatures, so that ΔF and ΔH are frequently close in value even at relatively high temperatures. In a comprehensive series of experiments on galvanic cells, Richards[3] showed, furthermore, that as the temperature decreases, ΔF approaches ΔH more and more closely, in the manner indi-

[3] T. W. Richards, Z. physik. Chem., **42**, 129 (1902).

cated in Fig. 1.   Although these results are really only fragmentary evidence, they did furnish the clues which led Nernst to the first formulation of the third law of thermodynamics.

A. Nernst heat theorem.   The trend of ΔF and ΔH toward each other may be expressed in mathematical form as follows:

$$\lim_{T \to 0} (\Delta F - \Delta H) = 0. \tag{11.20}$$

It is evident from a rearrangement of equation (11.1),

$$\Delta F - \Delta H = T\,\Delta S, \tag{11.21}$$

that the relation expressed by equation (11.20) may follow simply because as T approaches zero, so must $T\,\Delta S$, as long as ΔS is finite.   Nernst, however, made the additional assumption, based on the appearance of Richards' curves (Fig. 1), that the limiting value of ΔS was not only finite but actually zero for all condensed systems:

$$\lim_{T \to 0} \Delta S = \lim_{T \to 0} \left(\frac{\partial \Delta F}{\partial T} \right)_P = 0. \tag{11.22}$$

In essence, this assumption amounts to saying not only that ΔF approaches ΔH as T approaches $0°K$, but also that the ΔF curve (Fig. 1) approaches a horizontal limiting tangent.

It is conceivable, of course, that equation (11.20) may be valid and yet that expression (11.22) will not be fulfilled.   This situation would occur if the limiting value of ΔS were some finite number but not zero, and would correspond graphically to the curve illustrated in Fig. 2, where ΔF and ΔH approach each other at absolute zero but where the limiting slope of ΔF is finite.   In fact, Richards suggested that some of his data be extrapolated to give a graph such as shown in Fig. 2.   Thus from the data available to Nernst there was no assurance of the validity of equation (11.22).   Nevertheless, numerous experiments since then have confirmed this postulate, if it is limited to perfect crystalline systems.   *Apparent* exceptions have been accounted for satisfactorily.   The term "perfect" implies that we are dealing with a single pure substance.   There are other restrictions implied by this term, but they will be discussed later.

B. Planck's formulation.   In Nernst's statement of the third law no comment is made on the value of the absolute entropy at $0°K$, although it must be finite or zero if $\Delta S_{0°K}$ is to be finite for a reaction involving condensed phases.   Planck[4] extended Nernst's assumption by making the additional postulate that *the absolute value of the entropy of a pure solid or a pure liquid approaches zero at $0°\ K$:*

$$\lim_{T \to 0} S = 0. \tag{11.23}$$

---

[4] M. Planck, *Thermodynamik*, Veit & Co., Leipzig, 3d Ed., 1911, page 279.

If equation (11.23) is assumed, Nernst's theorem, equation (11.22), follows immediately for pure solids and liquids. Planck's formulation is also consistent with the treatment of entropy which is introduced in statistical mechanics.

Planck's statement asserts that $S_{0°K}$ is zero only for *pure solids and pure liquids*, whereas Nernst assumed that his theorem was applicable to *all condensed phases*, including solutions. According to Planck, solutions at $0°K$ have a positive entropy equal to the entropy of mixing.[5]

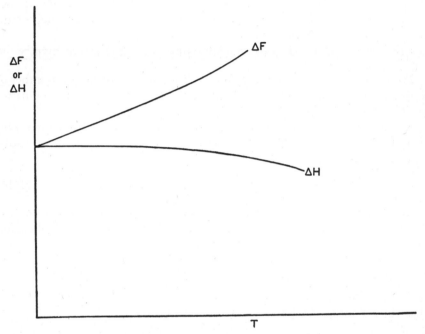

Fig. 2. A conceivable limiting approach of ΔF and of ΔH as the temperature approaches absolute zero.

C. Statement of Lewis and Randall. Lewis and Gibson[6] also emphasized the positive entropy of solutions at $0°K$ and in addition pointed out that supercooled liquids, such as glasses, even when composed of a single element (for example, sulfur), probably retain positive entropies even as the temperature approaches absolute zero. For these reasons Lewis and Randall[7] proposed the following statement of the third law of thermodynamics.

[5] An equation for the entropy of mixing in ideal solutions is derived in Chapter 17.
[6] G. N. Lewis and G. E. Gibson, *J. Am. Chem. Soc.*, **42**, 1529 (1920).
[7] G. N. Lewis and M. Randall, *Thermodynamics*, McGraw-Hill Book Company, Inc., New York, 1923, page 448.

If the entropy of each element in some crystalline state be taken as zero at the absolute zero of temperature: every substance has a finite positive entropy, but at the absolute zero of temperature the entropy may become zero, and does so become in the case of perfect crystalline substances.

It is this statement which we shall adopt as our working form of the third law of thermodynamics. From a practical standpoint of the chemist, that is, in making free-energy calculations, this statement is the most convenient formulation. Nevertheless, it should be realized that from a theoretical point of view more elegant formulations have been suggested.[8]

III. THERMODYNAMIC PROPERTIES AT ABSOLUTE ZERO

From the third law of thermodynamics it is possible to derive a number of limiting relationships which must be valid for thermodynamic quantities at the absolute zero of temperature.

A. Equivalence of F and H. It follows immediately from the statement of the third law and the definition of free energy that for any substance

$$F_{0°K} = H_{0°K} - TS_{0°K} = H_{0°K}. \tag{11.24}$$

B. ΔC_p in a chemical transformation. Starting with the expression (11.22), we can see readily from the Gibbs-Helmholtz relation [equation (8.124)],

$$\Delta F = \Delta H + T \left(\frac{\partial \Delta F}{\partial T} \right)_P, \qquad \overset{-\Delta S}{}$$

that the following relation must also be valid:

$$\lim_{T \to 0} \left(\frac{\partial \Delta F}{\partial T} \right)_P = \lim_{T \to 0} \frac{\Delta F - \Delta H}{T} = 0. \tag{11.25}$$

A glance at the expression $\lim_{T \to 0} \dfrac{\Delta F - \Delta H}{T}$ will show that as it stands it is an indeterminate form, since both $(\Delta F - \Delta H)$ and T approach zero. To resolve an indeterminate expression, we may apply the mathematical rule of differentiating numerator and denominator, respectively, with respect to the independent variable, T. Carrying out this procedure, we obtain from equation (11.25)

 L'Hôpital's Rule

---

[8] O. Stern, *Ann. Physik*, **49**, 823 (1916).   E. D. Eastman and R. T. Milner, *J. Chem. Phys.*, **1**, 444 (1933).   R. H. Fowler and E. A. Guggenheim, *Statistical Thermodynamics*, The Macmillan Company, New York, 1939, page 224.   P. C. Cross and H. C. Eckstrom, *J. Chem. Phys.*, **10**, 287 (1942).

$$\lim_{T \to 0} \frac{(\partial \, \Delta F / \partial T)_P - (\partial \, \Delta H / \partial T)_P}{1} = 0 \qquad (11.26)$$

and
$$\lim_{T \to 0} \left(\frac{\partial \, \Delta F}{\partial T} \right)_P = \lim_{T \to 0} \left(\frac{\partial \, \Delta H}{\partial T} \right)_P = \lim_{T \to 0} \Delta C_p. \qquad (11.27)$$

In view of our fundamental postulate, equation (11.22), it follows that

$$\lim_{T \to 0} \Delta C_p = 0. \qquad (11.28)$$

Many investigators have shown that ΔC_p does approach zero as T approaches absolute zero.   Nevertheless, these results in themselves do not constitute experimental evidence for the third law, because as long as the limiting value of $(\partial \, \Delta F / \partial T)_P$ is not infinite, it can be shown by a procedure corresponding to equations (11.26) and (11.27) that equation (11.28) must be valid.

C. Limiting values of C_p and C_v. The third law asserts that the entropy of any substance (referred to the corresponding elements) must be finite or zero at absolute zero.   In view of the finite values observed for ΔS at higher temperatures, it follows that the entropy of a substance must be finite at all (finite) temperatures.

In Chapter 7 it was shown that for a reversible isobaric temperature change in a substance

$$dS_P = \frac{DQ_P}{T} = \frac{C_p \, dT}{T}. \qquad (11.29)$$

This differential equation may be integrated, *at constant pressure*, to give

$$S = \int_0^T \frac{C_p \, dT}{T} + S_0, \qquad (11.30)$$

where S_0 represents an integration constant.   Since S must be finite at all temperatures, it follows that

$$\lim_{T \to 0} C_p = 0. \qquad (11.31)$$

If C_p had a finite value at $T = 0$, it is obvious that the integral in equation (11.30) would not converge, since T in the denominator goes to zero, and hence that S would not be finite.

By a completely analogous procedure, we can show that

$$\lim_{T \to 0} C_v = 0. \qquad (11.32)$$

D. Temperature coefficients of pressure and volume. In view of the statement that

$$\lim_{T \to 0} S = 0, \qquad (11.23)$$

it follows that in the limit of absolute zero of temperature the entropy of a perfect crystalline substance must be independent of changes in pressure or volume.[9]   Thus

$$\lim_{T \to 0} \left(\frac{\partial S}{\partial P} \right)_T = 0 \tag{11.33}$$

and

$$\lim_{T \to 0} \left(\frac{\partial S}{\partial V} \right)_T = 0. \tag{11.34}$$

In Chapter 8 we showed that

$$\left(\frac{\partial S}{\partial P} \right)_T = - \left(\frac{\partial V}{\partial T} \right)_P \tag{11.35}$$

and

$$\left(\frac{\partial S}{\partial V} \right)_T = \left(\frac{\partial P}{\partial T} \right)_V. \tag{11.36}$$

Combining equations (11.33) and (11.35), we obtain the expression

$$\lim_{T \to 0} \left(\frac{\partial V}{\partial T} \right)_P = 0. \tag{11.37}$$

Similarly, equations (11.34) and (11.36) lead to

$$\lim_{T \to 0} \left(\frac{\partial P}{\partial T} \right)_V = 0. \tag{11.38}$$

In other words, the temperature gradients of the pressure and volume vanish as absolute zero is approached.

IV.   TABLES OF ENTROPIES AT 298°K

In the statement which we have adopted for the third law, it is assumed (arbitrarily) that the entropy of each element in some crystalline state may be taken as zero at $0°K$.   For every perfect crystalline substance, then, the entropy is also zero at $0°K$.   Consequently, we have set S_0, the integration constant in equation (11.30), equal to zero.   Thus we may write

$$S = \int_0^T \frac{C_p \, dT}{T}. \tag{11.39}$$

Evidently, then, we can evaluate the entropy of a perfect crystalline solid at any specified temperature by integration of the heat-capacity data. The entropy so obtained is frequently called the "absolute" entropy and is

---

[9] One may generalize this statement to all the variables, x_1, x_2, ... x_i, which may determine the state of a substance at the absolute zero and hence conclude that $\lim_{T \to 0} (\partial S / \partial x_i)_T = 0$.

indicated as S_T^o.   In no sense, however, is S_T^o truly an absolute entropy, for we must always remember that behind equation (11.39) lies the assumption that the entropy of each element in some state at $0\,°K$ may be taken as zero.   It is clearly recognized that this assumption is merely one of convenience, since entropies associated with the nucleus, for example, are not zero.

Should we be interested in the "absolute" entropy of a substance at a temperature at which it is no longer a solid, it is necessary merely to add the entropy of transformation to a liquid or gas and the subsequent entropies of warming.   The same procedure would apply to a solid which exists in different crystalline forms as the temperature is raised.   The procedure may be illustrated best by an outline of some sample calculations.

A.   Typical calculations.

1.   *For solid or liquid.*   For either of these final states it is necessary to have heat-capacity data for the solid down to temperatures approaching absolute zero.   The integration indicated by equation (11.39) is then carried out in two steps.   From approximately $20\,°K$ up, graphical methods may be used.   Below $20\,°K$, however, few data are available.   It is customary, therefore, to rely on the Debye equation in this region.

a.   *Use of Debye equation at very low temperatures.*   It is generally assumed that the Debye equation expresses the behavior of the heat capacity adequately below about $20\,°K$.[10]   This relation [equation (4.40)],

$$C_p \simeq C_v = 464.5\,\frac{T^3}{\theta^3}\ \text{cal mole}^{-1}\,\text{deg}^{-1}, \tag{11.40}$$

contains but one constant, θ, which may be determined from one value of C_p in the region near $20\,°K$ or lower.   The integral for the absolute entropy then becomes

$$S = \int_0^T \frac{kT^3}{T}\,dT = \int_0^T kT^2\,dT, \tag{11.41}$$

where k represents $464.5/\theta^3$.

b.   *Absolute entropy of methylammonium chloride.*   Heat capacities for this solid in its various crystalline modifications have been determined very precisely down to $12\,°K$;[11] some of these data are summarized in Fig. 3.   In this case, where there are three crystalline forms between $0°$ and $298\,°K$, one can calculate the absolute entropy by integrating equation (11.39) for each allotrope in the temperature region in which it is most

---

[10] Deviations from the T^3 law and their significance have been discussed recently by K. Clusius and L. Schachinger, *Z. Naturforschung*, 2a, 90 (1947), and by G. L. Pickard and F. E. Simon, *Proc. Phys. Soc.*, 61, 1 (1948).

[11] J. G. Aston and C. W. Ziemer, *J. Am. Chem. Soc.*, 68, 1405 (1946).

stable, and by adding to the integrals thus obtained the two entropies of transition. The details as they are carried out in actual practice, at a pres-

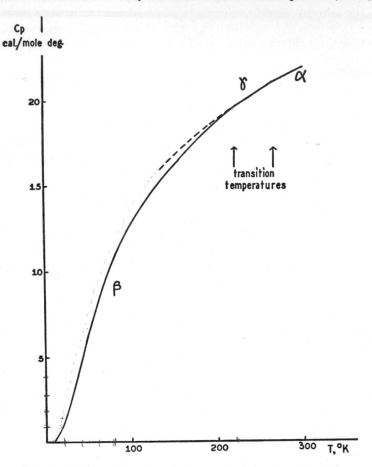

Fig. 3. Heat capacities of methylammonium chloride. Symbols α, β and γ indicate the three allotropic forms of the solid.

sure of one atmosphere, may be summarized as follows:

(a) CH_3NH_3Cl (s, 0°K, beta form) $=$ CH_3NH_3Cl (s, 12.04°K, beta form);

$$\Delta S_1 = \int_{0°K}^{12.04°K} \frac{C_p\, dT}{T} = 0.067 \text{ cal mole}^{-1} \text{ deg}^{-1}, \text{ Debye equation with } \theta = 200.5$$

(b) CH_3NH_3Cl (s, 12.04°, beta form) $=$ CH_3NH_3Cl (s, 220.4°, beta form);

$$\Delta S_2 = \int_{12.04°}^{220.4°} \frac{C_p\, dT}{T} = 22.326, \text{ graphical integration}$$

(c) CH_3NH_3Cl (s, 220.4°, beta form) $=$ CH_3NH_3Cl (s, 220.4° gamma form);

$$\Delta S_3 = \frac{\Delta H}{T} = \frac{425.2}{220.4} = 1.929$$

(d) CH_3NH_3Cl (s, 220.4°, gamma form) $=$ CH_3NH_3Cl (s, 264.5°, gamma form);

$$\Delta S_4 = \int_{220.4°}^{264.5°} \frac{C_p' \; dT}{T} = 3.690, \text{ graphical integration}$$

(e) CH_3NH_3Cl (s, 264.5°, gamma form) $=$ CH_3NH_3Cl (s, 264.5°, alpha form);

$$\Delta S_5 = \frac{\Delta H'}{T'} = \frac{673.6}{264.5} = 2.547$$

(f) CH_3NH_3Cl (s, 264.5°, alpha form) $=$ CH_3NH_3Cl (s, 298.16°, alpha form);

$$\Delta S_6 = \int_{264.5°}^{298.16°} \frac{C_p'' \; dT}{T} = 2.555, \text{ graphical integration}$$

Addition of Steps (a) to (f) gives

(g) CH_3NH_3Cl (s, 0°K) $=$ CH_3NH_3Cl (s, 298.16°K);

$$\Delta S° = \sum_1^6 \Delta S_i = 33.114 \text{ cal mole}^{-1} \text{ deg}^{-1}.$$

Thus for methylammonium chloride, S_{298}^o is 33.114 cal mole^{-1} deg^{-1} (or entropy units, E.U.).

2. *For a gas.* The procedure for the calculation of the absolute entropy of a gas in its standard state is substantially the same as that for a solid or liquid, except that there is one major correction factor necessary for the gas but not for the condensed phases. Since the absolute entropy is calculated from calorimetric heat-capacity data at one atmosphere pressure, the result obtained refers, of course, to such a pressure. For most real gases, however, the entropy at one atmosphere pressure is not the same as the entropy of the gas in its standard state. Frequently it will be necessary, therefore, to apply a correction to the observed entropy to account for deviations from ideal gas behavior.

a. *Correction for gas imperfection.* It is perhaps easiest to visualize the derivation of this correction by reference to Fig. 4. From thermal data one obviously obtains the entropy indicated by line b. However, it is the entropy indicated by line d which we need for our tables; that is, we want to find the entropy the substance would have if it behaved as an ideal gas at the given temperature and one atmosphere pressure.

At this stage in our discussion of thermodynamic concepts there is no obvious direct method of going from b to d. Since entropy is a thermodynamic property, however, we can use any roundabout path to calculate ΔS, so long as the initial and final states are b and d, respectively.

If we know S_{298} for the real gas at one atmosphere pressure, then we know ΔS for the reaction

$$A \text{ (s, 0°K)} = A \text{ (g, } P = 1, 298.16°); \quad \Delta S = S_{298.16} \text{ for } A. \quad (11.42)$$

To this equation we can add the following two transformations,

$$A \text{ (g, } P = 1, \text{ real, } 298.16°) = A \text{ (g, } P = P^*, \text{ real, } 298.16°) \quad (11.43)$$

$$A \text{ (g, } P = P^*, \text{ ideal, } 298.16°) = A \text{ (g, } P = 1, \text{ ideal, } 298.16°), \quad (11.44)$$

for each of which ΔS can be calculated. If P^* is made sufficiently low, or better yet, if P^* approaches zero, then the real gas approaches the ideal gas in properties such as the entropy. Hence the right-hand side of equation (11.43) is the same state as the left-hand side of equation (11.44), and addition of equations (11.43) and (11.44) gives

$$A \; (\text{g}, P = 1, \text{real}, 298.16°) = A \; (\text{g}, P = 1, \text{ideal}, 298.16°). \quad (11.45)$$

The sum of the entropy changes for reactions (11.42) and (11.45) gives us $S_{298.16}$ for the gas.

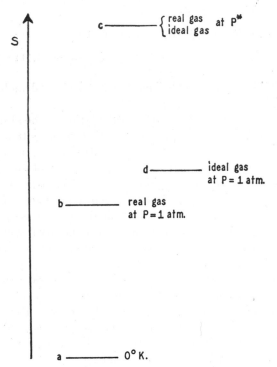

Fig. 4. Schematic representation of various entropies in connection with correction for gas imperfection.

There remains, then, the problem of calculation of the entropy changes for equations (11.43) and (11.44). Since the temperature is fixed, these quantities may be calculated from the relation

$$\left(\frac{\partial S}{\partial P}\right)_T = - \left(\frac{\partial V}{\partial T}\right)_P, \quad (11.35)$$

or

$$dS = - \left(\frac{\partial V}{\partial T}\right)_P dP. \quad (11.46)$$

Thus, for the transformation indicated in equation (11.43),

$$S\,(P = P^*) - S\,(P = 1) = -\int_1^{P^*} \left(\frac{\partial V}{\partial T}\right)_P dP = +\int_{P^*}^1 \left(\frac{\partial V}{\partial T}\right)_P dP$$

(11.47)

and for that in equation (11.44)

$$S_i\,(P = 1) - S_i\,(P = P^*) = -\int_{P^*}^1 \left(\frac{\partial V}{\partial T}\right)_P dP = -\int_{P^*}^1 \frac{R}{P}\, dP,$$

(11.48)

where the subscript i refers to the ideal gas. If P^* is allowed to approach zero pressure, then

$$S_i(P = P^*) = S\,(P = P^*).$$ (11.49)

Hence the correction for gas imperfection becomes

$$S_i\,(P = 1) - S\,(P = 1) = \int_{P^*}^1 \left(\frac{\partial V}{\partial T}\right)_P dP - \int_{P^*}^1 \frac{R}{P}\, dP$$ (11.50)

$$= \int_{P^*}^1 \left[\left(\frac{\partial V}{\partial T}\right)_P - \frac{R}{P}\right] dP.$$ (11.51)

Equation (11.51) may be integrated if the value of $(\partial V/\partial T)_P$ is known for the gas under consideration. Generally this coefficient is obtained from the Berthelot equation of state (Chapter 6),

$$Pv = RT\left[1 + \frac{9}{128}\frac{P}{P_c}\frac{T_c}{T}\left(1 - 6\frac{T_c^2}{T^2}\right)\right].$$ (11.52)

Differentiation gives readily

$$\left(\frac{\partial V}{\partial T}\right)_P = \frac{R}{P} + \frac{9}{128}R\frac{1}{P_c}T_c\left(12\frac{T_c^2}{T^3}\right) = \frac{R}{P}\left[1 + \frac{27}{32}\frac{P}{P_c}\frac{T_c^3}{T^3}\right].$$ (11.53)

The equation for the correction becomes, therefore,

$$(S_i - S)_{P=1\,\text{atm}} = \int_{P^*}^1 \frac{27}{32}\frac{R}{P_c}\frac{T_c^3}{T^3}\, dP$$ (11.54)

$$= \frac{27}{32}R\frac{P}{P_c}\frac{T_c^3}{T^3}\Bigg]_{P^*}^{P=1}.$$ (11.55)

If we let P^* approach zero,

$$(S_i - S)_{P=1\,\text{atm}} = \frac{27}{32}\frac{R}{P_c}\frac{T_c^3}{T^3}.$$ (11.56)

Thus from a knowledge of the critical constants of the gas it is possible now to evaluate the correction in the entropy for deviations from ideal behavior, if the Berthelot equation is applicable.[12]

b. *Entropy of gaseous cyclopropane at its boiling point.* Heat capacities for cyclopropane have been measured down to temperatures approaching absolute zero by Ruhrwein and Powell.[13] Their calculation of the absolute entropy of the gas at the boiling point, $240.30°K$, may be summarized as follows:

(a) C_3H_6 (s, $0°K$) $= C_3H_6$ (s, $15°K$); $\Delta S_1 = 0.243$ cal mole^{-1} deg^{-1}, Debye equation with $\theta = 130$

(b) C_3H_6 (s, $15°K$) $= C_3H_6$ (s, $145.54°K$); $\Delta S_2 = 15.733$, graphical integration

(c) C_3H_6 (s, $145.54°K$) $= C_3H_6$ (l, $145.54°K$); $\Delta S_3 = \dfrac{\Delta H_{fusion}}{T} = 8.939$

(d) C_3H_6 (l, $145.54°K$) $= C_3H_6$ (l, $240.30°K$); $\Delta S_4 = 9.176$, graphical integration

(e) C_3H_6 (l, $240.30°K$) $= C_3H_6$ (real gas, $240.30°K$); $\Delta S_5 = \dfrac{\Delta H_{vaporization}}{T} = 19.946$

Summing steps from (a) to (e), we obtain

(f) C_3H_6 (s, $0°K$) $= C_3H_6$ (real gas, $240.30°K$);

$$\Delta S = \sum_1^5 \Delta S_i = 54.04 \text{ cal mole}^{-1} \text{ deg}^{-1}.$$

The correction for deviations from ideality remains to be inserted. The critical constants were taken as $375°K$ and 50 atm. Hence from equation (11.56) we obtain

(g) C_3H_6 (real gas, $240.30°K$) $= C_3H_6$ (ideal gas, $240.30°K$);

$$\Delta S_6 = 0.13 \text{ cal mole}^{-1} \text{ deg}^{-1}.$$

Therefore the absolute entropy, $S^\circ_{240.30}$, of cyclopropane in the ideal gas (that is, standard) state is 54.17 E.U.

The correction for gas imperfection may often seem small, and there may be a tendency to neglect it. We should keep in mind, therefore, that an error of 0.1 E.U. affects the free energy by about 30 cal near room temperature, since ΔS would be multiplied by T. An error of 30 cal would change an equilibrium constant of 1.00 to 1.05, a difference of 5 per cent.

B. **Apparent exceptions to the third law.** There are several cases in which calculations of the entropy change of a reaction from values of the absolute entropy obtained from thermal data and the third law are in

---

[12] It has been pointed out recently [J. O. Halford, *J. Chem. Phys.*, **17**, 111, 405 (1949)] that the Berthelot equation may be inadequate for the calculation of entropy corrections due to gas imperfection, especially for vapors such as water and ethyl alcohol.

[13] R. A. Ruhrwein and T. M. Powell, *J. Am. Chem. Soc.*, **68**, 1063 (1946).

disagreement with values calculated directly from measurements of $\Delta H°$ and determinations of $\Delta F°$ from experimental equilibrium constants. For example, for the reaction

$$H_2 \text{ (g)} + \tfrac{1}{2}O_2 \text{ (g)} = H_2O \text{ (l)}, \tag{11.57}$$

$$\Delta S° \text{ (thermal data)} = -36.7^{14} \text{ cal mole}^{-1} \text{ deg}^{-1}, \tag{11.58}$$

$$\Delta S° \text{ (equilibrium)} = \frac{\Delta H° - \Delta F°}{T} = -39.1 \text{ cal mole}^{-1} \text{ deg}^{-1}. \tag{11.59}$$

Thus there is a large discrepancy between the two entropy values.

A satisfactory accounting of this discrepancy was not available until the development of statistical thermodynamics, with its methods of calculating entropies from spectroscopic data, and the discovery of the existence of *ortho-* and *para*-hydrogen. It was then found that the major portion of the deviation observed between equations (11.58) and (11.59) is due to the failure to obtain a true equilibrium between these two forms of H_2 molecules, differing in their nuclear spins, during thermal measurements at very low temperatures (Fig. 5). If true equilibrium were established at all times, more *para*-hydrogen would be formed as the temperature is lowered, and at 0°K all the hydrogen molecules would be in the *para* form and the entropy would be zero. In practice, measurements are actually made on a mixture of *ortho:para* of 3:1. This mixture at 0°K has a positive entropy. If the hydrogen were in contact with an appropriate catalyst for *ortho-para* conversion, an equilibrium mixture would be obtained and the entropy could be calculated correctly from the integral giving the area under the equilibrium curve in Fig. 5.

With the development of statistical thermodynamics and the calculation of the absolute entropies of many substances from spectroscopic data, several other substances in addition to hydrogen have been found to exhibit values of molal entropies in disagreement with those calculated from thermal data alone[15] (Table 1). Here again the discrepancies may be accounted for on the assumption that even near absolute zero not all the molecules are in the same state and that true equilibrium has not been attained. In the cases of CO, $COCl_2$, N_2O, and NO, the close similarity in the sizes of the atoms makes different orientations possible in the crystals,

---

[14] This figure is obtained when values of $S°_{298}$ for H_2 and H_2O are those determined from calorimetric data alone. Most recent critical tables list values of $S°_{2\,98}$ corrected for the effects discussed later in this section. If the latter values are used, $\Delta S°$ from the third law turns out to be -39.0 cal mole^{-1} deg^{-1}.

[15] W. F. Giauque and H. L. Johnston, *J. Am. Chem. Soc.*, **50**, 3221 (1928). H. L. Johnston and W. F. Giauque, *ibid.*, **51**, 3194 (1929). W. F. Giauque, *ibid.*, **52**, 4816 (1930). J. O. Clayton and W. F. Giauque, *ibid.*, **54**, 2610 (1932). W. F. Giauque and M. F. Ashley, *Phys. Rev.*, **43**, 81 (1933). R. W. Blue and W. F. Giauque, *J. Am. Chem. Soc.*, **57**, 991 (1935). W. F. Giauque and J. W. Stout, *ibid.*, **58**, 1144 (1936). W. F. Giauque and W. M. Jones, *ibid.*, **70**, 120 (1948). L. Pauling, *Phys. Rev.*, **36**, 430 (1930). L. Pauling, *J. Am. Chem. Soc.*, **57**, 2680 (1935).

TABLE 1
MOLAL ENTROPIES

Substance	Temperature, °K	$S°$ (spectro-scopic)	$S°$ (calori-metric)	Deviation
CO	298.1	47.313	46.2	1.11
$COCl_2$	280.6	68.26	66.63	1.63
H_2O	298.1	45.10	44.28	0.82
N_2O	298.1	52.581	51.44	1.14
NO	121.4	43.75	43.0	0.75

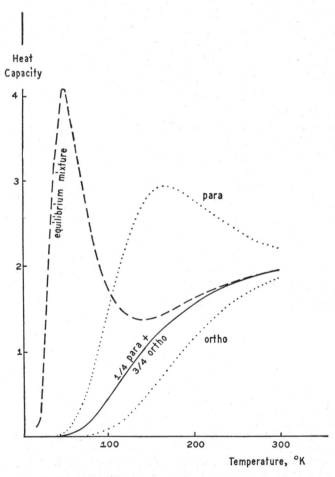

Fig. 5. Heat capacities (excluding translation) for hydrogen gas as a function of temperature. [Based on data of W. F. Giauque, J. Am. Chem. Soc., **52**, 4816 (1930).]

TABLE 2

ENTROPIES AT 298.16°K

Substance	S°_{298}, cal mole^{-1} deg^{-1}	Substance	S°_{298}, cal mole^{-1} deg^{-1}
Elements*, †			
Al (s)...........	6.75	Pb (s)..........	15.51
Sb (s)..........	10.5	Li (s).........	6.70
A (g)..........	36.983	Mg (s)........	7.77
As (s)..........	8.4	Mn (s)........	7.61
Ba (s)..........	15.1	Hg (l).........	18.5
Be (s)..........	2.28	Mo (s)........	6.83
Bi (s)..........	13.6	Ne (g)........	34.948
B (s)...........	1.7	Ni (s).........	7.12
Br$_2$ (l).........	36.4	N$_2$ (g).........	45.767
Cd (s)..........	12.3	O$_2$ (g).........	49.003
Ca (s)..........	9.95	P (white)......	10.6
C (graphite).....	1.3609	K (s).........	15.2
Cl$_2$ (g)..........	53.31	Rn (g)........	42.10
Cr (s)..........	5.68	Se (gray)......	10.0
Co (s)..........	6.8	Si (s).........	4.47
Cu (s)..........	7.97	Ag (s)........	10.20
F$_2$ (g)..........	48.58	Na (s)........	12.2
Ge (s)..........	10.14	S (rhombic).....	7.62
He (g).........	30.126	Sn (white)......	12.3
H$_2$ (g).........	31.211	Te (s).........	11.88
D$_2$ (g)..........	34.602	Ti (s).........	6.6
I$_2$ (s)..........	27.9	V (s)..........	7.0
Fe (s)..........	6.47	Xe (g)........	40.53
Kr (g).........	39.19	Zn (s)........	9.95
		Zr (s)........	9.5
Inorganic Compounds*, †			
BaO (s)........	16.8	H$_2$O (g).......	45.106
BaCl$_2$·2H$_2$O (s)..	48.5	HBr (g).......	47.437
BaSO$_4$ (s).......	31.6	HCl (g).......	44.617
BF$_3$ (g)........	60.70	HF (g)........	41.47
Ca(OH)$_2$ (s).....	17.4	HI (g)........	49.314
CO (g).........	47.301	ICl (g)........	59.12
CO$_2$ (g)........	51.061	NO (g)........	50.339
CNCl (g).......	56.31	NaCl (s)......	17.3
CuO (s)........	10.4	SO$_2$ (g)........	59.40
H$_2$O (l).........	16.716	SO$_3$ (g)........	61.24
Organic Compounds†			
Methane (g).....	44.50	Acetylene (g)...	47.997
Ethane (g)......	54.85	Benzene (g).....	64.34
Propane (g).....	64.51	Toluene (g).....	76.42
n-Butane (g).....	74.10	o-Xylene (g)....	84.31
Ethylene (g).....	52.45	m-Xylene (g)...	85.49
Propylene (g)....	63.80	p-Xylene (g)....	84.23
1-Butene (g).....	73.48	Cyclohexane (g).	71.28

* K. K. Kelley, *U. S. Bureau of Mines Bulletin 434* (1941).

† National Bureau of Standards, *Tables of Selected Values of Chemical Thermodynamic Properties.*

whereas in the case of H_2O, hydrogen bonds maintain an irregularity in the distribution of molecules in the crystal. Because of these exceptional situations, it is advisable to use the formulation of the third law suggested by Eastman and Milner,[8] or to interpret the term "perfect crystal" as excluding situations in which several orientations of the molecules are simultaneously present.

C. Tabulation of entropy values. Several sources are available which have critical tabulations of absolute entropies:

International Critical Tables.
K. K. Kelley, *U. S. Bureau of Mines Bulletin 434* (1941).
Landolt-Börnstein, *Physikalisch-chemische Tabellen.*
National Bureau of Standards, *Tables of Selected Values of Chemical Thermodynamic Properties.*
National Bureau of Standards, American Petroleum Institute Research Project 44, *Selected Values of Properties of Hydrocarbons.*

The new tables of the National Bureau of Standards are being distributed as the data are compiled. When they are complete, they will supersede all previous tabulations.

A few typical values of absolute entropies have been assembled in Table 2. Data obtained from spectroscopic studies have been included even though the methods used in their calculation have not been discussed.

Exercises

1. Assuming that

$$\lim_{T \to 0} \left(\frac{\partial \Delta F}{\partial T} \right)_P = 0$$

for reactions involving perfect crystalline solids, prove that

$$\lim_{T \to 0} \left(\frac{\partial \Delta A}{\partial T} \right)_V = 0.$$

2. Prove that $\lim_{T \to 0} \Delta C_v = 0$.

3. Assume that the limiting slope, as T approaches zero, of a graph of ΔF vs. T is a finite value but not zero. Prove that ΔC_p for the reaction would still approach zero at 0°K.

4. It has been suggested recently [G. J. Janz, *Can. J. Research*, **25b**, 331 (1947)] that α-cyanopyridine might be prepared from cyanogen and butadiene by the reaction

$$C_4H_6 \text{ (g)} + C_2N_2 \text{ (g)} = \text{(s)} + H_2 \text{ (g)}.$$

Pertinent thermodynamic data are given in the table below. Would you consider it worth while, on a thermodynamic basis, to attempt to work out this reaction?

Substance	ΔHf°_{298} (cal mole^{-1})	S°_{298} (cal mole^{-1} deg^{-1})
Butadiene (g)	26,748	66.42
Cyanogen (g)	71,820	57.64
α-Cyanopyridine (s)	62,000	77.09
Hydrogen (g)	0	31.21

5. Methylammonium chloride exists in a number of crystalline forms, as is evident from Fig. 3. The thermodynamic properties of the β and γ forms have been investigated by Aston and Ziemer [J. Am. Chem. Soc., 68, 1405 (1946)] down to temperatures near 0°K, and some of their data are listed below. From the information here given, find the heat of transition from the β to the γ form at 220.4° K.

C_p for β at 12.0°K = 0.202 cal mole^{-1} deg^{-1},
$\int C_p \, d \ln T$ from 12.0 to 220.4°K = 22.326 cal mole^{-1} deg^{-1},
C_p for γ at 19.5°K = 1.426 cal mole^{-1} deg^{-1},
$\int C_p \, d \ln T$ from 19.5 to 220.4°K = 23.881 cal mole^{-1} deg^{-1},
Normal transition temperature (that is, at $P = 1$ atm) = 220.4°K.

6. The equilibrium constant, K, for the formation of a deuterium atom from two hydrogen atoms may be defined by the equation

$$2H = D; \quad K = \frac{p_D}{(p_H)^2}.$$

The equation for the temperature dependence of K is

$$\log_{10} K = 20.260 + \frac{3}{2} \log_{10} T - \frac{7.04 \times 10^9}{T}.$$

(a) Calculate K at a temperature of 10^8 degrees.
(b) Calculate ΔF° and ΔH° at the same temperature.
(c) What is the change in entropy for the conversion of atomic hydrogen into atomic deuterium at a temperature of 10^8 degrees?

CHAPTER 12

Determination of Standard Free Energies with the Aid of the Third Law of Thermodynamics

HAVING considered the details of the calculation of absolute entropies for standard conditions, let us now turn our attention to their primary use—the calculation of standard free energies.

The methods of estimating thermodynamic functions may be divided into two categories—precise and approximate. Where sufficient data are available, the precise methods are used. In the absence of adequate information, one may still obtain reasonable estimates of entropies, enthalpies, and free energies by having recourse to one or more of the semi-empirical approximate methods.

I. PRECISE METHODS

A. Enthalpy determination. As was mentioned in the preceding chapter, it is necessary in all cases to obtain information on the standard enthalpy change for a given reaction before the free-energy change can be calculated from the third law. ΔH_T°, at some temperature T, may be obtained from ΔHf° of each of the substances involved in the transformation. Data on the standard heats of formation are found tabulated in either of two ways. The obvious method is to list ΔHf° at some convenient temperature, such as $25°C$, or at a series of temperatures. A typical table has been included in Chapter 5. Values at temperatures not listed are generally calculated with the aid of heat-capacity equations.

Under the stimulus of statistical thermodynamics, another method of tabulation, using $H_T^\circ - H_0^\circ$ or $(H_T^\circ - H_0^\circ)/T$, where the subscripts refer to the absolute temperatures, has come into general use. This method of presentation, illustrated in Table 1, avoids the inclusion of empirical heat-capacity equations in the results and allows more ready comparison of data from different sources.

The procedure necessary to the calculation of ΔHf° at any temperature, T, from the new type of table requires a brief discussion. From the definition of the standard heat of formation of a compound, C, it is evident that reference is made to the reaction

$$A \quad + \quad B \quad + \ldots = \quad C; \, \Delta Hf_T^\circ. \quad (12.1)$$

$$\left\{ \begin{matrix} \text{element in} \\ \text{standard state} \\ \text{at temperature } T \end{matrix} \right\} \quad \left\{ \begin{matrix} \text{element in} \\ \text{standard state} \\ \text{at temperature } T \end{matrix} \right\} \quad \left\{ \begin{matrix} \text{compound in} \\ \text{standard state} \\ \text{at temperature } T \end{matrix} \right\}$$

Since the enthalpies are available in terms of the function $H_T^\circ - H_0^\circ$, it is apparent that the sum of the following equations gives the desired ΔHf_T°:

$$A\ (0°\text{K}) + B\ (0°\text{K}) + \ldots = C\ (0°\text{K}) ; \quad \Delta Hf_0^\circ \tag{12.2}$$

$$C\ (0°\text{K}) = C\ (T°\text{K}) ; \quad \Delta H^\circ = (H_T^\circ - H_0^\circ)_C \tag{12.3}$$

$$A\ (T°\text{K}) = A\ (0°\text{K}) ; \quad \Delta H^\circ = (H_0^\circ - H_T^\circ)_A \tag{12.4}$$
$$= -\,(H_T^\circ - H_0^\circ)_A$$

$$B\ (T°\text{K}) = B\ (0°\text{K}) ; \quad \Delta H^\circ = (H_0^\circ - H_T^\circ)_B \tag{12.5}$$
$$= -(H_T^\circ - H_0^\circ)_B$$

$$A\ (T°\text{K}) + B\ (T°\text{K}) + \ldots = C\ (T°\text{K}) ;$$
$$\Delta Hf_T^\circ = \Delta Hf_0^\circ + (H_T^\circ - H_0^\circ)_{\text{compound}} - \sum (H_T^\circ - H_0^\circ)_{\text{elements}} \tag{12.6}$$

TABLE 1*

ENTHALPY FUNCTION, $H_T^\circ - H_0^\circ$

Substance	ΔHf_0°, kcal mole^{-1}	$H_T^\circ - H_0^\circ$ in cal mole^{-1} at $T°$K					
		298.16	400	600	800	1000	1500
H_2 (g)	0	2023.81	2731.0	4,128.6	5,537.4	6,965.8	10,694.2
O_2 (g)	0	2069.78	2792.4	4,279.2	5,854.1	7,497.0	11,776.4
C (graphite)	0	251.56	502.6	1,198.1	2,081.7	3,074.6	5,814
CO (g)	−27.2019	2072.63	2783.8	4,209.5	5,699.8	7,256.5	11,358.8
CO_2 (g)	−93.9686	2238.11	3194.8	5,322.4	7,689.4	10,222	17,004
H_2O (g)	−57.107	2367.7	3194.0	4,882.2	6,689.6	8,608.0	13,848
Methane (g)	−15.987	2397	3323	5,549	8,321	11,560	21,130
Ethane (g)	−16.517	2856	4296	8,016	12,760	18,280	34,500
Propane (g)	−19.482	3512	5556	10,930	17,760	25,670	48,650
Ethylene (g)	14.522	2525	3711	6,732	10,480	14,760	27,100
Acetylene (g)	54.329	2391.5	3541.2	6,127	8,999	12,090	20,541
Benzene (g)	24.000	3401	5762	12,285	20,612	30,163	57,350
Toluene (g)	17.500	4306	7269	15,334	25,621	37,449	71,250
o-Xylene (g)	11.096	5576	9291	19,070	31,386	45,531	85,960
m-Xylene (g)	10.926	5325	8925	18,563	30,817	44,933	85,330
p-Xylene (g)	11.064	5358	8929	18,499	30,690	44,755	85,080

* Taken from National Bureau of Standards, *Tables of Selected Values of Chemical Thermodynamic Properties.*

Each of the quantities in equation (12.6) may be obtained from tables such as Table 1.

B. Entropy determination. Standard absolute entropies for many substances are available in tables, such as Table 2 of Chapter 11. Generally the values listed are for 25°C, but many of the original sources, such as the tables of the National Bureau of Standards, give enumerations for other temperatures also. If heat-capacity data are available, no difficulty is encountered in converting entropy values from one temperature to another by methods analogous to those outlined at the beginning of the preceding chapter. Similar procedures can be used to obtain absolute entropies from heat-capacity data, if the appropriate integrations have not been made in the literature.

It is perhaps trivial to point out that in a reaction such as that represented by equation (12.1), the standard entropy change, ΔS_T°, at the temperature T, is given by the expression

$$\Delta S_T^{\circ} = S_T^{\circ} \text{ (compound)} - \sum S_T^{\circ} \text{ (elements)}. \tag{12.7}$$

C. Change in standard free energy. When adequate enthalpy and entropy data are available, the calculation of ΔF_T° is simply a matter of substitution into the equation

$$\Delta F_T^{\circ} = \Delta H_T^{\circ} - T \Delta S_T^{\circ}. \tag{12.8}$$

Generally data on ΔH° and ΔS° are available at least at one temperature. The conversion of the free-energy data from one temperature to another can be carried out by the methods outlined in Chapter 8.

With the development of statistical methods and the use of spectroscopic information, an alternative method of presentation of free-energy data as a function of temperature has also come into use. This consists of tabulations of the function $(F_T^{\circ} - H_0^{\circ})/T$, as is illustrated in Table 2. This method of tabulation avoids the use of empirical equations, with their associated constants, and allows direct comparison of data from different sources. Although we shall not discuss the methods used for calculating this new function from experimental data, we shall find it simple to use tables of these functions to obtain the free-energy change in a specified reaction.

If, for example, we wish to know the free energy of formation of a specified compound at a given temperature, we can start with equation (12.2), that for the formation of the compound at $0°K$:

$$A \ (0°K) + B \ (0°K) + \ldots = C \ (0°K); \quad \Delta F_0^{\circ} = \Delta H f_0^{\circ}. \tag{12.9}$$

At $0°K$ the enthalpy of formation of the compound, $\Delta H f_0^{\circ}$, equals the free-energy change, since ΔS° is finite or zero and T is zero. If we add the following equations to (12.9), to bring the reaction up to the temperature T, we must also add the accompanying free-energy changes.

$$C \ (0°K) = C \ (T°K); \quad (\Delta F^{\circ})_C = (F_T^{\circ} - F_0^{\circ})_C = (F_T^{\circ} - H_0^{\circ})_C \tag{12.10}$$
$$B \ (T°K) = B \ (0°K); \quad (\Delta F^{\circ})_B = (F_0^{\circ} - F_T^{\circ})_B = -(F_T^{\circ} - F_0^{\circ})_B$$
$$= -(F_T^{\circ} - H_0^{\circ})_B \tag{12.11}$$
$$A \ (T°K) = A \ (0°K); \quad (\Delta F^{\circ})_A = -(F_T^{\circ} - H_0^{\circ})_A. \tag{12.12}$$

In each case we make use of the relation

$$F_0^{\circ} = H_0^{\circ} \tag{12.13}$$

at $0°K$ [see equation (11.24)]. The summation of equations (12.9)–(12.12) leads to the expression

$$A \ (T°K) + B \ (T°K) + \ldots = C \ (T°K);$$
$$\Delta F_T^{\circ} = \Delta H_0^{\circ} + (F_T^{\circ} - H_0^{\circ})_{\text{compound}} - \sum (F_T^{\circ} - H_0^{\circ})_{\text{elements}}. \tag{12.14}$$

TABLE 2*

FREE-ENERGY FUNCTION, $\dfrac{F_T^\circ - H_0^\circ}{T}$

| Substance | $(F_T^\circ - H_0^\circ)/T$ in cal mole^{-1} deg^{-1} at T°K | | | | | |
	298.16	400	600	800	1000	1500
H_2 (g)	− 24.423	− 26.422	− 29.203	− 31.186	− 32.738	− 35.590
O_2 (g)	− 42.061	− 44.112	− 46.968	− 49.044	− 50.697	− 53.808
C (graphite)	− 0.5172	− 0.8245	− 1.477	− 2.138	− 2.771	− 4.181
CO (g)	− 40.350	− 42.393	− 45.222	− 47.254	− 48.860	− 51.864
CO_2 (g)	− 43.555	− 45.828	− 49.238	− 51.895	− 54.109	− 58.481
H_2O (g)	− 37.172	− 39.508	− 42.768	− 45.131	− 47.018	− 50.622
Methane (g)	− 36.46	− 38.86	− 42.39	− 45.21	− 47.65	− 52.84
Ethane (g)	− 45.27	− 48.24	− 53.08	− 57.29	− 61.11	− 69.46
Propane (g)	− 52.73	− 56.48	− 62.93	− 68.74	− 74.10	− 85.86
Ethylene (g)	− 43.98	− 46.61	− 50.70	− 54.19	− 57.29	− 63.94
Acetylene (g)	− 39.976	− 42.451	− 46.313	− 49.400	− 52.005	− 57.231
Benzene (g)	− 52.93	− 56.69	− 63.70	− 70.34	− 76.57	− 90.45
Toluene (g)	− 61.98	− 66.74	− 75.52	− 83.79	− 91.53	− 108.75
o-Xylene (g)	− 65.61	− 71.74	− 82.81	− 93.01	− 102.46	− 123.30
m-Xylene (g)	− 67.63	− 73.50	− 84.20	− 94.18	− 103.48	− 124.11
p-Xylene (g)	− 66.26	− 72.15	− 82.83	− 92.76	− 102.02	− 122.59

* Taken from National Bureau of Standards, *Tables of Selected Values of Chemical Thermodynamic Properties*.

To reduce the range of numerical values, it is convenient to tabulate $(F_T^\circ - H_0^\circ)/T$ instead of $(F_T^\circ - H_0^\circ)$. Hence we have the following expression for the free energy of formation, $\Delta F f_T^\circ$, of any compound at some specified temperature, T:

$$\frac{\Delta F f_T^\circ}{T} = \left[\frac{\Delta H f_0^\circ}{T} + \left(\frac{F_T^\circ - H_0^\circ}{T}\right)\right]_{\text{compound}} - \Sigma \left(\frac{F_T^\circ - H_0^\circ}{T}\right)_{\text{elements}}. \quad (12.15)$$

Thus the free energy of formation of a compound can be evaluated from tables of this new function.

It is apparent, of course, that if the free energy of formation of each substance in a reaction is known, it is a simple matter to calculate the free-energy change in any reaction involving these substances, since

$$\Delta F_T^\circ = \Sigma \Delta F f_T^\circ \text{ (products)} - \Sigma \Delta F f_T^\circ \text{ (reactants)}. \quad (12.16)$$

II. APPROXIMATE METHODS

Precision data are available for relatively few compounds. In many situations, however, it is desirable to have some idea of the feasibility or impossibility of a given chemical transformation long before the necessary thermodynamic data may become available. To accomplish this purpose, several groups of investigators[1-4] have proposed empirical methods of

[1] G. S. Parks and H. M. Huffman, *The Free Energies of Some Organic Compounds*, Reinhold Publishing Corporation, New York, 1932.

[2] R. R. Wenner, *Thermochemical Calculations*, McGraw-Hill Book Company, Inc., New York, 1941.

[3] J. W. Andersen, G. H. Beyer, and K. M. Watson, *National Petroleum News, Tech. Sec.*, **36**, R476 (July 5, 1944).

[4] J. G. M. Bremner and G. D. Thomas, *Trans. Faraday Soc.*, **44**, 230 (1948).

correlation which allow one to make approximate estimates of the thermo-dynamic properties required to calculate free energies and equilibrium distributions. We shall consider in some detail only one of these procedures, that of Andersen, Beyer, and Watson.

A. Group-contribution method of Andersen, Beyer, and Watson. In common with several other systems, this method is based on the assumption that a given thermodynamic property, such as the entropy, of an organic substance may be resolved into contributions from each of the constituent groups in the molecule. With tables of such group contributions assembled from available experimental data, we can estimate the thermodynamic properties of any molecule by adding the contributions of the constituent groups.

Generally speaking, there are several alternative methods of associating groups into a specified molecule. In the Andersen-Beyer-Watson approach, the thermodynamic properties in *the ideal gaseous state* are estimated by consideration of a given compound as built up from a base group, such as one of those listed in Table 3, which has been modified by appropriate substitutions to yield the desired molecule. Thus, aliphatic hydrocarbons may be built up from methane by repeated substitutions of methyl groups for hydrogen atoms. Similarly, any amide may be viewed as a derivative of formamide, any primary amine as a derivative of methylamine.

As the next step in the process of building up larger and more complex molecules, we may consider the changes associated with primary substitutions of CH_3 groups in each of the nine base groups listed in Table 3. For the first such substitution on a single carbon atom, the changes in thermodynamic properties are listed in Table 4. For the cyclic base groups—cyclopentane, benzene, and naphthalene—several carbon atoms are available for successive primary substitutions (no more than one on each carbon atom); the magnitude of the contribution depends upon the number and position of the added methyl groups, as well as on the type of base ring. Benzene and naphthalene may be considered together, but cyclopentane forms a category of its own. In the latter ring system, *ortho* refers to adjacent carbons on the nucleus, *meta* to carbon atoms separated by a (minimum of a) single carbon within the ring. In the naphthenes formed by enlargement of the cyclopentane ring, we may also refer to a *para* substitution when the second replacement is on a carbon atom in the ring separated by (a minimum of) two carbons from the atom on which the first substitution was made.

Attention may be turned now to the effects of a second substitution on a single carbon atom of one of the base groups. These secondary replacements have to be treated in more detail because the changes in thermodynamic properties depend upon the nature of the carbon atom on which the replacement is being made, as well as upon that of the adjacent carbon

TABLE 3
BASE GROUP PROPERTIES

Group	$\Delta Hf^{\circ}_{298.1}$ (g), kcal mole^{-1}	$S^{\circ}_{298.1}$ (g), cal mole^{-1} deg^{-1}	Heat-Capacity Constants for Ideal Gas at $T^{\circ}K$		
			a	$b\,(10^3)$	$c\,(10^6)$
Methane	-17.9	44.5	3.42	17.85	-4.16
Cyclopentane	-21.4	70.7	2.62	82.67	-24.72
Benzene	18.1	64.4	0.23	77.83	-27.16
Naphthalene	35.4	80.7	3.15	109.40	-34.79
Methylamine	-7.1	57.7	4.02	30.72	-8.70
Dimethylamine	-7.8	65.2	3.92	48.31	-14.09
Trimethylamine	-10.9	....	3.93	65.85	-19.48
Dimethyl ether	-46.0	63.7	6.42	39.64	-11.45
Formamide	-49.5	.....	6.51	25.18	-7.47

TABLE 4
CONTRIBUTIONS OF PRIMARY CH$_3$ SUBSTITUTION GROUPS REPLACING HYDROGEN

Base Group	$\Delta(\Delta Hf^{\circ}_{298.1})$ (g), kcal mole^{-1}	$\Delta S^{\circ}_{298.1}$ (g), cal mole^{-1} deg^{-1}	Heat-Capacity Constants for Ideal Gas at $T^{\circ}K$		
			Δa	$\Delta b(10^3)$	$\Delta c(10^6)$
1. Methane	-2.2	10.4	-2.04	24.00	-9.67
2. Cyclopentane					
(a) Enlargement of ring	-9.3	0.7	-1.04	19.30	-5.79
(b) First substitution	-5.2	11.5	-0.07	18.57	-5.77
(c) Second substitution:					
ortho	-12.2	....			
meta	-8.4	....	-0.24	16.56	-5.05
para	-7.1	....			
(d) Third substitution	-7.0	....	....	....	....
3. Benzene and naphthalene					
(a) First substitution	-4.5	12.0	0.36	17.65	-5.88
(b) Second substitution:					
ortho	-6.3	8.1	5.20	6.02	1.18
meta	-6.5	9.2	1.72	14.18	-3.76
para	-8.0	7.8	1.28	14.57	-3.98
(c) Third substitution (*sym*)	....	8.0	0.57	16.51	-5.19
4. Methylamine	-5.7	....			
5. Dimethylamine	-6.3	....	-0.10	17.52	-5.35
6. Trimethylamine	-4.1	....			
7. Formamide Substitution on C atom	-9.0	....	6.11	-1.75	4.75

atom. For this reason, these carbon atoms are characterized by "type numbers," as follows:

Type Number	Nature
1	$-CH_3$
2	$-\overset{\mid}{C}H_2$
3	$-\overset{\mid}{C}H\underset{\mid}{}$
4	$-\overset{\mid}{\underset{\mid}{C}}-$
5	C in benzene or naphthalene ring

The thermodynamic changes associated with secondary methyl substitutions then may be tabulated as in Table 5. The number in the column headed "A" is the type number of the carbon atom on which the second methyl substitution is made, and that in the column headed "B" is the highest type number of an adjacent carbon atom.

In connection with these secondary methyl substitutions in Table 5, it is necessary to introduce two special categories for use in estimating changes in thermodynamic properties for esters and ethers. The last entry in Table 5 gives the changes accompanying the replacement of the H atom in the OH of a carboxyl group by a CH_3 group to form a methyl ester. Similarly, the next to the last entry refers to the replacement of an H atom on the methyl group of an OCH_3 group in either an ether or an ester to form the corresponding ethyl ether or ester, as the case may be.

The effects of introducing multiple bonds in a molecule are treated in a category by themselves. The appropriate corrections have been assembled in Table 6 and require no special comments, except perhaps an emphasis upon the *additional* contribution which must be introduced every time a pair of conjugated double bonds is formed by any of the preceding substitutions in this table.

Finally we shall consider the changes in properties accompanying the introduction of various functional groups in place of one or two of the methyl groups on a given carbon atom. These are listed in Table 7. The only likely source of confusion is perhaps in the formation of aldehydes or ketones. The figures given in the table refer to the changes accompanying the replacement of *two* methyl groups by a $=O$. Thus, when such a substitution leads to the formation of an aldehyde, the entropy change is -12.3 cal mole^{-1} deg^{-1}. Similarly, a loss of two methyl groups and gain of a $=O$ to give a ketone is accompanied by a change of -2.4 entropy units.

TABLE 5

SECONDARY METHYL SUBSTITUTIONS REPLACING HYDROGEN

A	B	$\Delta(\Delta Hf^{\circ}_{298.1})$ (g), kcal mole^{-1}	$\Delta S^{\circ}_{298.1}$ (g), cal mole^{-1} deg^{-1}	Heat-Capacity Constants for Ideal Gas at T°K		
				Δa	$\Delta b(10^3)$	$\Delta c(10^6)$
1	1	−4.5	9.8	−0.97	22.86	− 8.75
1	2	−5.2	9.2	1.11	18.47	− 6.85
1	3	−5.5	9.5	1.00	19.88	− 8.03
1	4	−5.0	11.0	1.39	17.12	− 5.88
1	5	−6.1	10.0	0.10	17.18	− 5.20
2	1	−6.6	5.8	1.89	17.60	− 6.21
2	2	−6.8	7.0	1.52	19.95	− 8.57
2	3	−6.8	6.3	1.01	19.69	− 7.83
2	4	−5.1	6.0	2.52	16.11	− 5.88
2	5	−5.8	2.7	0.01	17.42	− 5.33
3	1	−8.1	2.7	−0.96	27.47	−12.38
3	2	−8.0	4.8	−1.19	28.77	−12.71
3	3	−6.9	5.8	−3.27	30.96	−14.06
3	4	−5.7	1.7	−0.14	24.57	−10.27
3	5	−9.2	1.3	0.42	16.20	− 4.68
1 —O— in ester or ether		−7.0	14.4	−0.01	17.58	− 5.33
Substitution of H of an OH group to form ester		+9.5	16.7	0.44	16.63	− 4.95

TABLE 6

MULTIPLE-BOND CONTRIBUTIONS REPLACING SINGLE BONDS

A	Type of Bond	B	$\Delta(\Delta Hf^{\circ}_{298.1})$ (g), kcal mole^{-1}	$\Delta S^{\circ}_{298.1}$ (g), cal mole^{-1} deg^{-1}	Heat-Capacity Constants for Ideal Gas at T°K		
					Δa	$\Delta b(10^3)$	$\Delta c(10^6)$
1	=	1	32.8	− 2.1	1.33	−12.69	+ 4.77
1	=	2	30.0	0.8	1.56	−14.87	+ 5.57
1	=	3	28.2	2.2	0.63	−23.65	+13.10
2	=	2	28.0	− 0.9	0.40	−18.87	+ 9.89
2	=	2 *cis*	28.4	− 0.6	0.40	−18.87	+ 9.89
2	=	2 *trans*	27.5	− 1.2	0.40	−18.87	+ 9.89
2	=	3	26.7	1.6	0.63	−23.65	+13.10
3	=	3	25.5	....	−4.63	−17.84	+11.88
Additional correction for each pair of conjugated double bonds			−3.8	−10.4	Approximately zero		
1	≡	1	74.4	− 6.8	5.58	−31.19	+11.19
1	≡	2	69.1	− 7.8	6.42	−36.41	+14.53
2	≡	2	65.1	− 6.3	4.66	−36.10	+15.28
Correction for double bond adjacent to aromatic ring			−5.1	− 4.3	Approximately zero		

The procedure followed in the use of the tables of Andersen, Beyer, and Watson has been described for the estimation of standard entropies. These tables also include columns of base structure and group contributions for estimating $\Delta Hf^\circ_{298.1}$, the standard heat of formation of a compound, as well as Δa, Δb, and Δc, the constants in the heat-capacity equations described in Chapter 8, page 126. Thus it is possible to estimate $\Delta F^\circ_{298.1}$ by appropriate summations of group contributions to $\Delta H^\circ_{298.1}$ and $\Delta S^\circ_{298.1}$. Then, if information is required at some other temperature, the constants

TABLE 7

SUBSTITUTION GROUP CONTRIBUTIONS REPLACING CH₃ GROUP

Group	$\Delta(\Delta Hf^\circ_{298.1})$ (g), kcal mole^{-1}	$\Delta S^\circ_{298.1}$ (g), cal mole^{-1} deg^{-1}	Heat-Capacity Constants for Ideal Gas at T°K		
			Δa	$\Delta b(10^3)$	$\Delta c(10^6)$
—OH (aliphatic, meta, para)	−32.7	2.6	3.17	−14.86	5.59
—OH ortho	−47.7	.....	....	.....	.....
—NO₂	1.2	2.0	6.3	−19.53	10.36
—CN	39.0	13.1	3.64	−13.92	4.53
—Cl	0 for first Cl on a carbon; 4.5 for each additional	0	2.19	−18.85	6.26
—Br	10.0	3.0*	2.81	−19.41	6.33
—F	−35.0	− 1.0*	2.24	−23.61	11.79
—I	24.8	5.0*	2.73	−17.37	4.09
=O aldehyde	−12.9	−12.3	3.61	−55.72	22.72
=O ketone	−13.2	− 2.4	5.02	−66.08	30.21
—COOH	−87.0	15.4	8.50	−15.07	7.94
—SH	15.8	5.2	4.07	−24.96	12.37
—C₆H₅	32.3	21.7	−0.79	53.63	−19.21
—NH₂	12.3	− 4.8	1.26	− 7.32	2.23

* Add 1.0 to the calculated entropy contributions of halides for methyl derivatives; for example, methyl chloride = 44.4 (base) + 10.4 (primary CH₃) −0.0 (Cl substitution) + 1.0.

of the heat-capacity equations may be inserted into the appropriate equations for ΔF° as a function of temperature, and ΔF° may be evaluated at any desired temperature. The details of this procedure have been described adequately in Chapter 8 and hence will not be repeated at this point.

B. Typical problems in estimation of entropies. The use of these tables is illustrated best by consideration of several specific examples.

Example 1. Estimate the entropy, $S^\circ_{298.1}$, of *trans*-2-pentene (g).

Base group, H—C—H 44.5

Primary CH_3 substitution → CH_3—CH_3............... 10.4
Secondary CH_3 substitutions → CH_3—CH_2—CH_2—CH_2—CH_3

Type Numbers		
Carbon A	Carbon B	$\Delta S^\circ_{298.1}$
1	1	9.8
1	2	9.2
1	2	9.2

Introduction of double bond at 2-position

2	2 trans	−1.2

Summation of group contributions.................... 81.9 E.U.
Experimental value (NBS tables).................... 81.81 E.U.

Example 2. Estimate the entropy, $S^\circ_{298.1}$, of acetaldehyde (g).

Base group
$$H—\overset{\displaystyle H}{\underset{\displaystyle H}{C}}—H$$
.... 44.5

Primary CH_3 substitution → CH_3—CH_3............... 10.4

Secondary CH_3 substitutions →
$$CH_3—\overset{\displaystyle CH_3}{\underset{\displaystyle CH_3}{CH}}$$

Type Numbers		
Carbon A	Carbon B	$\Delta S^\circ_{298.1}$
1	1	9.8
2	1	5.8

Substitution of =O replacing 2 —CH_3

→
$$CH_3—\overset{\displaystyle H}{C}=O$$
.... −12.3

Summation of group contributions.................... 58.2 E.U.
Experimental value (NBS tables).................... 63.5 E.U.

These examples illustrate the procedure used in the Andersen-Beyer-Watson method. The first is one with unusually good agreement, the second an example of an uncommonly poor case. In general it is preferable to consider the group substitutions in the same order as has been used in the presentation of the tables. Experience shows that when more than one base group is possible, the one with the largest entropy should be chosen. Also, best agreement with experimental values, when they have been known, has been obtained by the use of the minimum number of substitutions necessary to construct the molecule. In cases where several alternate routes with the minimum number of substitutions are possible, the average of the different results should be used.

C. Other methods. Although the tables proposed by Parks and Huffman[5] are based on older data, they are often more convenient to use,

[5] G. S. Parks and H. M. Huffman, *The Free Energies of Some Organic Compounds,* Reinhold Publishing Corporation, New York, 1932.

since they are simpler and also because they have been worked out for the liquid and solid states, as well as for the gaseous phase. Several special methods of estimating entropies are described also by Wenner,[6] who has given a great deal of attention to inorganic, as well as to organic, compounds. Since these additional methods involve no new fundamental principles, they will not be described in any detail here. Nevertheless, the practicing chemist should have a general acquaintance with more than one method of estimating entropies, if he expects a high degree of success in predictions of the feasibility of new reactions.

D. Accuracy of the approximate methods. Free-energy changes and equilibrium constants calculated from the enthalpy and entropy values estimated by the group-contribution method are generally reliable only as to order of magnitude. Andersen, Beyer, and Watson, for example, have found that their estimated enthalpies and entropies usually differ from experimental values, when known, by less than 4.0 kcal mole^{-1} and 2.0 cal mole^{-1} deg^{-1}, respectively. If errors of this magnitude occurred cumulatively, the free-energy change would be incorrect by approximately 4.6 kcal mole^{-1} near 25°C. Such an error in the free energy corresponds to an uncertainty of several powers of 10 in an equilibrium constant. With few exceptions, such an error is an upper limit. Nevertheless, it must be emphasized that approximate methods of calculating these thermodynamic properties are reliable for estimating the feasibility of a projected reaction but are not adequate for calculating equilibrium compositions to better than the order of magnitude.

Exercises

1. The following problem illustrates the application of calculations involving the third law to a specific organic compound, n-heptane.

(a) From tables of the National Bureau of Standards of the enthalpy function, $H° - H_0°$, find $\Delta H°$ for the reaction (at 298.16°K)

7C (graphite) + 8H$_2$ (hypothetical ideal gas)
$$= n\text{-}C_7H_{16} \text{ (hypothetical ideal gas).} \quad (12.17)$$

(b) The following equation can be derived for the pressure coefficient of the enthalpy:[7]

$$\left(\frac{\partial H}{\partial P}\right)_T = V - T\left(\frac{\partial V}{\partial T}\right)_P. \quad (12.18)$$

Using the Berthelot equation of state, derive an expression to evaluate ($H_{ideal} - H_{real}$), that is, the correction in enthalpy for deviations from ideal behavior, at any pressure, P.

---

[6] R. R. Wenner, *Thermochemical Calculations*, McGraw-Hill Book Company, Inc., New York, 1941.

[7] See, for example, O. A. Hougen and K. M. Watson, *Chemical Process Principles*, Vol. II, *Thermodynamics*. John Wiley & Sons, Inc., New York, 1947, pages 460–463.

(c) Find the critical constants for n-heptane in the *International Critical Tables*. Calculate the change in enthalpy for the transformation (at 298.16°K)

n-C_7H_{16} (hypothetical ideal gas, $P = 1$ atm)

$$= n\text{-}C_7H_{16} \text{ (real gas, } P = p_{\text{vap}}), \quad (12.19)$$

where p_{vap} represents the pressure of the vapor in equilibrium with liquid n-heptane

(d) With the aid of the result in Part (c) plus National Bureau of Standards data on the vaporization of n-heptane, find $\Delta H^\circ_{298.16}$ for the reaction

$$7C \text{ (graphite)} + 8H_2 \text{ (g)} = n\text{-}C_7H_{16} \text{ (l)}. \quad (12.20)$$

(e) Calculate $S^\circ_{298.16}$ for liquid n-heptane from the heat-capacity data in the table below and from those for solid n-heptane given in Table 4 of Chapter 2. Integrate by means of the Debye equation to obtain the entropy up to 15.14°K and carry out a graphical integration (C_p/T vs. T) thereafter.

Temperature (°K)	C_p (cal mole^{-1} deg^{-1})
194.60	48.07
218.73	48.49
243.25	49.77
268.40	51.71
296.51	53.68

(f) Calculate ΔS° for reaction (12.20) at 298.16°K. Use National Bureau of Standards data on graphite and hydrogen.

(g) Calculate the entropy of vaporization of liquid n-heptane at 298.16°K.

(h) Find the entropy change for the following transformation at 298.16°K:

$$n\text{-}C_7H_{16} \text{ (real gas, } P = p_{\text{vap}}) = n\text{-}C_7H_{16} \text{ (hyp. ideal gas, } P = 1). \quad (12.21)$$

(i) By appropriate summation of the results of Parts (f), (g), and (h), calculate ΔS° for reaction (12.17) at 298.16°K.

(j) Calculate $\Delta F f^\circ_{298.16}$ for liquid n-heptane.

(k) Calculate $\Delta F f^\circ_{298.16}$ for gaseous n-heptane in the (hypothetical ideal gas) standard state.

(l) From tables of the National Bureau of Standards on the free-energy function $(F^\circ - H^\circ_0)/T$, calculate $\Delta F f^\circ_{298.16}$ for gaseous n-heptane in the (hypothetical ideal gas) standard state. Compare with the value obtained in (k).

(m) Estimate $S^\circ_{298.16}$ for n-heptane (gas) by the group-contribution method of Andersen, Beyer, and Watson. Compare with the result obtainable from the information in Parts (f) and (i).

(n) Estimate $S^\circ_{298.16}$ for liquid n-heptane from the rules of Parks and Huffman. Compare with the result obtained in Part (e).

2. (a) Using the group contribution method of Andersen, Beyer, and Watson estimate $S^\circ_{298.1}$ for 1,2-dibromoethane (gas).

(b) Calculate the entropy change when gaseous 1,2-dibromoethane is expanded from 1 atm to its vapor pressure in equilibrium with the liquid phase at 298.1°K. Neglect any deviations of the gas from ideal behavior. Appropriate data on vapor pressures have been assembled conveniently by D. R. Stull, *Ind. Eng. Chem.*, **39**, 517 (1947).

(c) Using the data in Stull's publication, find the heat of vaporization of 1,2-dibromoethane at 298.1°K.

(d) Find the entropy, $S^\circ_{298.1}$, for liquid 1,2-dibromoethane.

(e) Compare the estimate obtained in Part (d) with that obtainable from the rules of G. S. Parks and H. M. Huffman in *The Free Energies of Some Organic Compounds.*

(f)   Compare the estimates of Parts (d) and (e) with the value found by K. S. Pitzer, *J. Am. Chem. Soc.*, **62**, 331 (1940).

3.   Recent precision measurements of heats of formation and entropies are probably accurate to perhaps 60 cal mole^{-1} and 0.2 cal mole^{-1} deg^{-1}, respectively. Show that either one of these uncertainties corresponds to a change of 10 per cent in an equilibrium constant at 25°C.

4.   It has been suggested recently [G. J. Janz, *Can. J. Research*, **25B**, 331 (1947)] that 1,4-dicyano-2-butene might be prepared in the vapor phase from the reaction of cyanogen with butadiene.

(a)   Estimate $\Delta Hf°$ and $S°$ for the dicyanobutene at 25°C by the group contribution method.

(b)   With the aid of the following additional information, calculate the equilibrium constant for the suggested reaction.

Substance	$\Delta Hf°_{298.16}$ (cal mole^{-1})	$S°_{298.16}$ (cal mole^{-1} deg^{-1})
Butadiene (g)	26,748	66.42
Cyanogen (g)	71,820	57.64

THERMODYNAMICS OF SYSTEMS OF VARIABLE COMPOSITION

CHAPTER 13

Partial Molal Quantities

I. NEED FOR A NEW FUNCTION IN DEALING WITH SOLUTIONS

In considering changes in thermodynamic quantities associated with a given reaction, we find no difficulty in physical interpretation so long as we do not deal with solutions. Thus in the reaction

$$Ag\ (s) + \tfrac{1}{2}Cl_2\ (g) = AgCl\ (s) \tag{13.1}$$

the volume of the system may be written in detail as

$$V = V_{Ag} + V_{Cl_2} + V_{AgCl} + \text{constant}, \tag{13.2}$$

where the constant is inserted to take account of the reaction vessel. Consequently, the change in volume per mole of Ag consumed, $\partial V/\partial n$, is given by the equation

$$\frac{\partial V}{\partial n} = \frac{\partial V_{Ag}}{\partial n} + \frac{\partial V_{Cl_2}}{\partial n} + \frac{\partial V_{AgCl}}{\partial n}, \tag{13.3}$$

where n represents the moles of Ag. Each of the partial derivatives on the right-hand side of equation (13.3) is a molal volume, since each participant in the reaction of equation (13.1) is a pure substance.[1] Therefore we may write

$$\frac{\partial V}{\partial n} = -\mathrm{v}_{Ag} - \frac{1}{2}\mathrm{v}_{Cl_2} + \mathrm{v}_{AgCl} = \Delta\mathrm{v}. \tag{13.4}$$

Thus the change in volume of the system, $\Delta\mathrm{v}$, per mole of Ag consumed is evidently the volume per mole of AgCl minus the sum of the molal volume of Ag and one-half the molal volume of gaseous Cl_2. It is this $\Delta\mathrm{v}$ which we would use in thermodynamic calculations such as the calculation of the change in ΔF for reaction (13.1) with change in the pressure on the system.

A corresponding problem may arise also for a reaction involving one or more solutions. Thus we may need to know the effect of pressure on the electromotive force of the cell

$$H_2, HCl\ (m), AgCl, Ag, \tag{13.5}$$

---

[1] If it is not apparent that $\partial V/\partial n$ for a pure substance is also the molal volume, see Section II–D of this chapter.

where m represents the molality of the dissolved HCl. In principle, we would require again a knowledge of a volume change, but this time for the reaction

$$\tfrac{1}{2}H_2 \text{ (g)} + AgCl \text{ (s)} = HCl \text{ } (m) + Ag \text{ (s)}. \tag{13.6}$$

By analogy with our discussion of this problem for a reaction involving pure substances, we may write for reaction (13.6), or in other words for the cell represented by (13.5),

$$V = V_{H_2} + V_{AgCl} + V_{HCl \text{ solution}} + V_{Ag} + \text{constant}. \tag{13.7}$$

Consequently, the change in volume of the cell per mole of Ag formed is given by the equation

$$\frac{\partial V}{\partial n} = \frac{\partial V_{H_2}}{\partial n} + \frac{\partial V_{AgCl}}{\partial n} + \frac{\partial V_{HCl \text{ solution}}}{\partial n} + \frac{\partial V_{Ag}}{\partial n}. \tag{13.8}$$

Once again we recognize that the partial derivative for each of the pure substances corresponds to a molal volume, v_i. We may introduce a new symbol for the partial derivative of the volume of the solution:

$$\frac{\partial V_{HCl \text{ solution}}}{\partial n} \equiv \bar{v}_{HCl}. \tag{13.9}$$

Thus, in parallel with equation (13.4) for the reaction involving only pure substances, we may write the following equation for the reaction involving a solution:

$$\frac{\partial V}{\partial n} = -\frac{1}{2}v_{H_2} - v_{AgCl} + \bar{v}_{HCl} + v_{Ag} = \Delta\bar{v}. \tag{13.10}$$

The quantity represented by the symbol $\bar{v}$ is called the *partial molal volume*, since it represents a partial derivative of a volume. The computation of changes in other extensive thermodynamic properties (such as free energy and enthalpy) associated with reactions involving solutions requires partial derivatives of these quantities also. We must introduce *partial molal quantities*, therefore, as general functions for the thermodynamic treatment of solutions. It will be necessary then to examine the characteristics of these functions and to appreciate their physical significance.

II. DEFINITION OF PARTIAL MOLAL QUANTITIES

The character of this new type of function can be described best if one deals first with a property such as volume, which can be visualized readily in geometric terms.

A. Volume. Let us picture an experiment in which we start with an arbitrary volume, V_0, of a solvent and measure the volume, V, of the solution formed as we add successive quantities of solute. The data obtained might fit a linear relation, such as is illustrated in Fig. 1. In such a case, it is obvious that each mole of added solute increases the volume by the same amount. This characteristic is also expressed by the fact that the slope of the line in Fig. 1, $(\partial V/\partial n_2)_{n_1,T,P}$, is a constant. (Here n_2 represents moles of solute; n_1, moles of solvent; T, temperature; and P, pressure.) In other words, no matter what the concentration of solute (within the limits of the graph), the addition of further quantities of it increases the

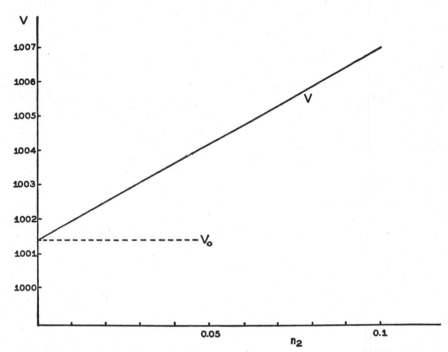

Fig. 1. Linear dependence of volume on concentration, for dilute solutions of glycol-amide in water. [Based on data of F. T. Gucker, Jr., W. L. Ford and C. E. Moser, *J. Phys. Chem.*, **43**, 153 (1939).] (Precise measurements at higher concentrations indicate that this figure does have some curvature, which is not detectable in dilute solutions).

volume by equal amounts per mole. Since $(\partial V/\partial n_2)_{n_1,T,P}$ is the increase in volume per mole of added solute, we shall represent it by the symbol $\bar{v}_2$.

$$\left(\frac{\partial V}{\partial n_2}\right)_{n_1,T,P} = \bar{v}_2. \qquad (13.11)$$

$\bar{v}_2$ may be thought of as the effective volume of one mole of solute in this solution.

More generally, the dependence of volume on n_2 will not fit a linear graph but one such as is represented in Fig. 2. In this case, it is obvious that the volume does not change by equal increments when equal quantities of solute are added successively to a fixed quantity of solvent. Let us consider, then, what must be the significance of the slope, $(\partial V/\partial n_2)_{n_1,T,P}$, represented by the dotted line in Fig. 2. It is evident from the principles of calculus that this slope must represent the change in volume per mole of added solute, n_2 (temperature, pressure, and moles of solvent, n_1, being

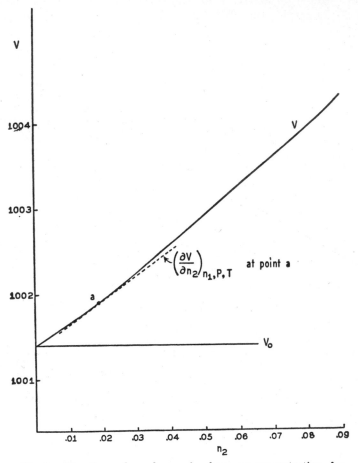

Fig. 2. Non-linear dependence of volume on concentration, for dilute solutions of sulfuric acid in water. [Based on data of I. M. Klotz and C. F. Eckert, *J. Am. Chem. Soc.*, **64**, 1878 (1942).]

maintained constant), at a fixed point on the curve, in other words at some specified value of n_2. The value of n_2 must be specified, since the slope obviously depends upon the position on the curve at which it is measured. In practice, this slope, which we may represent by $\bar{v}_2$, as in equation (13.11), refers to either one of the following two experiments:

(a) Measure the increase in total volume, V, of the solution when one mole of solute is added to a very large quantity (strictly speaking, an infinite quantity) of the solution.   Because very large quantities of solution are used, the addition of a mole of solute will not change the concentration of the solution appreciably.

(b)   Measure the change in total volume, V, of the solution when a very small quantity (strictly speaking, an infinitesimal amount) of solute is added to the solution.   Then calculate the change for one mole (that is, divide ΔV by Δn_2) as if there were no change in composition upon the addition of a full mole of solute.

These are two equivalent points of view concerning the meaning of $\bar{v}_2$. Clearly, though, $\bar{v}_2$ does not correspond to the actual change in volume when a full mole of solute is added to a limited quantity of solution, since $\bar{v}_2$ is the increase in V as we move up along the dotted line in Fig. 2, whereas the actual volume follows the solid line.

With these considerations of the experimental significance of $\bar{v}_2$, we are now in a position to consider the experimental meaning of the quantity $\Delta\bar{v}$ of equation (13.10).   First, however, we must emphasize one aspect of the notation in reaction (13.6) that may be overlooked.   It must be clearly understood that the reaction as written implies that the HCl is formed in a sufficiently large quantity of solution of m-molal concentration so that the addition of one mole of HCl does not change the concentration significantly.   Otherwise, of course, we could not specify a fixed concentration for the solution.

With this implication in mind we can see that the volume change, $\Delta\bar{v}$, of equation (13.10), associated with the consumption of one-half mole of pure hydrogen gas and one mole of pure silver chloride to produce one mole of hydrochloric acid in an m-molal solution and one mole of pure silver (all substances being at a specified temperature and pressure), must mean the following:

$$\Delta\bar{v} = \begin{bmatrix} \text{volume change of the solu-} \\ \text{tion when one mole of HCl} \\ \text{is dissolved in a sufficiently} \\ \text{large quantity of an } m\text{-} \\ \text{molal HCl solution so that} \\ \text{the concentration does not} \\ \text{change} \end{bmatrix} + \begin{bmatrix} \text{volume of one} \\ \text{mole of pure} \\ \text{Ag} \end{bmatrix} - \tfrac{1}{2}\begin{bmatrix} \text{volume of one} \\ \text{mole of pure} \\ \text{H}_2 \end{bmatrix}$$

$$- \begin{bmatrix} \text{volume of one} \\ \text{mole of pure} \\ \text{AgCl} \end{bmatrix}. \quad (13.12)$$

Thus $\bar{v}_{\text{HCl}}$ has a definite operational meaning which can be related, in turn, to the experimental quantity $\Delta\bar{v}$ for the chemical transformation as a whole.

There is, of course, no fundamental distinction as far as this treatment is concerned between a solute and a solvent. Thus for the solvent the partial molal volume is defined by the expression

$$\left(\frac{\partial V}{\partial n_1}\right)_{n_2, T, P} = \bar{v}_1. \tag{13.13}$$

Clearly this corresponds experimentally to the increase in volume of a solution upon the addition of one mole of pure solvent to a very large quantity of the solution with n_2, temperature, and pressure maintained constant.

B. Free energy. The extension of the notation for the partial molal quantity to other thermodynamic variables follows readily from the definitions introduced for the volume function. Thus for the free-energy function, F,

$$\bar{F}_2 = \left(\frac{\partial F}{\partial n_2}\right)_{n_1, T, P} \tag{13.14}$$

and

$$\bar{F}_1 = \left(\frac{\partial F}{\partial n_1}\right)_{n_2, T, P}. \tag{13.15}$$

$\bar{F}_2$ corresponds to the change in total free energy, F, of a solution when a mole of solute is added to an infinite amount of the solution.

The change in free energy in a reaction such as that represented by equation (13.6), which involves solutions, is given by the following expression:

$$\Delta\bar{F} = \bar{F}_{HCl} + F_{Ag} - \tfrac{1}{2}F_{H_2} - F_{AgCl}. \tag{13.16}$$

By the free-energy change for this reaction we mean:

$$\Delta\bar{F} = \begin{bmatrix} \text{free-energy change of the} \\ \text{solution when one mole of} \\ \text{HCl is dissolved in a suffi-} \\ \text{ciently large quantity of an} \\ \text{m-molal HCl solution so} \\ \text{that the concentration does} \\ \text{not change} \end{bmatrix} + \begin{bmatrix} \text{free energy of} \\ \text{one mole of} \\ \text{pure Ag} \end{bmatrix} - \tfrac{1}{2}\begin{bmatrix} \text{free energy of} \\ \text{one mole of} \\ \text{pure H}_2 \end{bmatrix}$$
$$- \begin{bmatrix} \text{free energy of} \\ \text{one mole of} \\ \text{pure AgCl} \end{bmatrix}. \tag{13.17}$$

The partial molal free energy has also been called the *chemical potential* by Gibbs. Since this term is used so frequently, we too shall refer to it often, designating it by the symbol μ:

$$\bar{F} = \mu. \tag{13.18}$$

C. General formulation. In view of the preceding discussion it is evident that we may define a partial molal quantity for any *extensive* thermodynamic property, G, by the expression

$$\bar{G}_i = \left(\frac{\partial G}{\partial n_i}\right)_{P, T, n_1, n_2, \ldots}. \tag{13.19}$$

$\bar{G}_1$ refers to the increase in total G when one mole of component 1 is added to an infinite amount of a solution at fixed temperature and pressure and with the number of moles, n_2, n_3, $\ldots$, of all other components kept fixed.

D. Partial molal quantities for pure phase. If a system is a single pure phase, a graph of volume (or any extensive thermodynamic property) gives a straight line passing through the origin[2] (Fig. 3). Obviously, then, the slope of this line is given by the following relation:

$$\frac{\partial V}{\partial n} = \bar{v} = \frac{V}{n} = \overset{\bullet}{v}. \qquad (13.20)$$

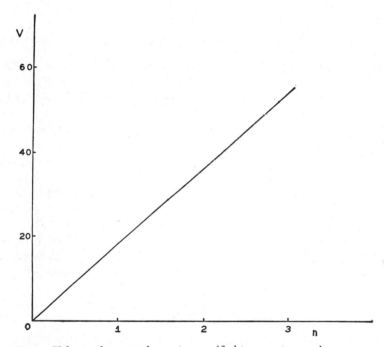

Fig. 3. Volume of a pure phase at a specified temperature and pressure. Data are for water at 25 °C and 1 atmosphere pressure.

In other words, the partial molal volume, $\bar{v}$, is identical with $\overset{\bullet}{v}$, the volume per mole of the pure phase.

If we are dealing with any extensive thermodynamic property, G, of a pure phase, its graph will be linear[2] also, and hence we can see readily that in general

$$\bar{G} = \overset{\bullet}{G}, \qquad (13.21)$$

where $\overset{\bullet}{G}$ refers to the molal value of the property G for a pure phase.

[2] We have neglected surface energies and forces when we make this statement. In most cases, however, surface forces are negligibly small in ordinary chemical problems.

III. FUNDAMENTAL EQUATIONS OF PARTIAL MOLAL QUANTITIES

Before we can describe conveniently the derivation of partial molal quantities from experimental data, it will be necessary to consider a number of fundamental relations between these quantities.

A. Integrated equation. A few moments' reflection will suffice to show that extensive thermodynamic properties are homogeneous functions of degree 1; for in Chapter 2 we defined a homogeneous function, $f(x,y,z,\ldots)$, as one in which replacement of each independent variable by an arbitrary parameter, k, times the variable merely multiplies the function by k^n, where n is the degree of homogeneity:

$$f(kx,ky,kz,\ldots) = k^n f(x,y,z,\ldots). \tag{13.22}$$

If n is 1, as in extensive thermodynamic properties, equation (13.22) expresses the condition that multiplication of each variable by k also multiplies $f(x,y,z,\ldots)$ by k. This condition is applicable (at fixed P and T) to extensive thermodynamic properties. For example, let us consider the total volume, V, of a system. If we double the number of moles of each component, at fixed temperature and pressure, we automatically double the volume. Or if we multiply the number of moles of each component by a constant, k, we automatically increase the volume by the factor k. This characteristic of the volume function is also an attribute of the free-energy function, the entropy function, the enthalpy, and all other extensive thermodynamic properties.[2]

1. *Volume.* If we have a homogeneous function, as defined by equation (13.22), we have shown in Chapter 2 that Euler's theorem is applicable. For a function of two variables, $f(x,y)$, this may be expressed as

$$x \left(\frac{\partial f}{\partial x}\right)_y + y \left(\frac{\partial f}{\partial y}\right)_x = n\, f(x,y). \tag{13.23}$$

Turning to the volume function, which we shall consider, for simplicity, for the case of a two-component system, at fixed pressure and temperature, we may write

$$V = f(n_1, n_2). \tag{13.24}$$

By applying equation (13.23), keeping in mind that the degree of homogeneity of the volume function is 1, we obtain

$$n_1 \left(\frac{\partial V}{\partial n_1}\right)_{n_2} + n_2 \left(\frac{\partial V}{\partial n_2}\right)_{n_1} = V. \tag{13.25}$$

The partial derivatives in equation (13.25) have been defined previously by equations (13.11) and (13.13), so that we may write

$$n_1 \bar{v}_1 + n_2 \bar{v}_2 = V. \tag{13.26}$$

2. *General case.* Similarly, for any extensive thermodynamic property, G, at constant pressure and temperature, we may write for a two-component system

$$G = f(n_1, n_2). \tag{13.27}$$

Consequently, from Euler's theorem

$$n_1 \left(\frac{\partial G}{\partial n_1} \right)_{n_2} + n_2 \left(\frac{\partial G}{\partial n_2} \right)_{n_1} = G; \tag{13.28}$$

and in view of equation (13.19),

$$n_1 \bar{G}_1 + n_2 \bar{G}_2 = G. \tag{13.29}$$

B. Equations derived from the integrated equation.

1. It is evident that differentiation of equation (13.29) leads to

$$dG = n_1 \, d\bar{G}_1 + \bar{G}_1 \, dn_1 + n_2 \, d\bar{G}_2 + \bar{G}_2 \, dn_2. \tag{13.30}$$

But we have pointed out that for any extensive thermodynamic property at constant pressure and temperature we may write equation (13.27) for a two-component system. When G is a function of two variables, $f(n_1, n_2)$, the methods of calculus give the following equation for the total differential:

$$\left. \begin{aligned} dG &= \left(\frac{\partial G}{\partial n_1} \right)_{n_2} dn_1 + \left(\frac{\partial G}{\partial n_2} \right)_{n_1} dn_2 \\ &= \bar{G}_1 \, dn_1 + \bar{G}_2 \, dn_2. \end{aligned} \right\} \tag{13.31}$$

In view of equation (13.30) for dG, we may obtain the following from equation (13.31):

$$n_1 \, d\bar{G}_1 + \bar{G}_1 \, dn_1 + n_2 \, d\bar{G}_2 + \bar{G}_2 \, dn_2 = \bar{G}_1 \, dn_1 + \bar{G}_2 \, dn_2, \tag{13.32}$$

or

$$n_1 \, d\bar{G}_1 + n_2 \, d\bar{G}_2 = 0. \tag{13.33}$$

Equation (13.33) is one of the most useful relations between partial molal quantities. It should be emphasized that this equation is valid only at constant pressure and temperature.

2. If we divide through by dn_1 in equation (13.33), we obtain

$$n_1 \frac{d\bar{G}_1}{dn_1} + n_2 \frac{d\bar{G}_2}{dn_1} = 0, \tag{13.34}$$

and if we write explicitly the restriction of constant pressure and temperature,

$$n_1 \left(\frac{\partial \bar{G}_1}{\partial n_1} \right)_{P,T} + n_2 \left(\frac{\partial \bar{G}_2}{\partial n_1} \right)_{P,T} = 0. \tag{13.35}$$

This equation is very useful in deriving certain relations between the partial molal quantity for a solute and that for the solvent. An analogous pair of equations can be written, of course, with dn_2 in place of dn_1.

3. Equation (13.33) may be used also to derive a very important expression for finding the partial molal quantity for one component when that for the other is known. Rearrangement of equation (13.33) leads to

$$d\bar{G}_1 = -\frac{n_2}{n_1} d\bar{G}_2. \tag{13.36}$$

This equation may be integrated from the infinitely dilute solution ($n_2 = 0$) to any finite concentration to give

$$\int_{\frac{n_2}{n_1} = 0}^{\frac{n_2}{n_1}} d\bar{G}_1 = -\int_{\frac{n_2}{n_1} = 0}^{\frac{n_2}{n_1}} \frac{n_2}{n_1} d\bar{G}_2, \tag{13.37}$$

or

$$\bar{G}_1 - \bar{G}_1^{\circ} = -\int_{\frac{n_2}{n_1} = 0}^{\frac{n_2}{n_1}} \frac{n_2}{n_1} d\bar{G}_2, \tag{13.38}$$

where $\bar{G}_1^{\circ}$ represents the partial molal quantity of the solvent at infinite dilution, that is, pure solvent. (For example, for water at room temperature, $\bar{v}_1^{\circ} = v^{\bullet} \simeq 18$ cc mole^{-1}.) Thus if we have either an analytical or a graphical representation of $\bar{G}_2$, the corresponding type of integration of the right-hand side of equation (13.38) may be carried out.

C. Partial molal equations for one mole of solution. Although the equations for partial molal quantities reduced to one mole of solution are not fundamentally different from those just considered, they do form a set of convenient relations for use in this important special case.

If

$$n_1 + n_2 = 1, \tag{13.39}$$

then by dividing each term in equation (13.29) by ($n_1 + n_2$) we obtain

$$G = N_1\bar{G}_1 + N_2\bar{G}_2, \tag{13.40}$$

where $\quad N_1 = \dfrac{n_1}{n_1 + n_2} =$ mole fraction of component 1, $\tag{13.41}$

$$N_2 = \frac{n_2}{n_1 + n_2} = \text{mole fraction of component 2,} \tag{13.42}$$

and $\quad G = \dfrac{G}{n_1 + n_2} =$ value of G for one mole of solution. $\tag{13.43}$

Correspondingly, for one mole of solution equation (13.33) reduces to

$$N_1 d\bar{G}_1 + N_2 d\bar{G}_2 = 0, \tag{13.44}$$

which in turn can be converted into another which parallels equation (13.34):

$$N_1 \frac{\partial \bar{G}_1}{\partial N_1} + N_2 \frac{\partial \bar{G}_2}{\partial N_1} = 0. \tag{13.45}$$

Another variant of these equations may be introduced if we are dealing with mole fractions. In contrast to the general case where n_1 and n_2 are completely independent of each other, N_1 cannot vary without some accompanying change in N_2. Thus, since

$$N_1 + N_2 = 1, \tag{13.46}$$

it follows that $\qquad\qquad dN_1 = -dN_2. \tag{13.47}$

Making use of this equality, we can convert equation (13.45) to the following additional, very useful expression:

$$N_1 \frac{\partial \bar{G}_1}{\partial N_1} - N_2 \frac{\partial \bar{G}_2}{\partial N_2} = 0. \tag{13.48}$$

In both (13.45) and (13.48) the partial-derivative notation is intended to emphasize the constancy of P and T.

IV. METHODS OF CALCULATING PARTIAL MOLAL QUANTITIES

Special methods for calculating partial molal quantities from experimental data have been developed for almost every one of the thermodynamic properties. To illustrate the general procedures involved, several methods will be described at this point for obtaining partial molal volumes, and one general method used in many other cases will be outlined in detail. In subsequent chapters other specific approaches for special cases will be considered.

A. Partial molal volumes.

1. *Total volume known as a function of molality.*

a. *Graphically.* If a graphical representation can be made of the experimental data in terms of the total volume as a function of the molality (Fig. 4A), then the calculation of $\bar{v}_2$ and $\bar{v}_1$ is a very simple procedure. In Chapter 2 we described a procedure for carrying out graphical differentiation. Applied to the problem at hand, this method requires the calculation and tabulation of values of n_2, V, ΔV, Δn_2, and $\Delta V / \Delta n_2$. Since n_1 may be fixed (for example, at 1000 grams divided by the molecular weight of the solvent), $\bar{v}_2$ may be obtained from a smooth curve drawn through the chords of a plot of $\Delta V / \Delta n_2$ vs. n_2 (Fig. 4B).

When $\bar{v}_2$ is available as a function of the composition (that is, as a function of n_2 at fixed n_1), it is possible to calculate $\bar{v}_1$ by graphical methods which carry out the procedure implicit in equation (13.38), as applied to volumes:

$$\bar{v}_1 - \bar{v}_1^\circ = - \int_0^{\frac{n_2}{n_1}} \frac{n_2}{n_1} \, d\bar{v}_2. \qquad (13.49)$$

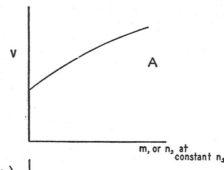

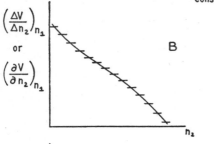

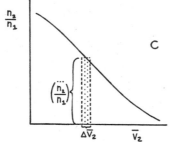

Fig. 4. Calculation of partial molal quantities by graphical means. A, graphical representation of total volume as a function of molality; B, chord-area plot for determination of $\bar{v}_2$; C, graphical integration for determination of $\bar{v}_1$.

In graphing, one is required to make a plot of n_2/n_1 vs. $\bar{v}_2$ (Fig. 4C) and then to carry out the necessary integration. As usual, if the chords are made sufficiently small,

$$\sum \left(\overline{\frac{n_2}{n_1}} \right) \Delta \bar{v}_2 = \int \frac{n_2}{n_1} \, d\bar{v}_2, \qquad (13.50)$$

where $(\overline{n_2/n_1})$ represents the average value of this ratio over the interval $\Delta \bar{v}_2$. Since $\bar{v}_1^\circ$ is the partial molal volume of solvent at infinite dilution, in other words of pure solvent, it can be calculated from the known molecular weight and density of the pure solvent.

b. *Analytically*. Frequently the total volume, V, can be expressed conveniently as an algebraic function of the composition. In the case of sodium chloride in water at 25 °C and one atmosphere pressure, for example, V (in cubic centimeters) may be expressed readily in terms of the following series in the molality, m:

$$V = 1001.38 + 16.6253m + 1.7738m^{3/2} + 0.1194m^2, \qquad (13.51)$$

or $\quad V = 1001.38 + 16.6253n_2 + 1.7738n_2^{3/2} + 0.1194n_2^2 \qquad (13.52)$

when 1000 grams of water is used. Obviously, $\bar{v}_2$ may be obtained by direct differentiation, since the quantity of solvent is fixed:

$$\bar{v}_2 = \left(\frac{\partial V}{\partial n_2}\right)_{n_1}$$

$$= 16.6253 + 2.6607 n_2^{1/2} + 0.2388 n_2 \qquad (13.53)$$

$$= 16.6253 + 2.6607 m^{1/2} + 0.2388 m. \qquad (13.54)$$

As in the graphical case, the partial molal volume of the solvent, $\bar{v}_1$, may be obtained readily by the integration illustrated in equation (13.49). To evaluate $d\bar{v}_2$, we need merely to differentiate equation (13.53):

$$d\bar{v}_2 = (1.3304 n_2^{-1/2} + 0.2388) dn_2. \qquad (13.55)$$

Since $n_1 = 1000/18.02 = 55.51$, it follows that

$$\bar{v}_1 - \bar{v}_1^{\circ} = -\int_0^{n_2} \frac{n_2}{n_1} d\bar{v}_2 = -\int \frac{n_2}{n_1} (1.3304 n_2^{-1/2} + 0.2388) dn_2 \qquad (13.56)$$

$$= -\frac{1}{n_1} \int (1.3304 n_2^{1/2} + 0.2388 n_2) dn_2 \qquad (13.57)$$

$$= -\frac{0.8869}{55.51} n_2^{3/2} - \frac{0.1194}{55.51} n_2^2. \qquad (13.58)$$

To complete the equation, we obtain $\bar{v}_1^{\circ}$ from the molecular weight (18.02) and density (0.99708 gram cm^{-3}) of water at 25 °C:

$$\bar{v}_1 = 18.08 - 0.015977 n_2^{3/2} - 0.002151 n_2^2; \qquad (13.59)$$

$$\bar{v}_1 = 18.08 - 0.015977 m^{3/2} - 0.002151 m^2. \qquad (13.60)$$

2. *Densities and composition known.* Most frequently, volume data for solutions are tabulated as density, D, vs. composition. It is essential, therefore, that we be able to convert such data into partial molal quantities.

Once again the procedure may be illustrated best by reference to an example, in this case the densities and weight-percentage concentrations of alcohol-water mixtures (Table 1, Columns 1 and 4) at 25 °C.

a. *Calculation of $\bar{v}_2$.* To obtain $\bar{v}_2$, that is, $(\partial V/\partial n_2)_{n_1}$, it is essential to keep in mind that we need values of V for a fixed quantity, n_1, of water, but for variable quantities, n_2, of alcohol. For this purpose, therefore, we convert the relative weights given in Column 1 to relative numbers of moles, that is, to n_2/n_1, Column 2. Obviously, the numbers in Column 2 are also the moles of alcohol accompanying one mole of water in each of the solutions listed in Column 1. From this information and the density (Column 4), we can calculate the volume in cubic centimeters which contains one mole of water; for the mass, M, of a solution containing n_2 moles of alcohol and 1 mole of water is

$$M = n_2 \times (\text{molecular weight of } C_2H_5OH)$$
$$+ 1 \times (\text{molecular weight of } H_2O), \qquad (13.61)$$

and hence the volume per mole of water is

$$V = \frac{M}{D} \text{ cm}^3 \text{ (mole } H_2O)^{-1}. \tag{13.62}$$

Numerical values for these volumes of solution containing a fixed quantity, one mole, of water are listed in Column 5. Obviously, the partial molal volumes may be determined graphically from a chord-area plot of the ratio of increments listed in Column 6 vs. n_2/n_1, since

$$\left(\frac{\Delta V}{\Delta n_2}\right)_{n_1} = \left(\frac{\Delta V}{\Delta(n_2/n_1)}\right)_{n_1 = 1}. \tag{13.63}$$

Such a graph is illustrated in Fig. 5. As in previous graphical differentiations, the partial derivative $(\partial V/\partial n_2)_{n_1}$ may be determined from the smooth curve drawn through the chords in Fig. 5 so as to balance areas.

TABLE 1

DENSITIES AND PARTIAL VOLUMES OF ALCOHOL-WATER MIXTURES

1	2	3	4	5	6	7	8
	$\dfrac{C_2H_5OH}{H_2O}$	$\dfrac{H_2O}{C_2H_5OH}$		V		V	
Wt.-% Alcohol	$=\dfrac{n_2}{n_1}$	$=\dfrac{n_1}{n_2}$	D	cm³ (mole $H_2O)^{-1}$	$\dfrac{\Delta V}{\Delta n_2}$	cm³ (mole $C_2H_5OH)^{-1}$	$\dfrac{\Delta V}{\Delta n_1}$
20	0.097769	10.228241	0.96639	23.3032		238.3510	
					53.59		18.064
25	0.130357	7.671181	0.95895	25.0496		192.1602	
					54.32		17.969
30	0.167603	5.966474	0.95067	27.0726		161.5282	
					55.10		17.839
35	0.210578	4.748826	0.94146	29.4403		139.8072	
					55.75		17.700
40	0.260716	3.835590	0.93148	32.2354		123.6419	
					56.31		17.555
45	0.319970	3.125296	0.92085	35.5719		111.1726	
					56.68		17.436
50	0.391074	2.556394	0.90985	39.6021		101.2650	
					57.03		17.300
55	0.477979	2.092140	0.89850	44.5582		93.2220	
					57.26		17.189
60	0.586611	1.704707	0.88699	50.7785		86.5624	
					57.50		17.048
65	0.726280	1.376878	0.87527	58.8096		80.9737	
					57.70		16.905
70	0.912564	1.095883	0.86340	69.5576		76.2236	

b. *Calculation of $\bar{v}_1$.* The partial molal volumes of the water in the alcohol solutions may be calculated by an analogous procedure. Obviously, in this case we wish to find V for a fixed quantity, let us say one mole, of solute, but variable quantities of water. The first step, then, consists of calculating the moles of water per mole of alcohol, n_1/n_2, from the data on percentage composition. Such values are listed in Column 3. From this information and the density, we can calculate the volumes of solutions containing one mole of alcohol but different quantities of water. These volumes are listed in Column 7. The partial molal volumes of the water, $\bar{v}_1$, may then be determined from a chord-area plot of the ratio of increments $\Delta V / \Delta n_1$ (listed in Column 8) vs. n_1/n_2, for

$$\left(\frac{\Delta V}{\Delta n_1}\right)_{n_2} = \left(\frac{\Delta V}{\Delta(n_1/n_2)}\right)_{n_1-1}.$$ (13.64)

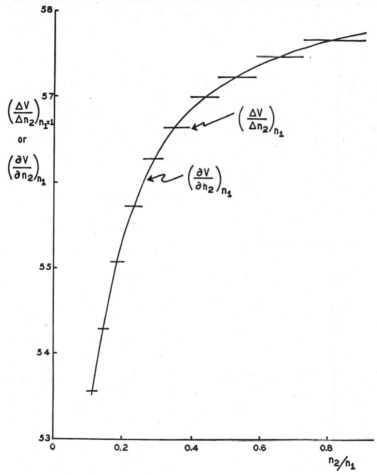

Fig. 5.   Graphical differentiation to obtain $\bar{v}_2$ in alcohol-water solutions.

An alternative method of determination of $\bar{v}_1$ is to apply equation (13.49), and hence to integrate graphically in a plot of n_2/n_1 vs. $\bar{v}_2$.

B.   **Partial molal quantities from apparent molal quantities.**   The apparent molal quantity is a very convenient function for calculating partial molal quantities, and hence we shall be interested in examining some of its properties.

1. *Definition of apparent molal quantity.*

a.   *Volume.*   As usual, we shall start with a consideration of the volume function, since it is simpler to visualize.   Referring to Fig. 6, as an example, we note that the volume of a solution may increase with added solute, as indicated by the solid curve.   In actual practice, of course, it is impossible to say which part of the volume of a solution belongs to the solvent and

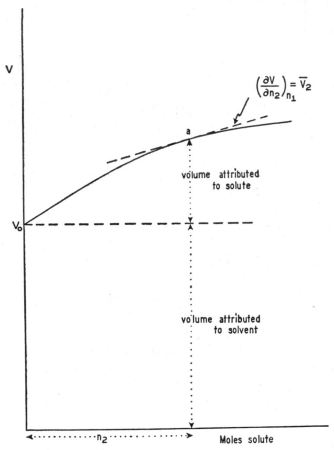

Fig. 6.   Significance of the apparent molal volume.

which to the solute.   Nevertheless, the total volume, V, at the point a may be (arbitrarily) considered to consist of two portions.   The first, V_0, that of the pure solvent, that is, at infinite dilution, may be attributed also to the solvent in the solution at some finite concentration of solute.   The difference $V - V_0$ can be attributed, then, to the solute, as is indicated in Fig. 6.   This "apparent" volume of the solute, $V - V_0$, is converted to a molal volume at point a if $V - V_0$ is divided by the moles of solute, n_2, present in the solution at this point. Hence ϕV_2, the *apparent molal volume* of the solute, is defined by the expression

$$\phi V_2 = \frac{\text{total volume} - \text{volume of pure solvent}}{\text{moles of solute}}, \quad (13.65)$$

or

$$\phi V_2 = \frac{V - n_1 \bar{v}_1^{\,\circ}}{n_2} = \frac{V - n_1 v_1^{\,\bullet}}{n_2}. \quad (13.66)$$

The corresponding quantity, the apparent molal volume of the solvent, ϕV_1, generally is not considered, because it would be logically absurd to consider the volume of the solute constant in the same solution in which it has been assumed already that the volume of the solvent is constant. Furthermore, the term $n_2 \bar{v}_2^{\,\circ}$, which would correspond to $n_1 \bar{v}_1^{\,\circ}$ in equation (13.66), cannot be evaluated readily, since in the case of the solute,

$$\bar{v}_2^{\,\circ} \neq v_2^{\,\bullet} \quad (13.67)$$

except by coincidence.

It can be shown readily that at infinite dilution the apparent molal volume and partial molal volume of the solute are identical, since by substituting equation (13.26) for V in equation (13.66), we obtain

$$\phi V_2 = \frac{V - n_1 \bar{v}_1^{\,\circ}}{n_2} = \frac{n_1 \bar{v}_1 + n_2 \bar{v}_2 - n_1 \bar{v}_1^{\,\circ}}{n_2}. \quad (13.68)$$

Obviously,

$$\lim_{n_2 \to 0} \bar{v}_1 = \bar{v}_1^{\,\circ} \quad (13.69)$$

and

$$\lim_{n_2 \to 0} \phi V_2 = \bar{v}_2^{\,\circ}. \quad (13.70)$$

b. *General case.* It is evident from the discussion for volume that the general apparent molal quantity, ϕG_2, is defined by the expression

$$\phi G_2 = \frac{G - n_1 \bar{G}_1^{\,\circ}}{n_2}. \quad (13.71)$$

Similarly, it can be shown readily that

$$\lim_{n_2 \to 0} \phi G_2 = \bar{G}_2^{\,\circ}. \quad (13.72)$$

2. *Equations relating $\bar{G}$'s to ϕG_2.* We are now in a position to derive the expressions relating the apparent molal quantity, which is most convenient for carrying out calculations from experimental data, to partial molal quantities.

It is obvious that equation (13.71) can be rearranged to give G explicitly.

$$G = n_2 \phi G_2 + n_1 \bar{G}_1^{\,\circ}. \quad (13.73)$$

In view of the fundamental definition of $\bar{G}_2$,

$$\bar{G}_2 = \left(\frac{\partial G}{\partial n_2}\right)_{n_1}, \tag{13.74}$$

we may obtain from equation (13.73) by differentiation

$$\bar{G}_2 = n_2 \left(\frac{\partial \phi G_2}{\partial n_2}\right)_{n_1} + \phi G_2. \tag{13.75}$$

The derivative of the second term in equation (13.73) vanishes, since n_1, the moles of solvent, and $\bar{G}_1^{\circ}$, the partial molal quantity of *pure* solvent, both are independent of n_2.

Similarly, for $\bar{G}_1$ we obtain

$$\bar{G}_1 = \left(\frac{\partial G}{\partial n_1}\right)_{n_2} = n_2 \left(\frac{\partial \phi G_2}{\partial n_1}\right)_{n_2} + \bar{G}_1^{\circ}. \tag{13.76}$$

3. *Calculation of $\bar{G}$'s.* The calculation of partial molal quantities from apparent molal quantities is generally used when ϕG_2 is expressible as an analytic function. Thus we might have an equation of the following form:

$$\phi G_2 = a + bm + cm^2, \tag{13.77}$$

where m is the molality and a, b, and c represent constants. For simplicity, we may consider first a series of solutions containing 1000 grams of solvent, so that

$$n_2 = m \tag{13.78}$$

and

$$dn_2 = dm. \tag{13.79}$$

It follows directly from equation (13.75) that

$$\bar{G}_2 = m \left(\frac{\partial \phi G_2}{\partial m}\right)_{n_1} + \phi G_2 \tag{13.80}$$

$$= m[b + 2cm] + a + bm + cm^2. \tag{13.81}$$

Therefore $\qquad \bar{G}_2 = a + 2bm + 3cm^2. \tag{13.82}$

There are several possible procedures for calculating $\bar{G}_1$. One could make use of equation (13.76), carrying out the required differentiation with the aid of the substitution

$$m = \frac{\text{moles solute}}{\text{kilograms solvent}} = \frac{n_2}{n_1 M_1/1000} = \frac{1000 n_2}{M_1 n_1}. \tag{13.83}$$

Probably a simpler method, however, is to make use of the general relation

$$G = n_1 \bar{G}_1 + n_2 \bar{G}_2 \tag{13.29}$$

as a substitution in equation (13.73) to obtain

$$n_1\bar{G}_1 + n_2\bar{G}_2 = n_2\phi G_2 + n_1\bar{G}_1^{\circ}. \tag{13.84}$$

Considering again solutions containing 1000 grams of solvent, we obtain

$$n_1\bar{G}_1 + m(a + 2bm + 3cm^2) = m(a + bm + cm^2) + n_1\bar{G}_1^{\circ}. \tag{13.85}$$

Hence
$$n_1(\bar{G}_1 - \bar{G}_1^{\circ}) = -bm^2 - 2cm^3. \tag{13.86}$$

Since
$$n_1 = \frac{1000}{M_1}, \tag{13.87}$$

it follows that
$$\bar{G}_1 = \bar{G}_1^{\circ} - \frac{bM_1}{1000}\, m^2 - \frac{2cM_1}{1000}\, m^3. \tag{13.88}$$

It should be emphasized that equations (13.82) and (13.88) [as well as (13.53) and (13.59)], though derived for solutions containing 1000 grams of solvent, are applicable also to any quantity of solution containing a different amount of solvent, since partial molal quantities are intensive quantities and do not depend upon the amount of matter present. In mathematical terms, since G is a homogeneous function of degree 1, a partial derivative of G must be a homogeneous function of degree zero, that is, a function which does not change when each of the independent variables, $n_1, n_2, \ldots$, is multiplied by a parameter, k. Multiplying by k corresponds to increasing the total quantity of solution by the same factor. If $\bar{G}$ does not change when both n_1 and n_2 are multiplied by k, it does not change when the volume of solution is increased by the same factor. On the other hand, $\bar{G}$ may change if the concentration of the solution is changed, since in this case the independent variables, n_1 and n_2, are not multiplied by the *same* k.

V. THERMODYNAMIC RELATIONSHIPS FOR PARTIAL MOLAL QUANTITIES

In general we may say that all of the thermodynamic expressions which we have previously derived for molal thermodynamic properties may be restated by replacement of the extensive quantity G by the partial molal quantity. Thus, if

$$F = H - TS, \tag{13.89}$$

we may immediately also write

$$\bar{F} = \bar{H} - T\bar{S}. \tag{13.90}$$

That such a transformation is valid may be demonstrated readily by the following steps. If we differentiate equation (13.89) with respect to n, the number of moles of any component,

$$\frac{dF}{dn} = \frac{dH}{dn} - T\frac{dS}{dn} - S\frac{dT}{dn}. \tag{13.91}$$

Since partial molal quantities are defined [equation (13.19)] for *fixed temperature and pressure*, equation (13.91) reduces to

$$\bar{F} = \bar{H} - T\bar{s}. \tag{13.92}$$

Similarly, when we have equations involving temperature coefficients, such as

$$\left(\frac{\partial H}{\partial T}\right)_P = C_p, \tag{13.93}$$

we may write

$$\left(\frac{\partial \bar{H}}{\partial T}\right)_P = \bar{c}_p. \tag{13.94}$$

That this procedure is valid may be seen readily as follows:

$$\frac{\partial}{\partial n}\frac{\partial H}{\partial T} = \frac{\partial}{\partial n} C_p. \tag{13.95}$$

But

$$\frac{\partial}{\partial n}\frac{\partial H}{\partial T} = \frac{\partial}{\partial T}\frac{\partial H}{\partial n} = \frac{\partial}{\partial T} \bar{H}. \tag{13.96}$$

Hence

$$\left(\frac{\partial \bar{H}}{\partial T}\right)_P = \bar{c}_p. \tag{13.97}$$

By a similar process it can be shown that the same restatements can be made for equations involving pressure coefficients. Thus, if

$$\left(\frac{\partial F}{\partial P}\right)_T = V, \tag{13.98}$$

then

$$\left(\frac{\partial \bar{F}}{\partial P}\right)_T = \bar{v}. \tag{13.99}$$

Thus we arrive at the conclusion that the thermodynamic expressions derived for the description of the behavior of pure substances are equally applicable to solutions, so long as extensive quantities are replaced by partial molal quantities.

Exercises

1. Show that $\partial G/\partial N_1$ is not identical with $\partial G/\partial n_1$ if n_2 is maintained constant in both cases.

2. (a) If V is the volume (in cubic centimeters) of a two-component solution containing 1000 grams of solvent, show that

$$\bar{v}_2 = \frac{M_2 - V\dfrac{\partial D}{\partial m}}{D},$$

where M_2 is the molecular weight of the solute, D is the density of the solution, and m is the molality of solute.

(b) Show that

$$\bar{v}_1 = \frac{M_1\left(1 + \frac{mV}{1000}\frac{\partial D}{\partial m}\right)}{D}.$$

3. If c is the concentration in moles of solute per liter of solution, D is the density of the solution, and D_0 is the density of pure solvent, show that:

(a)
$$\phi V_2 = \frac{1000}{c} - \frac{1}{D_0}\left[\frac{1000D}{c} - M_2\right];$$

(b)
$$\bar{v}_2 = \frac{M_2 - 1000\frac{\partial D}{\partial c}}{D - c\frac{\partial D}{\partial c}};$$

(c)
$$\bar{v}_1 = \frac{M_1}{D - c\frac{\partial D}{\partial c}}.$$

4. (a) Verify the calculations in Table 1 for the 20 and 25 weight-percentage solutions.

(b) Plot the volume per mole of water (Column 5) vs. n_2/n_1.

(c) Make a careful graph of $\Delta V/\Delta n_2$ vs. n_2/n_1. Draw a smooth curve through the chord plot.

(d) Make a careful graph of $\Delta V/\Delta n_1$ vs. n_1/n_2. Draw a smooth curve through the chord plot.

(e) Determine the partial molal volumes of water and of alcohol, respectively, for a 45 per cent solution.

(f) As a method of checking the calculations in Table 1, calculate the volume of 500 grams of a 45 per cent solution from the partial molal volumes, and compare the value obtained with that which can be calculated directly by using the density.

(g) Make a graph of $\bar{v}_1$ vs. n_1/n_2. Carry out an appropriate integration to find the difference between $\bar{v}_2$ in a 65 per cent solution and $\bar{v}_2$ in a 25 per cent solution.

(h) Find $\bar{v}_2$ for a 25 per cent solution from the chord-area graph in Part (c), and add it to the difference calculated in Part (g). Compare this sum with the value of $\bar{v}_2$ in a 65 per cent solution which can be read from the graph in Part (c).

5. Making use of the data in Exercise 4 and data on the density of pure H_2O (0.99708 at 25°) and of pure alcohol (0.78506 at 25°), compute the volume changes for the following processes:

(a) C_2H_6O (pure) $= C_2H_6O$ (in large quantity of aqueous solution of 45% alcohol).

(b) H_2O (pure) $= H_2O$ (in large quantity of aqueous solution of 45% alcohol).

6. (a) By an appropriate summation of the two processes in Exercise 5, compute the volume change for the process:

4.884 C_2H_6O (pure) + 15.264 H_2O (pure)

$$= \begin{cases} 4.884\ C_2H_6O \\ 15.264\ H_2O \end{cases} \begin{array}{l} \text{(in large quantity of aque-} \\ \text{ous solution of 45\% alco-} \\ \text{hol).} \end{array}$$

(b)   Making use of the answer in Exercise 4(f), compute the volume change for the process

4.884 C_2H_6O (pure) + 15.264 H_2O (pure) = 500 grams of aqueous solution of 45% alcohol.

(c)   Compare the answers in Parts (a) and (b).   They should be the same, within computational error.   Why?

(d)   Write word statements which emphasize the differences in meaning between the equations in 5(a), 5(b), 6(a), and 6(b).

7.   The specific heats and apparent molal heat capacities of aqueous solutions of glycolamide at 25°C [F. T. Gucker, Jr., W. L. Ford, and C. E. Moser, *J. Phys. Chem.*, **43**, 153 (1939); *ibid.*, **45**, 309 (1941)] are listed in the following table.

Molality	Specific Heat (25° calorie)	$\phi C_{p(2)}$
0.0000	(1.00000)	...
0.2014	0.99223	35.9
0.4107	0.98444	36.0
0.7905	0.97109	36.33
1.2890	0.95467	...
1.7632	0.94048	36.84
2.6537	0.91666	37.41
4.3696	0.87899	38.29
4.3697	0.87900	38.29
6.1124	0.84891	39.01

(a)   Calculate the value of $\phi C_{p(2)}$ of a 1.2890 molal solution.

(b)   Plot $\phi C_{p(2)}$ as a function of the molality.

(c)   Using the method of averages, obtain an equation for the apparent molal heat capacity as a function of the molality.

(d)   Using the method of least squares, find an equation for $\phi C_{p(2)}$ as a function of m.

(e)   Derive an equation for the partial molal heat capacity of glycolamide, based on the least square equation for $\phi C_{p(2)}$.

(f)   Determine a few numerical values of $\bar{c}_{p(2)}$ and plot them on the graph on which you have $\phi C_{p(2)}$.

(g)   Find the equation for the partial molal heat capacity of water in aqueous solutions of glycolamide.

8.   Starting with the relations for the corresponding extensive quantities, prove the validity of the following expressions:

(a) $\partial \bar{F}/\partial P = \bar{v}$;   (b) $\partial \bar{F}/\partial T = -\bar{s}$;   (c) $\bar{F} = \bar{H} + T(\partial \bar{F}/\partial T)_P$

CHAPTER 14

Enthalpy in Systems of Variable Composition

THE HEAT content or enthalpy of solutions must also be treated in terms of partial molal quantities, as are the volume and heat capacity. However, in contrast to volume or heat capacity, absolute values of the enthalpies of components of a solution cannot be determined. It is necessary, therefore, to give some consideration to methods of expressing partial molal enthalpies on a scale of relative values.

I. DEFINITIONS

A. Partial molal enthalpy. In conformance with our general notation for partial molal quantities, we shall represent the partial molal enthalpy of component i of a solution by the equation

$$\bar{H}_i = \left(\frac{\partial H}{\partial n_i}\right)_{P,T,n_j}. \tag{14.1}$$

This quantity $\bar{H}_i$ may be thought of as the increase in enthalpy of a solution when one mole of component i is added to a large enough volume of the solution of a given concentration so that the composition is not changed significantly.

B. Relative partial molal enthalpy. To appreciate the difficulty in any attempt to determine $\bar{H}$, and the consequent need for relative values of partial enthalpies, let us approach the problem from an experimental viewpoint. As an example we may consider a mixing process, at constant pressure and temperature, such as that represented by the following equation:

$$n_2\, NaCl + n_1\, H_2O = \text{solution of } n_1\, H_2O \text{ and } n_2\, NaCl. \tag{14.2}$$

In this mixing of n_2 moles of NaCl and n_1 moles of H_2O to give the solution indicated, a quantity of heat, Q, is absorbed which may be represented also by the symbol ΔH.[1] It should be apparent that the total enthalpy, H, of the solution formed on mixing is related to ΔH by the equation

$$H \text{ (solution)} = n_1 H_1^\bullet + n_2 H_2^\bullet + \Delta H, \tag{14.3}$$

---

[1] An unusually large-sized symbol has been used at this point so that ΔH may be clearly distinguished from any molal enthalpies.

where $H_1^{\bullet}$ and $H_2^{\bullet}$ represent the molal enthalpies of pure water and pure solid sodium chloride, respectively. From equation (14.3) it is evident that no absolute value can be assigned to H, since none has been given to either $H_1^{\bullet}$ or $H_2^{\bullet}$.

Nevertheless, to obtain a relation between $\bar{H}_2$ and the experimental quantity ΔH, we may differentiate H in equation (14.3) with respect to n_2, keeping n_1 constant:

$$\bar{H}_2 = \left(\frac{\partial H}{\partial n_2}\right)_{n_1} = H_2^{\bullet} + \left(\frac{\partial \Delta H}{\partial n_2}\right)_{n_1}. \tag{14.4}$$

Again we observe that $\bar{H}_2$ has no absolute value, since none has been assigned to $H_2^{\bullet}$. If it were necessary, we could assign an arbitrary value to $\bar{H}_2^{\bullet}$, but since we shall need only differences between values of $\bar{H}_2$, we shall not need to know absolute values. Thus, if we wish to know the difference between $\bar{H}_2$'s in solutions of 4-molal and 3-molal concentration, we can subtract the respective $\bar{H}_2$'s and obtain from equation (14.4)

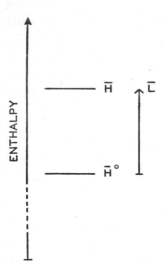

$$\bar{H}_2(4m) - \bar{H}_2(3m) = \left(\frac{\partial \Delta H}{\partial n_2}\right)_{n_1}'' - \left(\frac{\partial \Delta H}{\partial n_2}\right)_{n_1}', \tag{14.5}$$

where the double prime refers to the $4m$ solution and the single prime to the $3m$. Thus the difference between the two partial derivatives is the difference between the partial molal enthalpies.

Fig. 1.   Relative partial molal enthalpy.

In many thermodynamic problems it has proved convenient to choose some state as a reference state and to tabulate the difference between $\bar{H}$ in any state and $\bar{H}$ in the arbitrarily selected reference state. The most convenient choice for the reference state has usually been found to be the infinitely dilute solution. Without committing ourselves to this choice exclusively, we shall use it in most of our problems.

The relative values of partial molal enthalpies are used so frequently that it has become customary to use a special symbol, $\bar{L}$, to represent them. Thus $\bar{L}$, the *relative partial molal enthalpy*, is defined by the equation

$$\bar{L}_i = \bar{H}_i - \bar{H}_i^{\circ}, \tag{14.6}$$

where $\bar{H}_i^{\circ}$ is the partial molal enthalpy of component i in the infinitely dilute solution.

Relative partial molal enthalpies can also be visualized conveniently in terms of a diagram, such as Fig. 1. Although the absolute position of

$\bar{H}$ or $\bar{H}^\circ$ on the enthalpy scale cannot be specified, the difference between them can be determined.

To illustrate the relationships between $\bar{L}$, $\bar{H}$, and experimentally determined quantities, we shall examine a very simple (imaginary) example. Consider a solute X_2 whose heat of solution in 500 grams of water is given (in terms of the number of moles, n_2) by the equation

$$\Delta H = 40n_2 + 30n_2^2. \tag{14.7}$$

This heat of solution, ΔH, accompanies the process which may be expressed by the equation

$$n_2\, X_2 + 27.754\; H_2O = \text{solution of } n_2\, X_2 \text{ and } 27.754\; H_2O. \tag{14.8}$$

If we wish to calculate $\bar{H}_2$ first, we proceed to differentiate ΔH of equation (14.7) to obtain

$$\left(\frac{\partial \Delta H}{\partial n_2}\right)_{n_1} = 40 + 60n_2. \tag{14.9}$$

In view of equation (14.4) we may write an equation for $\bar{H}_2$ of the solute, X_2, as follows:

$$\bar{H}_2 = H_2^\bullet + 40 + 60n_2, \tag{14.10}$$

or
$$\bar{H}_2 - H_2^\bullet = 40 + 60n_2. \tag{14.11}$$

Either of these two equations may be used as it stands to evaluate differences in $\bar{H}_2$ between solutions of X_2 of different composition.

If we wish to calculate values of $\bar{L}_2$, we must first evaluate $\bar{H}_2^\circ$, the partial molal enthalpy in the reference state (infinite dilution). From equation (14.10) it follows that

$$\bar{H}_2^\circ = H_2^\bullet + 40. \tag{14.12}$$

Consequently $\bar{L}_2$ must be given by the expression

$$\bar{L}_2 = \bar{H}_2 - \bar{H}_2^\circ = 60n_2. \tag{14.13}$$

Some numerical results for this imaginary solute, X_2, calculated from the foregoing equations, have been assembled in Table 1. The values tabulated for the solutions of zero to 4-molal can be verified readily by reference to the appropriate equation among (14.7)–(14.13), and can be visualized easily by the preparation of a diagram similar to Fig. 1. A few additional remarks may be desirable, however, in connection with the last line in Table 1, referring to pure X_2. Columns 3 and 4 are blank, since no proc-

TABLE 1

RELATIVE PARTIAL MOLAL ENTHALPIES IN A TYPICAL PROBLEM

n_2	m	ΔH	$\left(\dfrac{\partial \Delta H}{\partial n_2}\right)_{n_1}$	$\bar{H}_2 - H_2^{\bullet}$	$\bar{L}_2 = \bar{H}_2 - \bar{H}_2^{\circ}$
0.0	0.0	0.0	40	40	0.0
0.1	0.2	4.3	46	46	6.0
0.5	1.0	27.5	70	70	30.0
1.0	2.0	70.0	100	100	60.0
2.0	4.0	200.0	160	160	120.0
Pure X_2		...	...	0.0	-40.0

ess of solution is being considered. Similarly, equations (14.11) and (14.13) are not applicable, since we are not referring to $\bar{H}_2$ or $\bar{L}_2$ of X_2 in solution but to pure X_2. It is possible, nevertheless, to make an entry in Column 5 for pure X_2, since $\bar{H}_2 - H_2^{\bullet}$ is merely $H_2^{\bullet} - H_2^{\bullet}$ and hence identically zero. Furthermore, $\bar{L}_2$ can also be specified, since it is $H_2^{\bullet} - \bar{H}_2^{\circ}$, which can be evaluated by rearrangement of equation (14.12) to the following:

$$H_2^{\bullet} - \bar{H}_2^{\circ} = -40. \qquad (14.14)$$

This example with an imaginary solute, X_2, should assist in establishing the significance of various enthalpy functions and should provide the basis for a discussion of actual procedures used to relate measured heats and $\bar{L}$'s. Before we consider the experimental problem any further, however, it is desirable to make a clear distinction between what are known as the *differential heat* and *integral heat* of solution, respectively.

II. DIFFERENTIAL HEAT OF SOLUTION

If one mole of a solute, for example, crystalline NaCl is dissolved in a solution of a given concentration and in a quantity so large that the additional NaCl causes no appreciable change in concentration, then the attendant heat effect is called the *differential heat of solution*. The process described is represented conventionally by the equation

$$\text{NaCl (s)} = \text{NaCl } (m = 1), \qquad (14.15)$$

if the solution is one-molal in concentration. It may be visualized more readily, however, from the following more detailed equation (for an aqueous solution):

$$\text{NaCl (s)} + \left\{ \begin{matrix} \text{solution of } n_1 \text{ H}_2\text{O and} \\ n_2' \text{ NaCl; } n_1 \text{ and } n_2' \\ \text{very large; } m = 1 \end{matrix} \right\} = \left\{ \begin{matrix} \text{solution of } n_1 \text{ H}_2\text{O and } n_2'' \\ \text{NaCl; } n_1 \text{ and } n_2'' \text{ very} \\ \text{large; } n_2'' = n_2' + 1; m = 1 \end{matrix} \right\}. \quad (14.16)$$

An equivalent viewpoint is to visualize the process as one in which an *infinitesimal* amount of solute is dissolved in a given quantity of solution, in which case there would also be no change in the concentration of the solution. This latter process makes the significance of the term "differential heat" more evident.

In either case, the term "differential heat" is used to make a clear distinction between the preceding processes and an alternative situation in which one mole of solute, for example, NaCl, is dissolved in a specified quantity, such as 1000 grams, of pure solvent to give a (one-molal) solution. This procedure may be represented by an equation of the form

$$\text{NaCl (s)} + 55.51\,H_2O\ (l) = \text{solution of NaCl and } 55.51\,H_2O; \quad (14.17)$$

the accompanying heat effect is called the *integral heat of solution*. A comparison of equations (14.16) and (14.17) indicates immediately that the differential and integral heats are not equal, in general.

An expression for the differential heat of solution in terms of relative partial enthalpies can be derived in any one of several ways. A glance at equation (14.16) shows that the solvent is substantially in the same state before and after the dissolution process, since a sufficiently large quantity of solution is used so that the concentration does not change significantly on the addition of solute. Thus

$$\bar{H}_{1(\text{initial})} = \bar{H}_{1(\text{final})}. \quad (14.18)$$

For the solute, however, the dissolution process involves a pronounced modification of state and hence, in general,

$$\overset{\bullet}{H}_2 \text{ (solid)} \neq \bar{H}_2 \text{ (solution)}. \quad (14.19)$$

Recognizing the validity of equations (14.18) and (14.19), we can proceed in a straightforward manner to calculate ΔH for the process represented by equation (14.16) or (14.15). We start with the relation

$$\Delta H = H_{\text{final}} - H_{\text{initial}}. \quad (14.20)$$

For any extensive thermodynamic property of a solution,

$$G = n_1 \bar{G}_1 + n_2 \bar{G}_2. \quad (14.21)$$

Hence $\quad \Delta H = [n_1 \bar{H}_1 + (n_2' + 1)\bar{H}_2] - [(n_1 \bar{H}_1 + n_2' \bar{H}_2) + \overset{\bullet}{H}_2]. \quad (14.22)$

Clearing parentheses, we obtain

$$\Delta H = \bar{H}_2 - \overset{\bullet}{H}_2. \quad (14.23)$$

Furthermore, we may add and subtract $\bar{H}_2^{\circ}$, the partial molal enthalpy of the solute in its reference state, and obtain an equation in terms of relative partial enthalpies:

$$\Delta H = \bar{H}_2 - \overset{\bullet}{H}_2 + \bar{H}_2^{\,\circ} - \bar{H}_2^{\,\circ} \tag{14.24}$$

$$= (\bar{H}_2 - \bar{H}_2^{\,\circ}) - (\overset{\bullet}{H}_2 - \bar{H}_2^{\,\circ}), \tag{14.25}$$

or $$\Delta H = \bar{L}_2 - \overset{\bullet}{L}_2. \tag{14.26}$$

The equivalence of equations (14.23) and (14.26) may be visualized also to advantage with a schematic diagram (Fig. 2). Equation (14.15) corresponds to putting one mole of NaCl into a large quantity of solution. The only heat effect, then, is due to the change in enthalpy of a mole of NaCl when it is converted from the crystalline state to the dissolved state

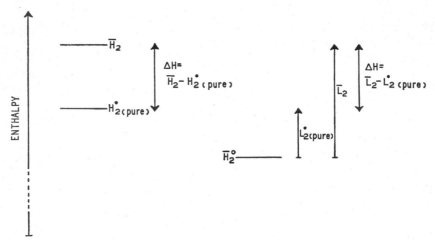

Fig. 2.   Diagrammatic representations of differential heat of solution.

at 1-molal concentration. The solution initially present undergoes no change in enthalpy, since it undergoes no significant change in state. Hence ΔH is given by the difference in levels (Fig. 2) between $\bar{H}_2$ and $\overset{\bullet}{H}_2$. If we use a reference state, $\bar{H}_2^{\,\circ}$, from which to express relative partial molal enthalpies, it is obvious from Fig. 2 that ΔH as given by $\bar{L}_2 - \overset{\bullet}{L}_2$ still has the same value.

Differential heats may refer to the solvent as well as to the solute.   Thus one might add one mole of water to a large quantity of 1-molal NaCl solution, with the amount of solution sufficiently large that no appreciable dilution occurs.[2] Such a process is represented conventionally by an equation analogous to (14.15):

$$H_2O \text{ (l)} = H_2O \text{ (solution, } m \text{ of NaCl is 1).} \tag{14.27}$$

---

[2] This process may be thought of as giving rise to a differential heat of solution by analogy with the preceding example, in which solute is added to a large quantity of solution.

Again the process may be visualized more readily by the following equivalent equation:

$$H_2O \; (l) \;\; | \;\; \begin{Bmatrix} \text{solution of } n_1' \text{ H}_2\text{O and} \\ n_2 \text{ NaCl; } n_1' \text{ and } n_2 \\ \text{very large; } m = 1 \end{Bmatrix} - \begin{Bmatrix} \text{solution of } n_1'' \text{ H}_2\text{O and } n_2 \\ \text{NaCl; } n_1'' \text{ and } n_2 \text{ very} \\ \text{large; } n_1'' = n_1' + 1; m = 1 \end{Bmatrix}. \quad (14.28)$$

By reasoning in a fashion similar to that used for the solute, we can show that

$$\Delta H = \bar{H}_1 - \overset{\bullet}{H}_1. \quad (14.29)$$

Since the reference state for the solvent is the pure solvent, that is, since

$$\bar{H}_1^\circ = \overset{\bullet}{H}_1, \quad (14.30)$$

it is evident that

$$\Delta H = \bar{H}_1 - \bar{H}_1^\circ = \bar{L}_1. \quad (14.31)$$

III. INTEGRAL HEAT OF SOLUTION

Let us now turn our attention to the relationship between ΔH and $\bar{L}$'s for a process such as that represented by equation (14.17), where two pure substances are mixed to form a solution and where the heat effect is referred to as the *integral heat of solution*. Recognizing the general validity of equations (14.20) and (14.21), we may write

$$\Delta H = H_{\text{final}} - H_{\text{initial}} = n_1 \bar{H}_1 + n_2 \bar{H}_2 - n_1 \overset{\bullet}{H}_1 - n_2 \overset{\bullet}{H}_2, \quad (14.32)$$

where $\overset{\bullet}{H}_1$ and $\overset{\bullet}{H}_2$ again refer to pure solvent and pure solute, respectively. In view of the identity of the reference state of the solvent with the pure solvent,

$$\Delta H = n_1(\bar{H}_1 - \bar{H}_1^\circ) + n_2(\bar{H}_2 - \overset{\bullet}{H}_2) \quad (14.33)$$

$$= n_1 \bar{L}_1 + n_2(\bar{H}_2 - \overset{\bullet}{H}_2). \quad (14.34)$$

If we insert $\bar{H}_2^\circ$, we obtain

$$\Delta H = n_1 \bar{L}_1 + n_2(\bar{H}_2 - \overset{\bullet}{H}_2 + \bar{H}_2^\circ - \bar{H}_2^\circ) \quad (14.35)$$

$$= n_1 \bar{L}_1 + n_2(\bar{H}_2 - \bar{H}_2^\circ) - n_2(\overset{\bullet}{H}_2 - \bar{H}_2^\circ), \quad (14.36)$$

or $\qquad \Delta H = n_1 \bar{L}_1 + n_2 \bar{L}_2 - n_2 \overset{\bullet}{L}_2. \quad (14.37)$

Thus it is possible to relate the integral heat of solution to relative partial molal enthalpies.

The derivation of equation (14.37) may also be considered profitably from another point of view. We may picture the formation of a solution from the pure components as taking place in three steps. For the process represented by equation (14.17), for example, we may write three separate steps whose sum is equation (14.17):

$$\text{NaCl (s)} + \begin{Bmatrix} \text{solution of } n_1 \\ \text{H}_2\text{O and } n_2 \\ \text{NaCl; } n_1 \text{ and} \\ n_2 \text{ very large;} \\ m = 1 \end{Bmatrix} = \begin{Bmatrix} \text{solution of } n_1 \\ \text{H}_2\text{O and } (n_2 + \\ 1) \text{ NaCl; } m = \\ 1 \end{Bmatrix} \quad ; \Delta H_A$$

(14.38)

$$55.51\,\text{H}_2\text{O (l)} + \begin{Bmatrix} \text{solution of } n_1 \\ \text{H}_2\text{O and } (n_2 + \\ 1) \text{ NaCl; } m = 1 \end{Bmatrix} = \begin{Bmatrix} \text{solution of } (n_1 \\ + 55.51) \text{ H}_2\text{O} \\ \text{and } (n_2 + 1) \\ \text{NaCl; } m = 1 \end{Bmatrix} \quad ; \Delta H_B$$

(14.39)

$$\begin{Bmatrix} \text{solution of } (n_1 \\ + 55.51) \text{ H}_2\text{O} \\ \text{and } (n_2 + 1) \\ \text{NaCl; } m = 1 \end{Bmatrix} = \begin{Bmatrix} \text{solution of } n_1 \\ \text{H}_2\text{O and } n_2 \\ \text{NaCl; } m = 1 \end{Bmatrix} + \begin{Bmatrix} \text{solution of} \\ 55.51 \text{ H}_2\text{O} \\ \text{and 1 NaCl;} \\ m = 1 \end{Bmatrix} ; \Delta H_C$$

(14.40)

$$\text{NaCl (s)} + 55.51\,\text{H}_2\text{O (l)} = \text{solution of NaCl and } 55.51\,\text{H}_2\text{O}. \qquad (14.17)$$

Since the enthalpy is a thermodynamic property, the sum of the ΔH's for equations (14.38) to (14.40) is ΔH for equation (14.17). It is merely necessary, therefore, to consider the individual heat effects.

From our preceding discussion of differential heats of solution it is evident that ΔH_A is merely the differential heat of solution of NaCl in a 1-molal solution; that is,

$$\Delta H_A = \bar{L}_2 - L_2^{\bullet}. \qquad (14.41)$$

Similarly, $$\Delta H_B = 55.51\bar{L}_1. \qquad (14.42)$$

Brief reflection is sufficient to show that ΔH_C must be zero,[3] since we are merely separating a solution into two parts which are identical in all the variables of state, that is, in temperature, pressure, and composition. Just as the mixing of two samples of a pure substance at the same temperature and pressure involves no thermodynamic changes, so are there no changes when two portions of a solution of a given concentration are mixed together at the same temperature and pressure. Consequently, the net ΔH for equations (14.38)–(14.40) is

$$\Delta H = \Delta H_A + \Delta H_B + \Delta H_C = \bar{L}_2 - L_2^{\bullet} + 55.51\bar{L}_1. \qquad (14.43)$$

Obviously, this is merely a special example of equation (14.37).

---

[3] The conclusion is based on the assumption that changes in surface energy are negligible. The separation of a solution into two parts would not occur, in general, without some change in surface area.

IV. INTEGRAL HEAT OF DILUTION

Another quantity of frequent interest is the heat accompanying the dilution of a given amount of solution with pure solvent, an example of which may be represented by the equation

$$\left\{\begin{matrix}\text{solution of } n_1 \text{ H}_2\text{O} \\ \text{and } n_2 \text{ NaCl}\end{matrix}\right\} + n_1' \text{ H}_2\text{O (l)} = \left\{\begin{matrix}\text{solution of } (n_1 + n_1') \\ \text{H}_2\text{O and } n_2 \text{ NaCl}\end{matrix}\right\}. \quad (14.44)$$

The calculation of the heat absorbed may be carried out in a fashion analogous to that described for the integral heat of solution:

$$\Delta H = H_{\text{final}} - H_{\text{initial}} \quad (14.45)$$

$$= (n_1 + n_1')\bar{\text{H}}_1' + n_2\bar{\text{H}}_2' - (n_1'\overset{\bullet}{\text{H}}_1 + n_1\bar{\text{H}}_1 + n_2\bar{\text{H}}_2). \quad (14.46)$$

The primed heats refer to the final solution.

Adding and subtracting $\bar{\text{H}}°$'s, we obtain the expression

$$\Delta H = (n_1 + n_1')\bar{\text{L}}_1' + n_2\bar{\text{L}}_2' - (n_1\bar{\text{L}}_1 + n_2\bar{\text{L}}_2). \quad (14.47)$$

Thus, once again, the process may be represented in terms of the relative partial molal heat contents of solute and of solvent.

V. DETERMINATION OF PARTIAL MOLAL ENTHALPIES FROM CALORIMETRIC MEASUREMENTS

It is evident from the preceding discussion that differential and integral heats may be calculated readily once $\bar{\text{L}}$'s are available. The task remains, therefore, of obtaining partial molal enthalpies from experimental calorimetric data.

A. From heats of solution. Let us consider a series of experiments in which variable quantities of pure solute are dissolved in a given quantity of pure solvent and the corresponding heats measured. These integral heats of mixing are related to the enthalpy of the solution as follows:

$$\Delta H = H_{\text{final}} - H_{\text{initial}} = H \text{ (solution)} - n_1\overset{\bullet}{\text{H}}_1 - n_2\overset{\bullet}{\text{H}}_2. \quad (14.48)$$

Again $\overset{\bullet}{\text{H}}$ refers to the pure component. Rearrangement of equation (14.48) gives the form previously introduced:

$$H \text{ (solution)} = n_1\overset{\bullet}{\text{H}}_1 + n_2\overset{\bullet}{\text{H}}_2 + \Delta H. \quad (14.3)$$

To find $\bar{\text{H}}_2$, we may differentiate H in equation (14.3) to obtain again

$$\bar{\text{H}}_2 = \overset{\bullet}{\text{H}}_2 + \left(\frac{\partial \Delta H}{\partial n_2}\right)_{n_1}. \quad (14.4)$$

Since we shall want $\bar{\text{L}}$'s rather than $\bar{\text{H}}$'s, we subtract $\bar{\text{H}}_2^\circ$ from both sides of equation (14.4) and thereby arrive at the expression

$$\bar{H}_2 - \bar{H}_2^\circ = \left(\frac{\partial \Delta H}{\partial n_2}\right)_{n_1} + \overset{\bullet}{H}_2 - \bar{H}_2^\circ, \tag{14.49}$$

or

$$\bar{L}_2 = \overset{\bullet}{L}_2 + \left(\frac{\partial \Delta H}{\partial n_2}\right)_{n_1}. \tag{14.50}$$

Obviously, if we knew $\overset{\bullet}{L}_2$, we could determine $\bar{L}_2$ by partial differentiation of the graphical or analytical relation between ΔH of mixing and n_2, since n_1 is maintained constant during the experiment.

The determination of $\overset{\bullet}{L}_2$ can be carried out by extrapolation of the experimental measurements. It is evident from equation (14.50) that since

$$\lim_{n_2 \to 0} \bar{L}_2 = 0, \tag{14.51}$$

$$\lim_{n_2 \to 0} \left(\frac{\partial \Delta H}{\partial n_2}\right)_{n_1} = -\overset{\bullet}{L}_2. \tag{14.52}$$

From the limiting value of the slope of a ΔH vs. n_2 relation it is thus possible to calculate $\overset{\bullet}{L}_2$. It follows, then, that $\bar{L}_2$ at any concentration can be evaluated from equation (14.50).

The calculation of $\bar{L}_1$ may be carried out in either of two ways. If data are available on the heats of solution of a fixed quantity of solute with variable quantities of solvent, then it is convenient to obtain from equation (14.3)

$$\bar{H}_1 = \left(\frac{\partial H}{\partial n_1}\right)_{n_2} = \left(\frac{\partial \Delta H}{\partial n_1}\right)_{n_2} + \overset{\bullet}{H}_1. \tag{14.53}$$

Hence

$$\bar{H}_1 - \overset{\bullet}{H}_1 = \bar{H}_1 - \bar{H}_1^\circ = \bar{L}_1 = \left(\frac{\partial \Delta H}{\partial n_1}\right)_{n_2}. \tag{14.54}$$

An alternative procedure is to make use of the differential relation between partial molal quantities, that is,

$$n_1 \, d\bar{G}_1 + n_2 \, d\bar{G}_2 = 0. \tag{14.55}$$

If we define L, the relative total enthalpy, by the equation

$$L = H - H_{\substack{\text{reference} \\ \text{state}}} \tag{14.56}$$

and if the infinitely dilute solution is taken as the reference state, it is evident that

$$L = H - H^\circ = n_1\bar{H}_1 + n_2\bar{H}_2 - n_1\bar{H}_1^\circ - n_2\bar{H}_2^\circ \tag{14.57}$$

$$= n_1\bar{L}_1 + n_2\bar{L}_2. \tag{14.58}$$

By the methods outlined in Chapter 13 on partial molal quantities, it can be shown readily that

$$n_1 \, d\bar{L}_1 + n_2 \, d\bar{L}_2 = 0. \tag{14.59}$$

Hence

$$\bar{L}_1 - \bar{L}_1^\circ = \bar{L}_1 = - \int_0^{n_2} \frac{n_2}{n_1} \, d\bar{L}_2. \tag{14.60}$$

Thus, partial molal enthalpies of the solvent may be evaluated from those of the solute.

The details of the actual procedure in the calculation of $\bar{L}$'s from experimental data on heats of solution may be illustrated by two examples, one using primarily a graphical approach, the other an analytical one.

1. *Graphical method.*   Lewis and Randall[4] have recalculated some data of Thomsen[5] on the heat absorbed when n_2 moles of gaseous HCl is dissolved in 1000 grams of H_2O.   A plot of ΔH as a function of n_2, at constant n_1, is shown in Fig. 3.   To find $\bar{L}_2$ it is necessary to calculate $(\partial \Delta H / \partial n_2)_{n_1}$ at various molalities of HCl.   These slopes have been evaluated by

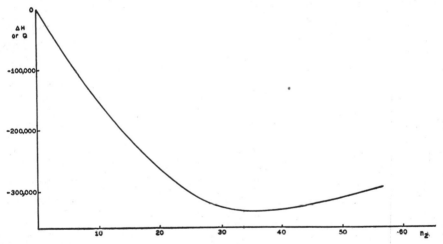

Fig. 3.   Heat absorbed upon solution of n_2 moles of gaseous HCl in 1000 grams of water. [Based on data from G. N. Lewis and M. Randall, *Thermodynamics*, page 96.]

the chord-area method and are assembled in Fig. 4.   The extrapolation required by equation (14.52) has also been carried out, and the value obtained for $-\bar{L}_2$ is $-17,300$ cal mole^{-1}.   Hence the relative partial molal

[4] G. N. Lewis and M. Randall, *Thermodynamics*, McGraw-Hill Book Company, Inc., New York, 1923, page 96.

[5] As an actual representation of the enthalpies of aqueous solutions of HCl, these data are inadequate, since they suggest an incorrect functional relationship between ΔH and m at low molalities.

heat contents of HCl in aqueous solutions may be expressed by the equation

$$\bar{L}_2 = \left(\frac{\partial \, \Delta H}{\partial n_2}\right)_{n_1} + 17{,}300,\qquad (14.61)$$

obtained by inserting the appropriate value for $\overset{\bullet}{L_2}$ into equation (14.50). Thus, for a 10-molal solution the slope may be read from Fig. 4, and $\bar{L}_2$ calculated:

$$\bar{L}_2 = -13{,}000 + 17{,}300 = 4300 \text{ cal mole}^{-1}.\qquad (14.62)$$

In this particular problem it is simple to put equation (14.61) into a more explicit analytical form, since the derivative $(\partial \, \Delta H/\partial n_2)_{n_1}$ in Fig. 4 is apparently linear. This linear graph can be represented analytically by the equation

$$\left(\frac{\partial \, \Delta H}{\partial n_2}\right)_{n_1} = -17{,}300 + 430 n_2.\qquad (14.63)$$

Hence equation (14.61) may be reduced to

$$\bar{L}_2 = 430 n_2 \text{ (cal mole}^{-1}),\qquad (14.64)$$

or
$$\bar{L}_2 = 430m \text{ (cal mole}^{-1})\qquad (14.65)$$

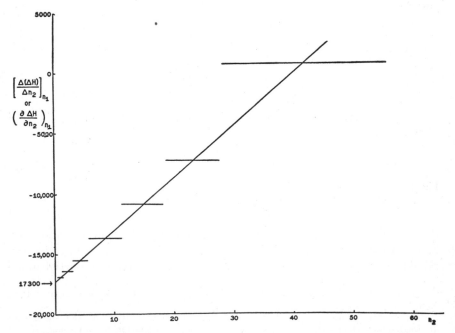

Fig. 4.   Chord-area plot of slopes of heat of solution curve in preceding figure.

since the quantity of solvent has been specified as 1000 grams.

Values of $\bar{L}_1$ for these aqueous HCl solutions may be determined by the integration suggested by equation (14.60). In view of the simplicity of the expression for $\bar{L}_2$, it is most convenient to carry out the integration by analytic means. It can be shown readily, then, that

$$\bar{L}_1 = -3.87m^2 \text{ (cal mole}^{-1}\text{)}. \tag{14.66}$$

2. *Analytical method.* Let us consider again the case of the imaginary solute X_2, in which the heat accompanying the solution of n_2 moles of X_2 in 1000 grams of water may be expressed by the relation

$$\Delta H = 40m + 30m^2. \tag{14.67}$$

To find $\bar{L}_2$, we proceed once again to find the limiting value of the partial derivative of ΔH, as suggested by equations (14.50) and (14.52):

$$\left(\frac{\partial \, \Delta H}{\partial n_2}\right)_{n_1} = \left(\frac{\partial \, \Delta H}{\partial m}\right)_{n_1 = 55.51} = 40 + 60m; \tag{14.68}$$

$$\lim_{m \to 0} \left(\frac{\partial \, \Delta H}{\partial m}\right)_{n_1} = -\overset{\bullet}{L_2} = 40. \tag{14.69}$$

Hence by substitution into equation (14.50), we obtain

$$\bar{L}_2 = 40 + 60m - 40, \tag{14.70}$$

or $$\bar{L}_2 = 60m. \tag{14.71}$$

The relative partial molal heat content of the solvent is obtained readily by appropriate substitutions into equation (14.60):

$$\bar{L}_1 = -\frac{1}{55.51} \int_0^m m(60dm) = -\frac{30}{55.51} m^2; \tag{14.72}$$

or $$\bar{L}_1 = -0.54m^2. \tag{14.73}$$

B. From apparent molal enthalpy.

1. *Relative apparent molal enthalpy.* It is sometimes convenient, particularly when dilution rather than solution measurements are made, to analyze calorimetric data in terms of the function known as the *apparent molal heat content.* From the general definition in Chapter 13 of an apparent molal quantity, it is evident that ϕH_2 is defined by the equation

$$\phi H_2 = \frac{H - n_1 \bar{H}_1^{\circ}}{n_2}. \tag{14.74}$$

Similarly, by the general procedure used to relate apparent with partial molal quantities, we can show that

$$n_2 \phi H_2 = H - n_1 \bar{H}_1^{\circ} \tag{14.75}$$

$$= n_1 \bar{H}_1 + n_2 \bar{H}_2 - n_1 \bar{H}_1^{\circ}. \tag{14.76}$$

Since from equation (13.72) it follows that

$$\lim_{m \to 0} \phi H_2 = \phi H_2^\circ = \bar{H}_2^\circ, \tag{14.77}$$

it is obvious that

$$n_2 \phi H_2 - n_2 \phi H_2^\circ = n_1 \bar{H}_1 + n_2 \bar{H}_2 - n_1 \bar{H}_1^\circ - n_2 \bar{H}_2^\circ \tag{14.78}$$

$$= n_1 \bar{L}_1 + n_2 \bar{L}_2. \tag{14.79}$$

By analogy with our preceding definitions of heat quantities, let us define the *relative* apparent molal heat content by the expression

$$\phi L_2 = \phi H_2 - \phi H_2^\circ. \tag{14.80}$$

It then follows that equation (14.79) may be reduced to

$$n_2 \phi L_2 = n_1 \bar{L}_1 + n_2 \bar{L}_2, \tag{14.81}$$

or

$$n_2 \phi L_2 = L. \tag{14.82}$$

The validity of equation (14.82) may be realized also from a graphical representation of the quantities involved. Although the absolute value of H° may not be known, it can be represented by an arbitrary point on a graph such as Fig. 5. Since L is given by the difference in heat content of the specified solution and that at zero molality, it is evident from the general definition of an apparent molal quantity that

$$\phi L_2 = \frac{L - L^\circ}{n_2} = \frac{L}{n_2}. \tag{14.83}$$

Equation (14.82) follows immediately.

2. $\bar{L}$'s *from* ϕL_2. The relationships between relative partial molal enthalpies and the relative apparent molal enthalpy are obtained readily from equations (14.81) and (14.82). From the latter it follows that

$$\bar{L}_2 = \left(\frac{\partial L}{\partial n_2} \right)_{n_1} = \frac{\partial}{\partial n_2} (n_2 \phi L_2), \tag{14.84}$$

or

$$\bar{L}_2 = \phi L_2 + n_2 \left(\frac{\partial \phi L_2}{\partial n_2} \right)_{n_1}. \tag{14.85}$$

Substitution of this expression into equation (14.81) gives a relation for $\bar{L}_1$:

$$n_1 \bar{L}_1 = n_2 \phi L_2 - n_2 \bar{L}_2$$

$$= n_2 \phi L_2 - n_2 \left[\phi L_2 + n_2 \left(\frac{\partial \phi L_2}{\partial n_2} \right)_{n_1} \right]; \tag{14.86}$$

$$\bar{L}_1 = - \frac{n_2^2}{n_1} \left(\frac{\partial \phi L_2}{\partial n_2} \right)_{n_1}. \tag{14.87}$$

Thus, to evaluate $\bar{L}_2$ and $\bar{L}_1$, we must obtain ϕL_2 from the experimental data.

3. ϕL_2 *and heat of dilution.*   For a dilution process, such as that represented by equation (14.44), the heat effect is expressible by equation (14.47). The first two terms of this equation give the total relative heat

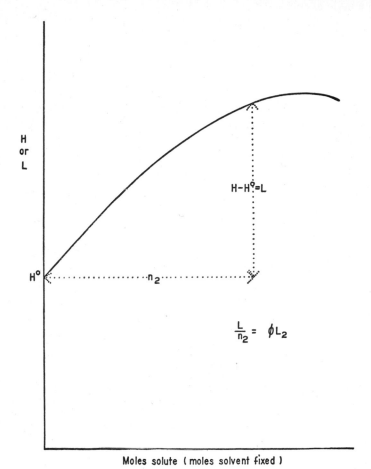

Fig. 5.   Relative apparent molal enthalpy.

content of the final solution, whereas the second two terms give the total relative heat content of the initial solution.   In other words,

$$L' = (n_1 + n_1')\bar{L}_1' + n_2\bar{L}_2' \qquad (14.88)$$

and

$$L = n_1\bar{L}_1 + n_2\bar{L}_2. \qquad (14.89)$$

Consequently, equation (14.47) may be reduced to

$$\Delta H_{\text{dilution}} = L' - L, \qquad (14.90)$$

which in view of equation (14.82) leads to

$$\Delta H_{\text{dilution}} = n_2 \phi L_2' - n_2 \phi L_2. \tag{14.91}$$

If the dilution process is carried out on a quantity of solution containing but one mole of solute, then it is evident that

$$\Delta H_{\substack{\text{dilution} \\ n_2 = 1}} = \phi L_2' - \phi L_2. \tag{14.92}$$

So far, we have an expression relating $\Delta H_{\text{dilution}}$, the quantity which may be measured experimentally, to the *difference* between two relative apparent molal heat contents. The problem of determining ϕL_2 from experimental data remains. As one might guess, however, an extrapolation to infinite dilution must be involved, just as it is in the case of heat of solution, described in Section V-**A**.

Let us consider a dilution process in which we start with a quantity of solution containing one mole of solute characterized by an apparent relative enthalpy, ϕL_2. This solution is diluted with pure solvent to produce another solution with an apparent relative enthalpy of $\phi L_2'$. If a series of such dilutions were carried out, each time with the same initial solution but with a larger quantity of added pure solvent, we would obtain a series of heats of dilution, each given by equation (14.92), but with $\phi L_2'$ referring to an increasingly dilute solution. An extrapolation of these measured heats of dilution to infinite dilution should lead to ϕL_2 of the initial solution, because if $\phi L_2'$ approaches zero at infinite dilution, as must follow from equation (14.80), then it also follows that

$$\lim_{m \to 0} \left(\Delta H_{\substack{\text{dilution} \\ n_2 = 1}} \right) = \lim_{m \to 0} (\phi L_2' - \phi L_2) \tag{14.93}$$

$$= -\phi L_2. \tag{14.94}$$

As an example of the use of this method, we may consider some data on the heats of dilution of aqueous solutions of hydrochloric acid.[6] The dilution experiments, and accompanying heats, may be summarized by the following equations:

(a) Solution of 1 HCl and 3 H_2O + 2 H_2O
    = solution of 1 HCl and 5 H_2O; $\Delta H = -1720$ cal.

(b) Solution of 1 HCl and 3 H_2O + 9 H_2O
    = solution of 1 HCl and 12 H_2O; $\Delta H = -3230$ cal.

(c) Solution of 1 HCl and 3 H_2O + 22 H_2O
    = solution of 1 HCl and 25 H_2O; $\Delta H = -3750$ cal.

(d) Solution of 1 HCl and 25 H_2O + 25 H_2O
    = solution of 1 HCl and 50 H_2O; $\Delta H = -247$ cal.

(e) Solution of 1 HCl and 25 H_2O + 75 H_2O
    = solution of 1 HCl and 100 H_2O; $\Delta H = -387$ cal.

[6] F. D. Rossini, *J. Res. Nat. Bureau Stds.*, **9**, 679 (1932).

(f) Solution of 1 HCl and 25 H_2O + 375 H_2O
$\qquad$ = solution of 1 HCl and 400 H_2O; ΔH = −549 cal.
(g) Solution of 1 HCl and 400 H_2O + 1200 H_2O
$\qquad$ = solution of 1 HCl and 1600 H_2O; ΔH = −91 cal.

From these data it is possible to tabulate values for the heats of dilution of the solution with 1 mole of HCl and 3 moles of H_2O to solutions with various final values of n_1. For example, the ΔH of dilution of the following process,

(h) Solution of 1 HCl and 3 H_2O + 397 H_2O
$\qquad$ = solution of 1 HCl and 400 H_2O; ΔH = −4299 cal,

is obtained by adding equations (c) and (f). By such arithmetic processes one may obtain the values assembled in Table 2.

TABLE 2

HEATS OF DILUTION OF HYDROCHLORIC ACID SOLUTIONS AT 25 °C
$HCl \cdot 3H_2O + n_1 H_2O$ (l) = $HCl \cdot (3 + n_1)H_2O$

n_1	$\dfrac{n_1 + 3}{n_2}$	m	$\Delta H =$ $\phi L_2' - \phi L_2(HCl \cdot 3H_2O)$ cal mole^{-1}
1597	1600	0.0347	−4390
397	400	0.140	−4299
97	100	0.555	−4137
47	50	1.110	−3997
22	25	2.220	−3750
9	12	4.626	−3230
2	5	11.10	−1720
0	3	18.50	0

TABLE 3

RELATIVE ENTHALPIES OF AQUEOUS SOLUTIONS OF HYDROCHLORIC ACID AT 25 °C

m	$\phi L_2',$ cal mole^{-1}	$\bar{L}_2,$ cal mole^{-1}	$\bar{L}_1,$ cal mole^{-1}
0.0347	90	131	−0.0253
0.140	181	262	−0.205
0.555	343	505	−1.62
1.110	483	771	−5.76
2.220	730	1136	−16.3
4.626	1250	2315	−88.8
11.10	2760	5580	−564.0
18.50	4480	8570	−1362.0

The extrapolation of the heats of dilution to infinite dilution is illustrated in Fig. 6. The square root of the molality, $m^{1/2}$, has been chosen as the abscissa in this case because the extrapolation is more convenient by this

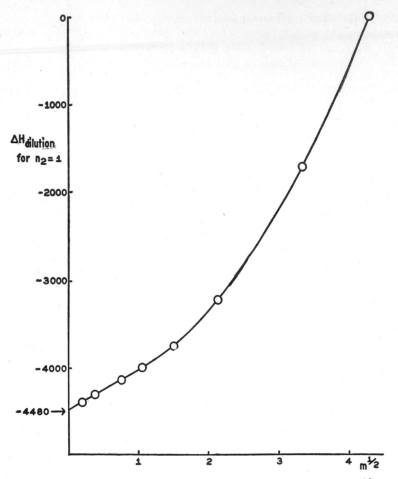

Fig. 6.   Extrapolation of data for heats of dilution of hydrochloric acid.

method for solutions of electrolytes.   For solutions of non-electrolytes, the molality itself can be used.

The extrapolated value of ϕL_2 for the solution with 1 mole of HCl and 3 moles of H_2O turns out to be $+4480$ cal.   From the data listed in the last column of Table 2, it is possible to evaluate $\phi L_2'$ for any of the other solutions.   The values calculated in this manner are listed in Table 3, together with the relative partial molal enthalpies which may be derived from them by the application of equations (14.85) and (14.87).

Exercises

1.   Gucker, Planck, and Pickard [*J. Am. Chem. Soc.*, **61**, 459 (1939)] have found that the following simple equation,

$$\phi L_2 = 128.9m,$$

expresses the relative apparent molal heat content of aqueous solutions of sucrose at 20°C. Find the expressions for $\bar{L}_2$ and $\bar{L}_1$, respectively, as a function of the molality.

2. If the heat of solution of HCl (g) in 25 moles of H_2O is $-17{,}150$ cal mole^{-1}, calculate $L_2^\bullet$ (g) for HCl.

3. (a) Calculate the differential heat of solution of one mole of HCl (g) in $HCl \cdot 25H_2O$, that is, in a solution of 1 mole of HCl and 25 moles of H_2O.

(b) Calculate the differential heat of solution of one mole of H_2O (l) in $HCl \cdot 25H_2O$.

(c) From Parts (a) and (b) calculate the integral heat of solution of HCl (g) in 25 moles of H_2O (l).

4. Prove that $\partial \bar{L}_1 / \partial T = \bar{c}_{p(1)} - \bar{c}_{p(1)}^\circ$.

5. With the aid of the data in Problem 1 and the following equation for the relative apparent molal heat content at 30°C,

$$\phi L_2 = 140.2m,$$

find expressions for $\bar{c}_{p(2)}$ and $\bar{c}_{p(1)}$ for aqueous sucrose solutions at 25°C. $\bar{c}_{p(2)}^\circ$ is 151.50 cal mole^{-1} deg^{-1}.

6. The heat absorbed when m moles of NaCl is dissolved in 1000 grams of H_2O is given by the expression

$$\Delta H = 923m + 476.1m^{3/2} - 726.1m^2 + 243.5m^{5/2}.$$

(a) Derive an expression for $\bar{L}_2$, and compute values of the relative partial molal enthalpy of NaCl in 0.01- and 0.1-molal solutions.

(b) Derive an expression for $\bar{L}_1$, and compute its value for 0.01- and 0.1-molal solutions.

(c) If the expression above for ΔH per 1000 grams of H_2O is divided by m, the number of moles of solute in 1000 grams of H_2O, one obtains the heat absorbed when 1 mole of NaCl is dissolved in enough water to give an m-molal solution. Show that this equation is

$$\Delta H \text{ (per mole of NaCl)} = 923 + 476.1m^{1/2} - 726.1m + 243.5m^{3/2}.$$

(d) Find $(\partial \Delta H / \partial n_1)_{n_2}$ from the preceding expression. Derive an equation for $\bar{L}_1$.

7. The following information on partial molal enthalpies is available [Gucker, Pickard, and Ford, *J. Am. Chem. Soc.*, **62**, 2698 (1940)] for glycine and its aqueous solutions at 25°C:

m	$\bar{L}_1$ (cal mole^{-1})	$\bar{L}_2$ (cal mole^{-1})
1.000	1.537	-165.5
3.33 (saturated)	...	-354
Glycine (pure)	...	-3765

(a) The heat of solution of an infinitesimal quantity of pure solid glycine in a saturated aqueous solution is 3411 cal mole^{-1}. Show that $L_2^\bullet$ (s) is -3765 cal mole^{-1}.

(b) Find ΔH per mole for the addition of an infinitesimal quantity of solid glycine to a 1-molal aqueous solution.

(c) Find ΔH per mole for the addition of an infinitesimal quantity of solid glycine to an infinitely dilute aqueous solution.

(d) Find ΔH for the addition of one mole of solid glycine to 1000 grams of pure water to form a 1-molal aqueous solution.

CHAPTER 15

Free Energy in Systems of Variable Composition.
The Fugacity Function

THE PROPERTIES of partial molal free energies can be deduced, of course, from the general formulation for all partial molal quantities which has been previously described. As in the case of enthalpies, absolute values of partial free energies cannot be determined. Hence it has been necessary once again to work with relative values. To appreciate the particular approach which is most preferred, it is desirable to consider first some general aspects of the free-energy function.

I. DEPENDENCE OF FREE ENERGY ON COMPOSITION

A. General equation for dF. We have considered previously the differential dF in *closed* systems, that is, systems in which matter does not enter or leave. When only pressure-volume work is possible, the expression for dF is

$$dF = -S\,dT + V\,dP. \qquad (15.1)$$

We wish to turn attention now to an *open* system, one in which the number of moles, n, of any one or more of the components may change. For generality we shall assume also that a chemical transformation can occur in this system, and we shall represent the reaction by the equation

$$A + B + \ldots = M + N + \ldots. \qquad (15.2)$$

Since the free energy of the system is then a function of $n_1, n_2, \ldots$, moles of the respective participants in the reaction, as well as of temperature and pressure, it is evident that the total differential is given by the expression

$$dF = \left(\frac{\partial F}{\partial T}\right)_{P,n} dT + \left(\frac{\partial F}{\partial P}\right)_{T,n} dP + \left(\frac{\partial F}{\partial n_1}\right)_{P,T,n_2,\ldots} dn_1$$
$$+ \left(\frac{\partial F}{\partial n_2}\right)_{P,T,n_1,n_3\ldots} dn_2 + \ldots. \qquad (15.3)$$

The first two partial derivatives in equation (15.3) may be evaluated by consideration of the *special* case in which there is no change in composition, that is, in which

$$dn_1 = dn_2 = \ldots = dn_i = 0. \qquad (15.4)$$

In this situation, equation (15.3) reduces to

$$dF = \left(\frac{\partial F}{\partial T}\right)_{P,n} dT + \left(\frac{\partial F}{\partial P}\right)_{T,n} dP. \tag{15.4a}$$

Furthermore, if there is no addition or removal of any component, then we have a closed system, and equation (15.1) is also valid. Hence it follows, as was shown also in Chapter 8, that

$$\left(\frac{\partial F}{\partial T}\right)_{P,n} = -S \tag{15.5}$$

and

$$\left(\frac{\partial F}{\partial P}\right)_{T,n} = V. \tag{15.6}$$

Therefore substitution into equation (15.3) leads to the general equation for an *open* system:

$$dF = -S\,dT + V\,dP + \left(\frac{\partial F}{\partial n_1}\right)_{T,P,n_2,\ldots} dn_1$$
$$+ \left(\frac{\partial F}{\partial n_2}\right)_{T,P,n_1,n_3,\ldots} dn_2 + \ldots. \tag{15.7}$$

In view of our general definition of a partial molal quantity (Chapter 13), it is evident that

$$\left(\frac{\partial F}{\partial n_1}\right)_{P,T,n_2,\ldots} = \bar{F}_1 \tag{15.8}$$

and

$$\left(\frac{\partial F}{\partial n_2}\right)_{P,T,n_1,n_3,\ldots} = \bar{F}_2, \tag{15.9}$$

so that equation (15.7) may be modified to read

$$dF = -S\,dT + V\,dP + \bar{F}_1\,dn_1 + \bar{F}_2\,dn_2 + \ldots. \tag{15.10}$$

Frequently one finds Gibbs's notation used in place of $\bar{F}$, whereupon equation (15.10) becomes

$$dF = -S\,dT + V\,dP + \mu_1\,dn_1 + \mu_2\,dn_2 + \ldots, \tag{15.11}$$

where μ is called the *chemical potential* and is equal to the partial molal free energy.

B. Condition of equilibrium in a closed system. We showed in Chapter 8 that in a closed system at constant pressure and temperature in which environmental pressure is the only restraint, equilibrium is obtained if

$$dF_{P,T} = 0. \tag{15.12}$$

For this special case the general equation, (15.10), obviously reduces to

$$\bar{F}_1 \, dn_1 + \bar{F}_2 \, dn_2 + \ldots = 0, \tag{15.13}$$

or
$$\sum \mu_i \, dn_i = 0. \tag{15.14}$$

C. Spontaneous process in closed system. We have shown previously that in a closed system at constant pressure and temperature where environmental pressure is the only restraint, a spontaneous process is possible if

$$dF_{P,T} < 0. \tag{15.15}$$

For this special case the general equation, (15.10), obviously reduces to

$$\bar{F}_1 \, dn_1 + \bar{F}_2 \, dn_2 + \ldots < 0, \tag{15.16}$$

or
$$\sum \mu_i \, dn_i < 0. \tag{15.17}$$

II. ESCAPING TENDENCY

A. Partial molal free energy and escaping tendency. G. N. Lewis proposed the very useful term "escaping tendency" to give a strong mechanical flavor to the concept of partial molal free energy, or chemical potential. The idea behind the term "escaping tendency" can be visualized best, perhaps, in terms of a concrete example.   Let us consider two solutions of iodine, in water and carbon tetrachloride, respectively, which have reached equilibrium at a fixed pressure and temperature (Fig. 1).   In this system at equilibrium, let us carry out a transfer of an infinitesimal quantity of iodine from the water to the carbon tetrachloride phase.   In view of equation (15.13), we may say that

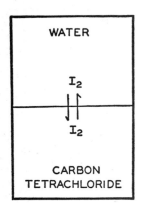

Fig. 1. Schematic diagram of distribution of iodine between water and carbon tetrachloride, at fixed temperature and pressure.

$$\bar{F}_{\substack{I_2 \text{ in} \\ H_2O}} \, dn_{\substack{I_2 \text{ in} \\ H_2O}} + \bar{F}_{\substack{I_2 \text{ in} \\ CCl_4}} \, dn_{\substack{I_2 \text{ in} \\ CCl_4}} = 0. \tag{15.18}$$

Obviously, in this closed system, any loss of iodine from the water phase is accompanied by an equivalent gain in the organic phase, so that

$$-dn_{\substack{I_2 \text{ in} \\ H_2O}} = dn_{\substack{I_2 \text{ in} \\ CCl_4}}. \tag{15.19}$$

Hence
$$\bar{F}_{\substack{I_2 \text{ in} \\ H_2O}} \, dn_{\substack{I_2 \text{ in} \\ H_2O}} + \bar{F}_{\substack{I_2 \text{ in} \\ CCl_4}} \, (-dn_{\substack{I_2 \text{ in} \\ H_2O}}) = 0. \tag{15.20}$$

It follows, then, that

$$\bar{F}_{\substack{I_2 \text{ in} \\ H_2O}} = \bar{F}_{\substack{I_2 \text{ in} \\ CCl_4}} \tag{15.21}$$

for this system in equilibrium at constant pressure and temperature. Thus, at equilibrium the chemical potential, or partial molal free energy, of the iodine is the same in all phases in which it is present. It is elucidative also to say that the escaping tendency of the iodine in the water must be the same as that of the iodine in the carbon tetrachloride.

It may be helpful to consider also the situation in which the iodine will diffuse spontaneously (at constant pressure and temperature) from the water into the carbon tetrachloride—a case in which the concentration in the former phase is greater than that which would exist in equilibrium with the organic phase. In view of equation (15.16), we may write

$$\bar{F}_{\substack{I_2\,in\\H_2O}}\, dn_{\substack{I_2\,in\\H_2O}} + \bar{F}_{\substack{I_2\,in\\CCl_4}}\, dn_{\substack{I_2\,in\\CCl_4}} < 0. \tag{15.22}$$

Even for the spontaneous diffusion of iodine, however, equation (15.19) must be valid in this closed system. Hence

$$\bar{F}_{\substack{I_2\,in\\H_2O}}\, dn_{\substack{I_2\,in\\H_2O}} + \bar{F}_{\substack{I_2\,in\\CCl_4}}\, (-dn_{\substack{I_2\,in\\H_2O}}) < 0, \tag{15.23}$$

or

$$(\bar{F}_{\substack{I_2\,in\\H_2O}} - \bar{F}_{\substack{I_2\,in\\CCl_4}})\, dn_{\substack{I_2\,in\\H_2O}} < 0. \tag{15.24}$$

Since the water loses iodine, it is evident that

$$dn_{\substack{I_2\,in\\H_2O}} < 0; \tag{15.25}$$

that is, dn is a negative number. In such a case, equation (15.24) can be valid only if the difference in partial molal free energies is a positive number. Thus

$$\bar{F}_{\substack{I_2\,in\\H_2O}} - \bar{F}_{\substack{I_2\,in\\CCl_4}} > 0, \tag{15.26}$$

or

$$\bar{F}_{\substack{I_2\,in\\H_2O}} > \bar{F}_{\substack{I_2\,in\\CCl_4}}. \tag{15.27}$$

We may say, then, that the escaping tendency of the iodine is greater in the water than in the carbon tetrachloride phase. In general, when the partial molal free energy, or chemical potential, of a given species is greater in one phase than in a second, we shall also say that the escaping tendency is greater in the former case than in the latter. The escaping tendency is thus a qualitative phrase, corresponding to the property given precisely by the partial molal free energy. The escaping tendency may be used, therefore, for comparative purposes in isothermal transformations.

A word of caution should be added. It is legitimate to speak of the escaping tendency of a given substance under different conditions (at constant temperature). However, in view of our basic correlation of this concept with $\bar{F}$'s, it is meaningless to compare escaping tendencies of different substances, since we cannot compare absolute values of partial molal free energies. For similar reasons we cannot compare escaping tendencies of a single substance at different temperatures.

The inability to specify absolute values of the partial molal free energy is one of the reasons that this thermodynamic quantity is not convenient for use as a quantitative measure of the escaping tendency. Instead a new function, the fugacity, has been proposed. The nature of this new function can be realized best by consideration first of the relation between pressure and $\bar{F}$.

B. Pressure and chemical potential.

1. *Gaseous state.* It is evident, of course, that if two isolated chambers contain a given gas at different pressures, P_1 and P_2, respectively, and if these chambers are connected by a porous plug, the gas from the chamber at the higher pressure will flow spontaneously into that at the lower pressure. Thus the escaping tendency of the gas at the higher pressure is greater than that at the lower. Similarly, the chemical potential is greater at the higher pressure.

2. *Liquid state.* In comparing the vapor pressures of a given substance in two different states, for example, pure water and a 1-molal aqueous solution of NaCl, respectively, we know that the vapor pressure of H_2O in

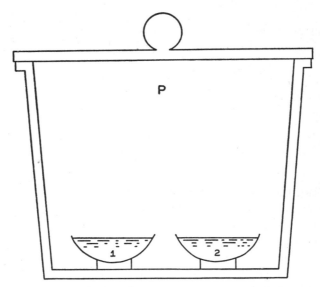

Fig. 2. Schematic diagram of experiment to correlate vapor pressure and escaping tendency in pure water (1) and an aqueous salt solution (2).

the former state is greater than that in the latter. By a procedure similar to that just described in the example with iodine, we can show that at constant temperature and fixed total pressure (Fig. 2)

$$\bar{F}_1^\circ > \bar{F}_1, \tag{15.28}$$

where $\bar{F}_1^{\circ}$ is the partial molal free energy of pure water and $\bar{F}_1$ that of water in the salt solution.   Furthermore, *if the vapor behaves as an ideal gas,* we can calculate this difference in free energy by the following three steps:

$$H_2O \; (l) \qquad\quad = \; H_2O \; (g, \; p^{\circ}_{\text{equil}}); \; \; \Delta F_{P,T} \; = \; 0 \tag{15.29}$$

$$H_2O \; (g, \; p^{\circ}_{\text{equil}}) \; = \; H_2O \; (g, \; p_{\text{equil}}); \; \; \Delta F \; = \; \int_{p^{\circ}}^{p} V \, dP \; = \; RT \ln \frac{p}{p^{\circ}} \tag{15.30}$$

$$H_2O \; (g, \; p_{\text{equil}}) \; = \; H_2O \; (\text{solution}); \; \; \Delta F_{P,T} \; = \; 0 \tag{15.31}$$

$$H_2O \; (l) \qquad\quad = \; H_2O \; (\text{solution}); \; \; \Delta F \quad = \; RT \ln \frac{p}{p^{\circ}} . \tag{15.32}$$

In these equations p_{equil} refers to the equilibrium vapor pressure of water above the solution and p°_{equil}, to that above pure water at the same temperature and same total pressure, P.

Thus the vapor pressure is an excellent quantitative measure of the escaping tendency of any liquid component, if the vapor behaves ideally.

3. *Solid state.* The situation here is analogous to that for the liquid state. If, for example, we visualize dishes 1 and 2 in Fig. 2 as containing rhombic and monoclinic sulfurs, respectively, it is evident that at 25°C the monoclinic will slowly evaporate and condense into dish 1 as the rhombic form, since the vapor pressure of the former is greater than that of the latter and hence the escaping tendency of the monoclinic exceeds that of the rhombic.   In fact, *if the vapor behaves as an ideal gas,* the free energy of transition can be calculated from the vapor pressures of the two allotropic forms by a procedure similar to that described in the preceding example for water.

Once again, then, it is evident that the vapor pressure is an excellent quantitative measure of the escaping tendency, if the vapors behave ideally. In the examples considered, ideal behavior may be a sufficiently good approximation to actual behavior, but in general such will not be the case. Since the integration

$$\Delta F \; = \; \int V \, dP \; = \; RT \ln \frac{P_2}{P_1} \tag{15.33}$$

is permissible for ideal gases but is not generally valid, we need some other function to allow calculations of free energies for nonideal gases and vapors.

III. THE FUGACITY FUNCTION

A. Definition. The definition of this new function has been developed with an eye on the convenience of equation (15.33), as well as with the recognition that this expression becomes more nearly correct, the lower the pressure.   Since the partial molal free energy, $\bar{F}$, is the fundamental measure of escaping tendency, the fugacity, f, has been defined by the expression

$$\bar{F} = RT \ln f + B(T), \qquad (15.34)$$

where $B(T)$ for a particular substance is a function of the temperature only. It has been necessary to introduce the term $B(T)$, since, otherwise, *absolute* values of f (as of $\bar{F}$) could not be determined. Equation (15.34) alone, however, does not allow the determination of absolute values of the fugacity, because it does not tell how $B(T)$ is to be fixed at any particular temperature. A further specification is therefore necessary:

$$\frac{f}{P} \to 1 \text{ as } P \to 0; \qquad (15.35)$$

that is, the fugacity is defined so as to approach the pressure as the pressure approaches zero. This behavior may be visualized readily by means of the illustration in Fig. 3.

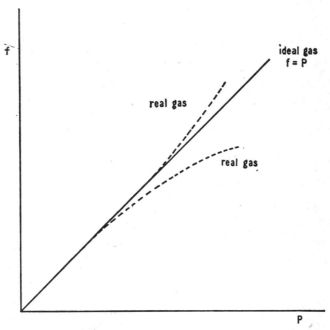

Fig. 3. Characteristics of the fugacity for ideal and real gases.

It is evident immediately from equations (15.34) and (15.35) that differences in free energy for an isothermal process are given by the equation

$$\Delta \bar{F} = RT \ln \frac{f_2}{f_1} \qquad (15.36)$$

and that $\Delta \bar{F}$ calculated from this relation approaches that given by equation (15.33) as the pressure approaches zero.

The definition of fugacity serves for the solid and liquid states as well as for the gaseous. Every substance has a finite vapor pressure and thus every liquid and solid is in equilibrium with its vapor. If the fugacity of the vapor of a particular component is known, then the fugacity of the same component in the solid or liquid state is known, since at equilibrium

$$\Delta \bar{\mathrm{F}}_{P,T} = 0 \qquad (15.37)$$

and therefore $\qquad f_{\text{state 2}} = f_{\text{state 1}}.$ (15.38)

If the constant, 1, in equation (15.35) is taken to be dimensionless, then the unit of fugacity is the same as that of pressure. Generally "atmosphere" is chosen.

B. Change of fugacity with pressure. The pressure dependence of the fugacity may be determined in a very straightforward manner. Since $\bar{\mathrm{F}}$ and f are related by equation (15.34) and $B(T)$ is a function of temperature only, it is evident that

$$\left(\frac{\partial \bar{\mathrm{F}}}{\partial P}\right)_T = RT\left(\frac{\partial \ln f}{\partial P}\right)_T. \qquad (15.39)$$

But the pressure dependence of the partial molal free energy is also given by the expression

$$\left(\frac{\partial \bar{\mathrm{F}}}{\partial P}\right)_T = \bar{\mathrm{v}}. \qquad (15.40)$$

Hence

$$\left(\frac{\partial \ln f}{\partial P}\right)_T = \frac{\bar{\mathrm{v}}}{RT}. \qquad (15.41)$$

To find the fugacity at one pressure from that at another, we may integrate equation (15.41), so long as the process is isothermal, and obtain

$$RT \ln \frac{f_2}{f_1} = \int_{P_1}^{P_2} \bar{\mathrm{v}} \, dP. \qquad (15.42)$$

C. Change of fugacity with temperature. Let us consider an isothermal process in which a substance (solid, liquid, or gas) is transformed from one state, A at a pressure P, to another, A^* at a different pressure P^*. Such a transformation may be represented as follows:

$$A(P) \rightarrow A^*(P^*). \qquad (15.43)$$

The free-energy change for such a transformation is given by the expression

$$\Delta F = \bar{\mathrm{F}}^* - \bar{\mathrm{F}} \qquad (15.44)$$

$$= RT \ln f^* + B(T) - RT \ln f - B(T). \qquad (15.45)$$

Hence $\qquad \dfrac{\bar{\mathrm{F}}^*}{T} - \dfrac{\bar{\mathrm{F}}}{T} = R \ln f^* - R \ln f.$ (15.46)

From equation (8.94) we know that the temperature derivatives of the terms on the left-hand side of equation (15.46) must be given by

$$\left(\frac{\partial(\bar{F}^*/T)}{\partial T}\right)_{P*} - \left(\frac{\partial(\bar{F}/T)}{\partial T}\right)_{P} = -\frac{\bar{H}^*}{T^2} + \frac{\bar{H}}{T^2}. \tag{15.47}$$

From equation (15.46) it also follows that the partial derivatives of the fugacities are given by

$$\left(\frac{\partial(\bar{F}^*/T)}{\partial T}\right)_{P*} - \left(\frac{\partial(\bar{F}/T)}{\partial T}\right)_{P} = R\left(\frac{\partial \ln f^*}{\partial T}\right)_{P*} - R\left(\frac{\partial \ln f}{\partial T}\right)_{P}. \tag{15.48}$$

Consequently,

$$\left(\frac{\partial \ln f^*}{\partial T}\right)_{P*} - \left(\frac{\partial \ln f}{\partial T}\right)_{P} = -\frac{\bar{H}^*}{RT^2} + \frac{\bar{H}}{RT^2}. \tag{15.49}$$

If P^* is taken to be a very low pressure, approaching zero, then as

$$P^* \to 0, \tag{15.50}$$

$$\frac{f^*}{P^*} \to 1. \tag{15.51}$$

It follows, then, that

$$\left(\frac{\partial \ln f^*}{\partial T}\right)_{P*} = \left(\frac{\partial \ln P^*}{\partial T}\right)_{P*} = 0. \tag{15.52}$$

Hence equation (15.49) reduces to

$$\left(\frac{\partial \ln f}{\partial T}\right)_{P} = \frac{\bar{H}^* - \bar{H}}{RT^2}. \tag{15.53}$$

$\bar{H}^*$ is the partial molal heat content of the given substance in State A^*, that is, the state of zero pressure. The difference $(\bar{H}^* - \bar{H})$ must be, therefore, the change in heat content or enthalpy when the substance is transformed from State A to State A^*, that is, to the state of zero pressure, or into infinite volume. This difference in enthalpy is sometimes called the *ideal heat of vaporization* from State A.

If State A is that of a gas at the pressure P, then $(\bar{H}^* - \bar{H})$ evidently corresponds to the enthalpy change when the gas expands isothermally from the pressure P into a vacuum. The pressure dependence of this heat would be given by the expression

$$\left(\frac{\partial(\bar{H}^* - \bar{H})}{\partial P}\right)_{T} = -\left(\frac{\partial\bar{H}}{\partial P}\right)_{T} = -\left(\frac{\partial H}{\partial P}\right)_{T}, \tag{15.54}$$

because $(\partial\bar{H}^*/\partial P)_T$ must be zero, since $\bar{H}^*$ is the partial molal enthalpy at a fixed (zero) pressure. In equation (15.54) we replace $\bar{H}$ by H, the molal enthalpy, since we are dealing with a pure gas.

From equation (6.69) we know that the pressure coefficient of the molal enthalpy of a gas is related to the Joule-Thomson coefficient, μ, by the equation

$$\left(\frac{\partial \mathrm{H}}{\partial P}\right)_T = -c_p\mu. \tag{15.55}$$

Combination of equations (15.54) and (15.55) leads to

$$\left(\frac{\partial(\bar{\mathrm{H}}^* - \bar{\mathrm{H}})}{\partial P}\right)_T = c_p\mu. \tag{15.56}$$

Because of this relation between $(\bar{\mathrm{H}}^* - \bar{\mathrm{H}})$ and μ, the former quantity is frequently referred to as the *Joule-Thomson heat*. It is obvious that the pressure coefficient of this Joule-Thomson heat can be calculated from the known values of the Joule-Thomson coefficient and the heat capacity of the gas. Similarly, since $(\bar{\mathrm{H}}^* - \bar{\mathrm{H}})$ is a derived function of the fugacity, a knowledge of the temperature dependence of the latter may be used to calculate the Joule-Thomson coefficient. An illustration of such a calculation will be presented in the following chapter.

Exercises

1. Consider a gas with the equation of state

$$Pv = RT + aP, \tag{15.57}$$

where a is a small *negative* number.

(a) Draw a rough sketch of a graph of PV vs. P for this gas. Include a dotted line for the corresponding graph of an ideal gas.

(b) Draw a dotted curve for a graph of V vs. P for an ideal gas. On this same graph draw a curve for v vs. P for a gas with the equation of state given by (15.57).

(c) As P approaches zero, what happens to the two curves in the graph in Part (b)?

(d) Rearrange equation (15.57) to read explicitly for v. As P approaches zero, what does v approach?

(e) Rearrange equation (15.57) into one for $[v - (RT/P)]$. As P approaches zero, what does the quantity in brackets approach?

(f) Draw a graph of $[v - (RT/P)]$ vs. P for the gas with the equation of state given by (15.57).

2. If the fugacity function is defined by equation (15.34), show that for a solution of two components

$$N_1\left(\frac{\partial \ln f_1}{\partial N_1}\right)_{P,T} = N_2\left(\frac{\partial \ln f_2}{\partial N_2}\right)_{P,T}.$$

CHAPTER 16

The Fugacity of Gases

FROM THE discussion in the preceding chapter it should be evident that the fugacity of a solid or liquid may be determined if that of its vapor in equilibrium is known. However, it is only in the gaseous state that direct measurements of absolute fugacities can be made. It behooves us, then, to consider methods of determining the fugacities of real gases.

I. CALCULATION OF THE FUGACITY OF A REAL GAS

Several methods have been developed for calculating fugacities from measurements of pressures and volumes of real gases.

A. Graphical method using the α function. A typical volume-pressure graph for a real gas is illustrated in Fig. 1, in conjunction with the volumes to be expected at corresponding pressures if the gas behaved ideally. From equation (15.42),

$$RT \ln \frac{f_2}{f_1} = \int_{P_1}^{P_2} \bar{v} \, dP, \qquad (16.1)$$

it is evident that the ratio of the fugacity f_2 at the pressure P_2 to the fugacity f_1 at the pressure P_1 can be obtained by graphical integration, as indicated by the stippled area in Fig. 1. However, as P_1 approaches zero, the area becomes infinite. Hence this direct method is not suitable for determining absolute values of the fugacities of a real gas.

In using equation (16.1), we have utilized only one part of the definition of fugacity. Let us consider also the second part, which in essence specifies the standard or reference state for the fugacity. While f approaches zero as P is reduced to zero, the ratio f/P approaches 1. It might seem promising, therefore, to carry out an integration of this ratio, rather than of f or P directly, since the absolute value of this function can be specified under certain conditions.[1]

Considering then the pressure coefficient of the ratio f/P, we obtain

$$\left(\frac{\partial \ln (f/P)}{\partial P} \right)_T = \left(\frac{\partial \ln f}{\partial P} \right)_T - \left(\frac{\partial \ln P}{\partial P} \right)_T. \qquad (16.2)$$

---

[1] It may be worth while to point out that this procedure of using the difference between two functions, either of which approaches infinity, is a very convenient one in many scientific problems. In the present case it leads to the recognition of the fact that though $RT \ln (f_2/f_1)$ and $RT \ln (P_2/P_1)$ both approach infinity as P_1 approaches zero, the difference function, $RT \ln [(f_2/P_2)/(f_1/P_1)]$, approaches a finite limit.

The pressure coefficient of $\ln f$ is given by

$$\left(\frac{\partial \ln f}{\partial P}\right)_T = \frac{\mathrm{v}}{RT}, \tag{16.3}$$

where v is the molal volume of the gas. Thus equation (16.2) becomes

$$\left(\frac{\partial \ln (f/P)}{\partial P}\right)_T = \frac{\mathrm{v}}{RT} - \frac{\partial \ln P}{\partial P} = \frac{\mathrm{v}}{RT} - \frac{1}{P}\frac{\partial P}{\partial P} \tag{16.4}$$

$$= \frac{\mathrm{v}}{RT} - \frac{1}{P} \tag{16.5}$$

$$= \frac{\mathrm{v}}{RT} - \frac{1}{P}\frac{RT}{RT} \tag{16.6}$$

$$= \frac{1}{RT}\left(\mathrm{v} - \frac{RT}{P}\right). \tag{16.7}$$

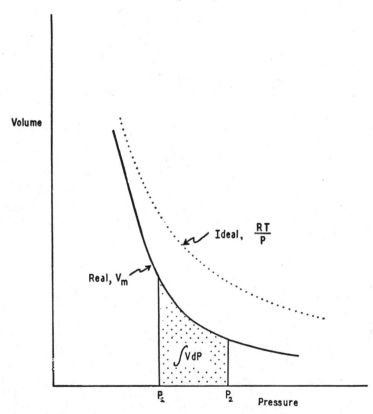

Fig. 1. Comparison of pressure-volume isotherms for a possible real gas and an ideal gas.

If we permit the quantity within the parentheses to be represented by $-\alpha$, that is, if

$$\alpha = \left(\frac{RT}{P} - \mathrm{v}\right), \tag{16.8}$$

we obtain

$$\left(\frac{\partial \ln (f/P)}{\partial P}\right)_T = -\frac{\alpha}{RT}. \tag{16.9}$$

Integration of this equation, for isothermal conditions, from zero pressure to some pressure P, gives

$$\int_0^P d \ln \frac{f}{P} = -\frac{1}{RT} \int_0^P \alpha \, dP; \tag{16.10}$$

$$\ln \frac{f}{P} - \ln \left(\frac{f}{P}\right)_{P=0} = -\frac{1}{RT} \int_0^P \alpha \, dP. \tag{16.11}$$

Since f/P approaches 1 as P approaches zero, the second term on the left side of equation (16.11) goes to zero. Hence

$$\ln f - \ln P = -\frac{1}{RT} \int_0^P \alpha \, dP, \tag{16.12}$$

or

$$\ln f = \ln P - \frac{1}{RT} \int_0^P \alpha \, dP. \tag{16.13}$$

To evaluate $\ln f$, it is thus necessary to integrate $\alpha \, dP$. v and RT/P each approach infinity as the pressure goes to zero. Nevertheless, the difference between them does not in general approach zero. Usually α can be measured over a range of pressures and an extrapolation made to zero pressure. A typical graph for α (for hydrogen gas) is illustrated in Fig. 2. The stippled area indicates the graphical evaluation of the integral in equation (16.13). Once this integral is known, the evaluation of f from equation (16.13) is simple.

B. Analytical methods.

1. *Based on van der Waals' equation of state.* In addition to graphical integration, we may integrate $d \ln f$ or $d \ln (f/P)$ by use of an equation of state, such as that of van der Waals. Integrating as in equation (16.11), we obtain

$$RT \ln \frac{f}{P} = -\int_0^P \alpha \, dP = \int_0^P \left(\mathrm{v} - \frac{RT}{P}\right) dP$$

$$= \int_0^P \mathrm{v} \, dP - \int_0^P RT \, d \ln P. \tag{16.14}$$

To evaluate the first integral, it is necessary to substitute for v or dP. A

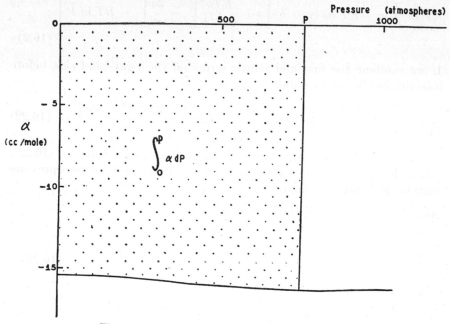

Fig. 2. The α function for hydrogen gas at 0 °C.

trial will show that it is necessary to substitute for dP. Thus, solving the van der Waals equation of state for P, we obtain

$$P = \frac{RT}{v - b} - \frac{a}{v^2};$$ (16.15)

$$dP = -\frac{RT}{(v - b)^2}\, dv + \frac{2a}{v^3}\, dv.$$ (16.16)

If we insert equation (16.16) into (16.14), we obtain

$$RT \ln \frac{f}{P} = \int_0^P v\left[-\frac{RT}{(v - b)^2} + \frac{2a}{v^3}\right] dv - RT \ln P\Big]_0^P$$ (16.17)

$$= -\int_0^P \frac{RTv}{(v - b)^2}\, dv + \int_0^P \frac{2a}{v^2}\, dv - RT \ln P\Big]_0^P$$ (16.18)

$$= -\int_0^P \frac{RT\,(v - b + b)}{(v - b)^2}\, dv + \int_0^P \frac{2a}{v^2}\, dv - RT \ln P\Big]_0^P$$ (16.19)

$$= -\int_0^P \frac{RT\,(v - b)}{(v - b)^2}\, dv - \int_0^P \frac{RTb}{(v - b)^2}\, dv$$
$$+ \int_0^P \frac{2a}{v^2}\, dv - RT \ln P\Big]_0^P;$$ (16.20)

$$RT \ln \frac{f}{P} = -RT \ln (v - b) \Big]_0^P + \frac{RTb}{v - b} \Big]_0^P - \frac{2a}{v} \Big]_0^P - RT \ln P \Big]_0^P.$$

$$\text{(16.21)}$$

If we combine the first and fourth terms on the right-hand side before inserting our limits, we obtain

$$RT \ln \frac{f}{P} = -RT \ln \{P(v - b)\} \Big]_0^P + \frac{RTb}{v - b} \Big]_0^P - \frac{2a}{v} \Big]_0^P. \quad \text{(16.22)}$$

There is no difficulty in inserting the upper limit, P, in equation (16.22). The values of the terms on the right at the lower limit of zero pressure may be determined as follows:

As
$$\left. \begin{array}{c} P \to 0, \\ v \to \infty, \\ \dfrac{RTb}{v - b} \to 0, \\ \dfrac{2a}{v} \to 0, \\ P(v - b) \to Pv \to RT. \end{array} \right\} \quad \text{(16.23)}$$

Making use of the results in (16.23), we can reduce equation (16.22) to the expression

$$RT \ln \frac{f}{P} = -RT \ln \{P(v - b)\} + \frac{RTb}{v - b} - \frac{2a}{v}$$
$$+ RT \ln RT + 0 + 0. \quad \text{(16.24)}$$

Hence

$$RT \ln f - RT \ln P = -RT \ln P + RT \ln \frac{RT}{v - b} + \frac{RTb}{v - b} - \frac{2a}{v}.$$

$$\text{(16.25)}$$

Finally, we obtain an expression explicit for $\ln f$:

$$\ln f = \ln \frac{RT}{v - b} + \frac{b}{v - b} - \frac{2a}{RTv}. \quad \text{(16.26)}$$

Thus the fugacity of a van der Waals gas may be evaluated from the constants a and b, at any given pressure or corresponding molal volume, v.

2. *An approximate method.* It frequently happens that α is roughly constant, particularly at relatively low pressures. A good example is hydrogen gas (Fig. 2). In such a case it is, of course, a simple matter to integrate equation (16.10) analytically and obtain

$$RT \ln \frac{f}{P} = -\alpha P. \tag{16.27}$$

This equation, in turn, may be converted into several other useful approximate forms. For example,

$$\ln \frac{f}{P} = -\frac{\alpha P}{RT}. \tag{16.28}$$

Therefore

$$\frac{f}{P} = e^{-\alpha P/RT}. \tag{16.29}$$

The exponential in equation (16.29) may be expanded to give

$$\frac{f}{P} = 1 - \frac{\alpha P}{RT} + \frac{1}{2!}\left(\frac{\alpha P}{RT}\right)^2 - \cdots \tag{16.30}$$

If we neglect all terms of higher power than (αP), since they are negligible in an alternating series whose nth term approaches zero, we obtain

$$\frac{f}{P} = 1 - \frac{\alpha P}{RT} = \frac{RT - [(RT/P) - v]P}{RT}. \tag{16.31}$$

Therefore

$$\frac{f}{P} = \frac{Pv}{RT}. \tag{16.32}$$

Another relation may be obtained by recognition of the fact that an *ideal* pressure, P_i, may be defined from the observed molal volume, v, and the ideal gas law:

$$P_i = \frac{RT}{v}. \tag{16.33}$$

With this relation equation (16.32) becomes

$$\frac{f}{P} = \frac{P}{P_i}. \tag{16.34}$$

Thus the fugacity may be estimated from the observed pressure, P, and the ideal pressure calculated from the observed volume.

C. **Universal fugacity coefficients.** It has been observed that the behavior of most pure gases can be represented adequately by a single chart of the compressibility factor, Z, which is defined as

$$Z = \frac{Pv}{RT}, \tag{16.35}$$

the reduced temperature, T_r, and the reduced pressure, P_r. It is evident, then, that the integration of equation (16.10), in which α is a function of v, could be carried out to give another chart of f/P, the *fugacity coefficient*,[2] as a function of T_r and P_r. Again a single chart should be applicable to all pure gases, within the precision to which the compressibility-factor chart is valid. Several investigators have prepared such charts, a typical example of which is illustrated in Fig. 3. With the critical constants of a gas it is a simple matter to calculate T_r and P_r, to read f/P from the chart, and then to calculate the fugacity from the expression

$$f = \frac{f}{P} \times P. \tag{16.36}$$

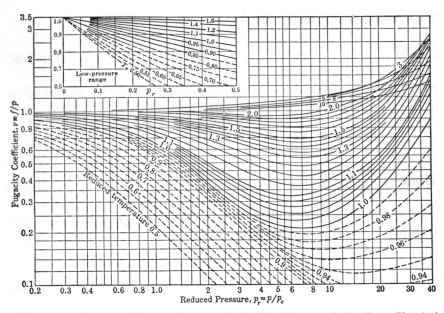

Fig. 3. Fugacity coefficients of gases. [Reproduced by permission from *Chemical Process Principles, Part II, Thermodynamics* by O. A. Hougen and K. M. Watson, published by John Wiley and Sons, Inc., 1947; based on data taken from B. W. Gamson and K. M. Watson, *National Petroleum News, Tech. Section*, **36**, R623 (Sept. 6, 1944).]

II. JOULE-THOMSON EFFECT FOR A VAN DER WAALS GAS

In the preceding chapter it was pointed out that the partial derivatives of an explicit fugacity function are related to the Joule-Thomson coefficient, μ. Let us, then, consider an example of the calculation of μ from a fugacity equation. We shall restrict our discussion to relatively low pressures and to a gas which obeys van der Waals' equation.

A. Limiting value of α for a van der Waals gas. The fugacity function will be obtained, as usual, by an integration of equation (16.10).

---

[2] This ratio is also referred to as an *activity coefficient*. It is frequently represented by the symbol γ.

Since we are limiting our considerations to low pressures, let us examine the value of α at such pressures.

Using the van der Waals equation in the form

$$v = \frac{RT}{P + (a/v^2)} + b \tag{16.37}$$

to represent the molal volume in equation (16.8), we obtain

$$\alpha = \frac{RT}{P} - \left[\frac{RT}{P + (a/v^2)} + b \right]. \tag{16.38}$$

This equation may be transformed to the following:

$$\alpha = \frac{RT}{P} - \left[\frac{RT}{P \left(1 + \dfrac{a}{Pv^2} \right)} + b \right]. \tag{16.39}$$

The terms within the parentheses may be expanded as follows:

$$\frac{1}{1 + (a/Pv^2)} = 1 - \frac{a}{Pv^2} + \left(\frac{a}{Pv^2} \right)^2 - \cdots. \tag{16.40}$$

We shall neglect all terms of higher power than 1. [In a series which alternates in sign and whose nth term approaches zero, such as (16.40), all terms beyond the last one used will not exceed the latter.] Equation (16.39) may be converted, then, into the expression

$$\alpha = \frac{RT}{P} - \left[\frac{RT}{P} \left(1 - \frac{a}{Pv^2} \right) + b \right], \tag{16.41}$$

which in turn may be reduced to

$$\alpha = \frac{RT}{P^2 v^2} a - b. \tag{16.42}$$

At low pressures, $$Pv = RT, \tag{16.43}$$

and equation (16.42) becomes

$$\alpha = \frac{a}{RT} - b. \tag{16.44}$$

B. **Fugacity at low pressures.** Having obtained a simple expression for α, we may now carry out the integration specified in equation (16.10).

$$\int_0^P d \ln \frac{f}{P} = -\frac{1}{RT} \int_0^P \left(\frac{a}{RT} - b \right) dP. \tag{16.45}$$

Therefore $$\ln \frac{f}{P} = -\frac{1}{RT} \left(\frac{a}{RT} - b \right) P, \tag{16.46}$$

or
$$\ln f = \ln P - \frac{aP}{R^2T^2} + \frac{bP}{RT}.$$ (16.47)

C. Enthalpy of a van der Waals gas. Since the temperature coefficient of $\ln f$ is related to the Joule-Thomson heat, we proceed to evaluate the latter from equation (16.47) as follows:

$$\left(\frac{\partial \ln f}{\partial T}\right)_P = \frac{\bar{H}^* - \bar{H}}{RT^2} = \frac{2aP}{R^2T^3} - \frac{bP}{RT^2}.$$ (16.48)

Therefore
$$\bar{H}^* - \bar{H} = \left(\frac{2a}{RT} - b\right)P,$$ (16.49)

or
$$\bar{H} = H = \bar{H}^* - \left(\frac{2a}{RT} - b\right)P.$$ (16.50)[3]

D. Joule-Thomson coefficient. To find μ, we need merely to obtain the pressure coefficient of the molal enthalpy, H, from equation (16.50):

$$\left(\frac{\partial H}{\partial P}\right)_T = -\left(\frac{2a}{RT} - b\right).$$ (16.51)

Since from equation (6.69)

$$\left(\frac{\partial H}{\partial P}\right)_T = -c_p\mu,$$ (16.52)

it follows that
$$\mu = \frac{1}{c_p}\left(\frac{2a}{RT} - b\right).$$ (16.53)

Thus we find μ expressed in terms of the constants a and b of the gas.

For any given gas, it follows from equation (16.53) that at high temperatures

$$\frac{2a}{RT} < b.$$ (16.54)

Therefore
$$\mu < 0;$$ (16.55)

that is, μ is negative and there is a temperature increase on expansion. In contrast, at low temperatures

$$\frac{2a}{RT} > b.$$ (16.56)

Therefore
$$\mu > 0;$$ (16.57)

that is, μ is positive and there is a temperature drop on expansion. Finally, at some temperature, T_i, called the *inversion temperature*,

---

[3] It should be kept in mind that this equation is valid only in the low-pressure region where α is essentially constant. In practice this region corresponds, in general, to pressures below 10 atmospheres.

$$\frac{2a}{RT_i} = b. \tag{16.58}$$

Therefore
$$\mu = 0; \tag{16.59}$$

that is, there is no change in temperature on expansion.

III. SOLUTIONS OF REAL GASES

The thermodynamics of solutions of real gases is generally based on the assumption that the mixtures follow the laws of ideal solutions, even though each gas cannot be described in terms of the ideal gas laws. As Lewis and Randall[4] pointed out: "It seems reasonable to suppose that the solution of a given pair of substances will be more nearly perfect when the density of the solution is less, or, in other words, when the average distance between the molecules is greater. We shall therefore expect gaseous solutions to be much more nearly perfect than corresponding liquid solutions, and since we find even among liquids numerous cases in which there is a close approach to the perfect solution, it is likely that almost any gaseous solution may be regarded as nearly perfect." These considerations may be expressed in mathematical form in terms of a simple assumption.

A. Rule for solutions of real gases.[5] We shall assume, as has been suggested by Lewis and Randall, that the fugacity, f_i, of a constituent of a gaseous solution may be expressed by the equation

$$f_i = N_i f_i^{\bullet}, \tag{16.60}$$

where $f_i^{\bullet}$ is the fugacity of the pure gas at the same temperature and at the same total pressure, P, and N_i is the mole fraction of the particular component.

B. Fugacity coefficients in gaseous solution. We may define the fugacity coefficient, γ_i, of a constituent of a gaseous solution by the expression

$$\gamma_i = \frac{f_i}{N_i P}. \tag{16.61}$$

In a gaseous solution for which equation (16.60) is valid, it follows that the fugacity coefficient, γ_i, is also given by the equation

$$\gamma_i = \frac{N_i f_i^{\bullet}}{N_i P} = \gamma_i^{\bullet}, \tag{16.62}$$

---

[4] G. N. Lewis and M. Randall, *Thermodynamics*, McGraw-Hill Book Company, Inc., New York, 1923, page 226.

[5] More detailed treatments of the thermodynamic properties of gaseous solutions have been described recently by J. A. Beattie, *Chem. Rev.*, **44**, 141 (1949), and by O. Redlich and J. N. S. Kwong, *ibid.*, **44**, 233 (1949).

where $\gamma_i^\bullet$ is the fugacity coefficient of the pure constituent at the total pressure P. Thus we see that γ_i is the same as $\gamma_i^\bullet$; that is, the fugacity coefficient of the constituent in an ideal gaseous solution at a total pressure P is the same as that of the pure gas at the pressure P (and the same temperature). Therefore it is necessary merely to have information on the fugacity coefficients of pure gases in order to obtain values in gaseous solutions, if the very approximate rule of Lewis and Randall is used.

C. **Equilibrium constant and free-energy change for reactions involving real gases.** If we represent the equilibrium among several gases by the equation

$$A \text{ (g)} + B \text{ (g)} = C \text{ (g)} + D \text{ (g)}, \qquad (16.63)$$

then by a procedure analogous to that used for ideal gases, we can show that the standard free-energy change may be calculated from the relation

$$\Delta F^\circ = -RT \ln K_f. \qquad (16.64)$$

In this equation K_f is given by the ratio

$$K_f = \frac{f_C f_D}{f_A f_B} \qquad (16.65)$$

and is actually the thermodynamic equilibrium constant, K. To obtain ΔF°, then, it is necessary to calculate the equilibrium constant in terms of fugacities.

In view of equation (16.61), the thermodynamic equilibrium constant may also be expressed by the following relation:

$$K = K_f = \frac{[(N_C P)\gamma_C][(N_D P)\gamma_D]}{[(N_A P)\gamma_A][(N_B P)\gamma_B]}. \qquad (16.66)$$

Each product of a mole fraction and the total pressure P may be defined as the partial pressure, p_i. Consequently, equation (16.66) may be rearranged to

$$K = \frac{p_C p_D}{p_A p_B} \frac{\gamma_C \gamma_D}{\gamma_A \gamma_B}. \qquad (16.67)$$

From equation (16.62) we know that the fugacity coefficient of each constituent in this gaseous solution may be replaced by the corresponding $\gamma^\bullet$ for the pure gas at the same total pressure P. It follows, then, that equation (16.67) may be converted into

$$K = \frac{p_C p_D}{p_A p_B} \frac{\gamma_C^\bullet \gamma_D^\bullet}{\gamma_A^\bullet \gamma_B^\bullet} = K_p K_\gamma, \qquad (16.68)$$

where K_p represents the familiar pressure equilibrium constant and K_γ is the ratio of the fugacity coefficients for the respective pure gases at the specified *total* pressure. Approximate values of K_γ may be obtained from tables and graphs of fugacity coefficients of *pure* gases, as in Fig. 3, and K_p may be calculated by the methods described in elementary physical chemistry.

Exercises

1. A gas obeys the equation of state

$$Pv = RT + AP,$$

where A is a constant. Find the following:
   (a) An expression for $\ln f$.
   (b) An expression for γ.
   (c) An expression for the Joule-Thomson heat.
   (d) An expression for the Joule-Thomson coefficient.

2. For hydrogen at 0°C, Amagat has prepared the following table of data. Find the fugacity of hydrogen at 1000 atm by the graphical method using the α function.

P	$\dfrac{Pv}{RT}$	P	$\dfrac{Pv}{RT}$
100	1.069	600	1.431
200	1.138	700	1.504
300	1.209	800	1.577
400	1.283	900	1.649
500	1.356	1000	1.720

3. Proceeding in a manner analogous to that used for ideal gases, prove the following equation for real gases:

$$\Delta F° = -RT \ln K_f. \tag{16.64}$$

4. R. H. Ewell [*Ind. Eng. Chem.*, **32**, 147 (1940)] has suggested the following reaction as a method of production of hydrogen cyanide:

$$N_2 (g) + C_2H_2 (g) = 2HCN (g). \tag{16.69}$$

   (a) From data in tables of the National Bureau of Standards, calculate $\Delta F°_{298}$ for this reaction.
   (b) By methods discussed previously, it is possible to calculate $\Delta F°$ at 300°C. The value obtained should be 6186 cal.
   (c) Calculate K_f for reaction (16.69) at 300°C.
   (d) By referring to the International Critical Tables, find the critical temperatures and pressures for the gases in reaction (16.69). Tabulate these values. Tabulate also the reduced temperatures for 300°C and the reduced pressures for a total pressure of 5 and of 200 atm, respectively.
   (e) Referring to Fig. 3, find γ's for the gases in reaction (16.69) at total pressures of 5 and 200 atm, respectively. Tabulate these values.
   (f) Calculate K_γ at 5 and 200 atm, respectively, and add to the table in Part (e).
   (g) Calculate K_p at 5 and 200 atm, respectively, and add to the table in Part (e).
   (h) If one starts with an equimolar mixture of N_2 and C_2H_2, what fraction of C_2H_2 is converted to HCN at 5 atm total pressure? At 200 atm total pressure?
   (i) What is the effect of increasing pressure on the yield of HCN?
   (j) According to Le Chatelier's rule, what should be the effect of increasing pressure on the yield of HCN?

CHAPTER 17

The Ideal Solution

W E CAN make use of the thermodynamic functions which we have developed for solutions by applying them first to the simplest type of solution—the ideal solution.

I. DEFINITION

An ideal solution is one in which each component obeys Raoult's law at all temperatures and pressures. By *Raoult's law* we mean the following statement: *the fugacity, f, of any component is equal to its mole fraction, N, multiplied by its fugacity in the pure state, $f^{\bullet}$*;

$$f = Nf^{\bullet}. \tag{17.1}$$

If for simplicity we consider a solution of only two components, then this solution is ideal if

$$f_1 = N_1 f_1^{\bullet} \tag{17.2}$$

and

$$f_2 = N_2 f_2^{\bullet}. \tag{17.3}$$

The dependence of the fugacities, in such a solution, upon the mole fractions is illustrated in Fig. 1.

II. SOME CONSEQUENCES OF THE DEFINITION

The restrictions, equations (17.2) and (17.3), which define an ideal solution (of two components) lead immediately to certain consequences which must be properties of such a solution. It is not necessary, therefore, to specify these properties explicitly in the definition.

A. Volume changes. *There is no change in volume upon mixing pure components which form an ideal solution.* That this statement is valid for an ideal solution can be shown readily as follows. If we convert equation (17.2) into a logarithmic form, we obtain

$$\ln f_1 = \ln f_1^{\bullet} + \ln N_1. \tag{17.4}$$

At any fixed mole fraction

$$\left(\frac{\partial \ln f_1}{\partial P} \right)_T = \left(\frac{\partial \ln f_1^{\bullet}}{\partial P} \right)_T. \tag{17.5}$$

But from equation (15.41) we have

$$\left(\frac{\partial \ln f}{\partial P}\right)_T = \frac{\bar{v}}{RT}. \tag{17.6}$$

Hence equation (17.5) may be reduced to

$$\bar{v}_1 = \bar{v}_1^{\bullet} = v_1, \tag{17.7}$$

where v_1 represents the molal volume of pure Component 1. The same argument can be applied to Component 2, so that we may write

$$\bar{v}_2 = \bar{v}_2^{\bullet} = v_2. \tag{17.8}$$

Thus the partial molal volume of each component in solution is equal to the molal volume of the corresponding pure substance.

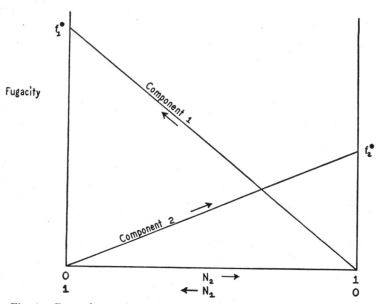

Fig. 1.   Dependence of fugacities on composition in an ideal solution.

Before the pure components are mixed, the total volume, V_{initial}, must be given by the sum of the products of the number of moles of each pure component multiplied by its molal volume:

$$V_{\text{initial}} = n_1 v_1 + n_2 v_2. \tag{17.9}$$

When the solution is formed, the total volume, V_{final}, must be given by [equation (13.26)]

$$V_{\text{final}} = n_1 \bar{v}_1 + n_2 \bar{v}_2. \tag{17.10}$$

In view of equations (17.7) and (17.8), it is evident that the volume change on mixing is zero:

$$\Delta V = V_{\text{final}} - V_{\text{initial}} = n_1\bar{v}_1 + n_2\bar{v}_2 - (n_1v_1 + n_2v_2) = 0. \quad (17.11)$$

B. Heat effects. *There is no heat evolved upon mixing pure components which form an ideal solution.* The validity of this statement can be proved readily from consideration of the temperature coefficient of the fugacity. Again from Raoult's law in logarithmic form, equation (17.4), it follows, at fixed mole fraction, that

$$\left(\frac{\partial \ln f_1}{\partial T}\right)_P = \left(\frac{\partial \ln f_1^{\bullet}}{\partial T}\right)_P. \quad (17.12)$$

From equation (15.53) we see that

$$\left(\frac{\partial \ln f}{\partial T}\right)_P = \frac{\text{H}^* - \bar{\text{H}}}{RT^2}, \quad (17.13)$$

where H* refers to the molal heat content of the substance as a vapor as P approaches zero. Thus equation (17.12) may be reduced to

$$\text{H}_1^* - \bar{\text{H}}_1 = \text{H}_1^* - \bar{\text{H}}_1^{\bullet}, \quad (17.14)$$

or

$$\bar{\text{H}}_1 = \bar{\text{H}}_1^{\bullet} = \text{H}_1, \quad (17.15)$$

where H_1 is the molal enthalpy of pure Component 1. Obviously, again the same argument applies to Component 2, so that we may write

$$\bar{\text{H}}_2 = \bar{\text{H}}_2^{\bullet} = \text{H}_2. \quad (17.16)$$

Thus before mixing, the enthalpy, H_{initial}, is given by

$$H_{\text{initial}} = n_1\text{H}_1 + n_2\text{H}_2 \quad (17.17)$$

and after the formation of the solution,

$$H_{\text{final}} = n_1\bar{\text{H}}_1 + n_2\bar{\text{H}}_2. \quad (17.18)$$

In view of the ideality of the solution, expressions (17.15) and (17.16) are valid, so that ΔH, the heat effect for a constant pressure process, is zero:

$$\Delta H = H_{\text{final}} - H_{\text{initial}}$$
$$= n_1\bar{\text{H}}_1 + n_2\bar{\text{H}}_2 - (n_1\text{H}_1 + n_2\text{H}_2) = 0. \quad (17.19)$$

III. THERMODYNAMICS OF TRANSFER OF COMPONENT FROM ONE SOLUTION TO ANOTHER

In preparation for a consideration of the thermodynamic changes accompanying the formation of a solution, let us derive the relations for the changes accompanying the isothermal, isobaric transfer of a component from one solution to another of different concentration.

A. Free-energy change. We may visualize this transfer by reference to Fig. 2, in which we have pictured two solutions, A and A', of very large volumes, so that the removal, or addition, of a mole of Component 1 will not change the composition significantly. The solutions differ in composition, N_1 and N_1' being the respective mole fractions of Component 1. Hence the fugacity of Component 1 must differ in the two solutions, as is indicated in Fig. 2, even though $f_1^\bullet$, the fugacity of pure 1, is the same for both solutions (since P and T are fixed). Similarly, the partial molal free

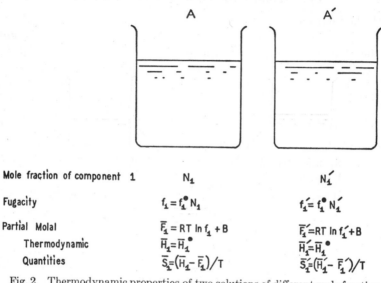

Mole fraction of component 1	N_1	N_1'
Fugacity	$f_1 = f_1^\bullet N_1$	$f_1' = f_1^\bullet N_1'$
Partial Molal Thermodynamic Quantities	$\bar{F}_1 = RT \ln f_1 + B$ $\bar{H}_1 = \bar{H}_1^\bullet$ $\bar{S}_1 = (\bar{H}_1 - \bar{F}_1)/T$	$\bar{F}_1' = RT \ln f_1' + B$ $\bar{H}_1' = \bar{H}_1^\bullet$ $\bar{S}_1' = (\bar{H}_1' - \bar{F}_1')/T$

Fig. 2. Thermodynamic properties of two solutions of different mole fraction, N_1 and N_1', respectively. A mole of Component 1 is to be transferred from solution A to A'.

energies of Component 1 in the two solutions must differ, even though the constant B is the same in both cases (since T is constant). The change in free energy on transferring a mole of Component 1 from Solution A to Solution A' is given, therefore, by the expression

$$\Delta\bar{F} = \bar{F}_1' - \bar{F}_1 = RT \ln f_1' - RT \ln f_1 = RT \ln \frac{f_1'}{f_1}$$

$$= RT \ln \frac{f_1^\bullet N_1'}{f_1^\bullet N_1} = RT \ln \frac{N_1'}{N_1}. \tag{17.20}$$

B. Enthalpy change. From equation (17.15) we know that the partial molal enthalpy of a component of an ideal solution is the same *at any concentration* as that of the pure component. Hence the enthalpy change on transferring a component from a solution at one concentration to that at another must be given by

$$\Delta\bar{H} = \bar{H}_1' - \bar{H}_1 = 0. \tag{17.21}$$

C. Entropy change. For any isothermal process

$$\Delta \bar{F} = \Delta \bar{H} - T \Delta \bar{S}. \tag{17.22}$$

In the process under consideration, $\Delta \bar{H}$ is zero. Hence

$$\Delta \bar{S} = - \frac{\Delta \bar{F}}{T} = - R \ln \frac{N_1'}{N_1}. \tag{17.23}$$

IV. THERMODYNAMICS OF MIXING

With the information at hand it is a simple matter to calculate the thermodynamic changes accompanying the formation of a mole of solution from the mixing of two pure components.

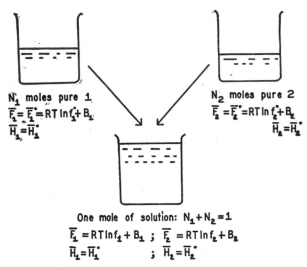

One mole of solution: $N_1 + N_2 = 1$

$\bar{F}_1 = RT \ln f_1 + B_1$; $\bar{F}_2 = RT \ln f_2 + B_2$

$\bar{H}_1 = \bar{H}_1^{\bullet}$; $\bar{H}_2 = \bar{H}_2^{\bullet}$

Fig. 3. Thermodynamics of formation of one mole of solution from pure components.

A. Free energy of mixing. We may visualize the formation of one mole of solution as it is illustrated in Fig. 3. Since the free energy is a thermodynamic property, the value of ΔF is independent of the path used to go from one state to another. Consequently, to determine ΔF all that we need is the value of the total free energy of the solution and that of both pure components. The former can be obtained from the equations developed in Chapter 13 and is given by the expression

$$F \text{ (solution)} = N_1 \bar{F}_1 + N_2 \bar{F}_2 \tag{17.24}$$

$$= N_1 (RT \ln f_1 + B_1) + N_2 (RT \ln f_2 + B_2). \tag{17.25}$$

Since the free energy is an extensive quantity, the total free energy for N_1 moles of pure Component 1 and N_2 moles of pure Component 2 is given by

$$F \text{ (pure components)} = N_1 \bar{F}_1^\bullet + N_2 \bar{F}_2^\bullet \qquad (17.26)$$

$$= N_1(RT \ln f_1^\bullet + B_1) + N_2(RT \ln f_2^\bullet + B_2). \quad (17.27)$$

The free-energy change, ΔF, on mixing becomes

$$\Delta F = F \text{ (solution)} - F \text{ (pure components)}$$

$$= N_1(RT \ln f_1 + B_1) + N_2(RT \ln f_2 + B_2)$$
$$- [N_1(RT \ln f_1^\bullet + B_1) + N_2(RT \ln f_2^\bullet + B_2)], \quad (17.28)$$

or $\quad \Delta F = N_1 RT \ln \dfrac{f_1}{f_1^\bullet} + N_2 RT \ln \dfrac{f_2}{f_2^\bullet}. \qquad (17.29)$

In view of equations (17.2) and (17.3), the expression in (17.29) may be reduced to

$$\Delta F = N_1 RT \ln N_1 + N_2 RT \ln N_2. \qquad (17.30)$$

Thus we have the free energy of formation of one mole of solution from the pure components.

B. Enthalpy of mixing. From arguments similar to those presented for the free-energy change, it is evident that the heat of mixing, ΔH, is zero, since in view of equations (17.15) and (17.16)

$$\Delta H = H \text{ (solution)} - H \text{ (pure components)} = N_1 \bar{H}_1 + N_2 \bar{H}_2$$
$$- (N_1 \bar{H}_1^\bullet + N_2 \bar{H}_2^\bullet)$$
$$= 0. \qquad (17.31)$$

C. Entropy of mixing. Once again we make use of equation (17.22) for an isothermal process. Since ΔH is zero, the entropy change, ΔS, when two pure components are mixed to form one mole of solution is given by

$$\Delta S = -\frac{\Delta F}{T} = -N_1 R \ln N_1 - N_2 R \ln N_2. \qquad (17.32)$$

It should be emphasized that the entropy of mixing is a positive quantity, despite the negative signs in equation (17.32). Since $N_1 < 1$ and $N_2 < 1$, their logarithms are both negative and hence ΔS is positive.

V. EQUILIBRIUM BETWEEN A PURE SOLID AND AN IDEAL LIQUID SOLUTION

It happens occasionally, as in solutions of naphthalene in benzene, that an ideal solution does not exist at all possible compositions because one of the components (in this case naphthalene) separates out as a solid at some specific mole fraction. If the dissolved solute forms an ideal solution with the solvent, several interesting solubility relations can be derived readily.

A. Fundamental relations. In discussions of this type of equilibrium between solid, solute, and ideal solution, it will be helpful to deal in terms of a specific example, such as naphthalene as solute and benzene as solvent.

A few moments' reflection will show that *dissolved* naphthalene must possess some of the properties of the liquid state.[1] Clearly it is not in the solid state, because the molecules do not occupy fixed positions. Similarly, it is not gaseous, since it does not occupy uniformly any volume made available by a containing vessel. Evidently, then, if the solution is an ideal one, the fugacity, f_2, of the dissolved solute obeys the expression

$$f_2 = N_2 f_2^{\bullet}, \tag{17.33}$$

where $f_2^{\bullet}$ is the fugacity of *pure, supercooled liquid naphthalene*. This supercooled liquid naphthalene is characterized also by a molal volume and enthalpy, $\bar{v}_2$ and $\bar{H}_2$, respectively, which are equal to the corresponding quantities for the dissolved solute.

Above a specified concentration (at a given temperature and pressure) no more solid naphthalene will dissolve; that is, the solution becomes saturated. If solid naphthalene is in equilibrium with the solution,

$$\text{naphthalene (s)} = \text{naphthalene (satd soln)}, \tag{17.34}$$

it follows that

$$f_s = f_2 \text{ (satd soln)}, \tag{17.35}$$

where f_s represents the fugacity of the pure solid. Clearly,

$$f_s < f_2^{\bullet}, \tag{17.36}$$

since we know that the supercooled liquid naphthalene can be transformed spontaneously into the solid; in other words, the escaping tendency of the supercooled liquid must be greater than that of the solid. The same conclusion can be reached by substituting expression (17.33) for f_2 in equation (17.35):

$$f_s = N_2 \text{ (satd soln) } f_2^{\bullet}. \tag{17.37}$$

Since $N_2 < 1$ in a saturated solution at ordinary temperatures, it is obvious that equation (17.36) is valid.

---

[1] In addition to the argument to be presented in this paragraph in the text, it is apparent that dissolved naphthalene possesses the properties of a liquid also from the following considerations. For the reaction

$$\text{naphthalene (s)} = \text{naphthalene (solution)},$$

the ΔH observed equals the ΔH_{fusion} of naphthalene. Furthermore, for the reaction

$$\text{naphthalene (supercooled liquid)} = \text{naphthalene (solution)},$$

the heat effect is zero. Obviously, then, dissolved naphthalene behaves as if it were a liquid.

An interesting consequence may be derived also from equation (17.37). By rearrangement we can obtain

$$N_2 \text{ (satd soln)} = \frac{f_s}{f_2^{\bullet}}. \qquad (17.38)$$

In this equation, N_2 represents the mole fraction of naphthalene *in the saturated solution* in benzene. It is determined only by f_s and $f_2^{\bullet}$, the fugacity of pure solid and supercooled liquid naphthalene, respectively. Nowhere in equation (17.38) does any property of the solvent, benzene, appear. Apparently, then, N_2 (satd soln) would be the same no matter what the solvent, so long as the solution is an ideal one. Thus we arrive at the conclusion that the solubility of naphthalene (in terms of mole fraction) is the same in all solvents with which it forms an ideal solution. Furthermore, though equations (17.33) to (17.38) were derived with reference to naphthalene, there is nothing in their derivation which restricts their application to this solid alone. Hence we may conclude that the solubility (in terms of mole fraction) of any specified solid is the same in all solvents with which it forms an ideal solution.

B. Change of solubility with pressure. If a solid is in equilibrium with the solute in an ideal solution under isothermal conditions,

$$\text{solid} \rightleftharpoons \text{solute in solution,} \qquad (17.39)$$

then
$$f_s = f_2, \qquad (17.40)$$

and
$$\ln f_s = \ln f_2. \qquad (17.41)$$

If the pressure is changed, the solubility may vary; but if equilibrium is maintained, then

$$d \ln f_s = d \ln f_2. \qquad (17.42)$$

In general, for either constituent of a two-component system we may write

$$\ln f = \phi(T,P,N),$$

since the escaping tendency depends upon temperature, pressure, and the mole fraction of the given constituent. Hence

$$d \ln f = \left(\frac{\partial \ln f}{\partial T}\right)_{P,N} dT + \left(\frac{\partial \ln f}{\partial P}\right)_{T,N} dP + \left(\frac{\partial \ln f}{\partial N}\right)_{T,P} dN. \quad (17.43)$$

For the pure solid solute, only the second term in equation (17.43) need be considered, since T and N are fixed. Thus

$$d \ln f_s = \left(\frac{\partial \ln f_s}{\partial P}\right)_{T,N} dP. \qquad (17.44)$$

For the dissolved solute, the first term in equation (17.43) vanishes, since the temperature is fixed. Thus

$$d \ln f_2 = \left(\frac{\partial \ln f_2}{\partial P}\right)_{T,N_2} dP + \left(\frac{\partial \ln f_2}{\partial N_2}\right)_{T,P} dN_2. \qquad (17.45)$$

Since in general

$$\left(\frac{\partial \ln f}{\partial P}\right)_T = \frac{\bar{v}}{RT}, \qquad (17.46)$$

we may write, in view of equations (17.42), (17.44), and (17.45),

$$\frac{v_s}{RT} dP = \frac{\bar{v}_2}{RT} dP + \left(\frac{\partial \ln f_2}{\partial N_2}\right)_{T,P} dN_2. \qquad (17.47)$$

The last term in equation (17.47) may be evaluated explicitly if we recognize the ideality of the solution:

$$f_2 = N_2 f_2^\bullet, \qquad (17.33)$$

$$\ln f_2 = \ln N_2 + \ln f_2^\bullet, \qquad (17.48)$$

$$d \ln f_2 = d \ln N_2 \qquad \text{at fixed } P \text{ and } T, \quad (17.49)$$

$$d \ln f_2 = \frac{dN_2}{N_2}, \qquad (17.50)$$

and thus $\qquad \left(\dfrac{\partial \ln f_2}{\partial N_2}\right)_{P,T} = \dfrac{1}{N_2}. \qquad (17.51)$

Substitution of equation (17.51) into (17.47) and rearrangement of terms leads to

$$\frac{v_s - \bar{v}_2}{RT} dP = \frac{dN_2}{N_2} = d \ln N_2, \qquad (17.52)$$

or $\qquad \left(\dfrac{\partial \ln N_2}{\partial P}\right)_T = \dfrac{v_s - \bar{v}_2}{RT}. \qquad (17.53)$

C. Change of solubility with temperature. The procedure for deriving the temperature coefficient of the solubility of a solute in an ideal solution parallels that just used for the pressure coefficient. Again we recognize the applicability of equations (17.39), (17.40), and (17.41) and the validity of equation (17.42) if equilibrium is maintained with variation in temperature. In this case, however, the second term in equation (17.43) vanishes for both solid and dissolved solute; the third term is also zero for the pure solid. Hence, in view of equation (17.42), we may write

$$\left(\frac{\partial \ln f_s}{\partial T}\right)_{P,N} dT = \left(\frac{\partial \ln f_2}{\partial T}\right)_{P,N_2} dT + \left(\frac{\partial \ln f_2}{\partial N_2}\right)_{P,T} dN_2. \qquad (17.54)$$

From previous considerations [equation (15.53)] we have

$$\left(\frac{\partial \ln f}{\partial T}\right)_P = \frac{\text{H}^* - \bar{\text{H}}}{RT^2}.$$ (17.13)

Consequently,

$$\frac{\text{H}^* - \text{H}_s}{RT^2}\, dT = \frac{\text{H}^* - \bar{\text{H}}_2}{RT^2}\, dT + \frac{1}{N_2}\, dN_2.$$ (17.55)

On rearrangement we obtain

$$\left(\frac{\partial \ln N_2}{\partial T}\right)_P = \frac{\bar{\text{H}}_2 - \text{H}_s}{RT^2}.$$ (17.56)

In view of the fact that the solution is ideal,

$$\bar{\text{H}}_2 = \bar{\text{H}}_2^{\bullet},$$ (17.16)

where $\bar{\text{H}}_2^{\bullet}$ represents the molal enthalpy of pure supercooled liquid solute. Obviously, then, at constant pressure

$$\bar{\text{H}}_2 - \text{H}_s = \bar{\text{H}}_2^{\bullet} - \text{H}_s = \Delta\text{H}_{\text{fusion of solute}}.$$ (17.57)

Hence

$$\left(\frac{\partial \ln N_2}{\partial T}\right)_P = \frac{\Delta\text{H}_{\text{fusion solute}}}{RT^2}.$$ (17.58)

From this temperature coefficient of the solubility we can also derive an equation of interest in ideal solutions. Assuming ΔH to be constant and integrating equation (17.58) over definite limits, we obtain

$$\ln \frac{N_2}{N_2'} = \int_{T'}^{T} \frac{\Delta\text{H}}{RT^2}\, dT = -\frac{\Delta\text{H}}{R}\left[\frac{1}{T} - \frac{1}{T'}\right].$$ (17.59)

We may let T' be the melting point of the solute. From equation (17.37) it can be seen that N_2' must be unity at the melting point:

$$N_2' = 1 \text{ at } T' = T_{\text{m.p.}};$$ (17.60)

in other words, the solute dissolves in the solvent in all proportions. At the melting point the solid solute can be in equilibrium with the liquid and hence the latter need no longer be supercooled. Thus we obtain the following expression, from equation (17.59), for the solubility of a solute in an ideal solution:

$$\ln N_2 = -\frac{\Delta\text{H}}{R}\left[\frac{1}{T} - \frac{1}{T_{\text{m.p.}}}\right].$$ (17.61)

Thus if we know the heat of fusion and the melting point of a solute, we can predict its solubility at various temperatures in all solvents with which it forms an ideal solution. In practice, the condition of ideality is likely to be fulfilled between substances of similar structure, and hence equation (17.61) makes a very good approximation formula.

Exercises

1. (a) Calculate ΔF, ΔH, and ΔS (per mole) at 298°K for the addition of an infinitesimal quantity of pure benzene to one mole of an ideal solution of benzene and toluene in which the mole fraction of the latter is 0.6.

(b) Calculate ΔF, ΔH, and ΔS (per mole) at 298°K for the mixing of 0.4 mole of pure benzene with 0.6 mole of pure toluene.

2. Calculate the entropy of "unmixing" of U^{235} and U^{238} from a sample of pure uranium from natural sources. The former isotope occurs to the extent of 0.7 per cent in the natural (ideal) mixture.

CHAPTER 18

The Dilute Solution: Non-Electrolytes

W E SHALL proceed in our discussion of solutions from ideal to nonideal solutions, limiting ourselves at first to non-electrolytes. For dilute solutions of non-electrolytes, several limiting laws have been found which describe the behavior of these systems with increasing precision as infinite dilution is approached. Although these laws may be stated on purely independent bases, one of the triumphs of thermodynamics is to demonstrate their fundamental interdependence. If we take any one of them as an empirical rule, all the others may be derived readily from thermodynamic principles already elaborated. It is necessary, however, to take one as an empirical fact. We shall illustrate this interdependence by assuming Henry's law as an empirical principle and deriving the other laws.

I. HENRY'S LAW

For dilute solutions of a nondissociating solute at constant temperature, Henry's law states that the fugacity of the solute is proportional to its mole fraction; that is,

$$f_2 = kN_2. \tag{18.1}$$

As solutions become more and more dilute, this law holds more and more accurately. For any solution of finite concentration there will be some deviation, which may or may not be within the limits of experimental error.

At first glance the difference between Henry's law and Raoult's law may not be evident, since both postulate a proportionality between the fugacity of the solute and its mole fraction. The difference lies in the nature of the proportionality constant. From Raoult's law this constant is $f_2^\bullet$, the fugacity of the pure solute. In Henry's law, however,

$$k \neq f_2^\bullet \tag{18.2}$$

in general.

This distinction may be made clearer by a graphical illustration (Fig. 1). A typical fugacity-mole fraction curve is shown by the solid line. If the solute formed an ideal solution with the solvent, the fugacity of the solute would be represented by the broken line in Fig. 1. It is evident that the actual behavior of the solute does not approach Raoult's law, except, as we shall soon prove, when its mole fraction approaches 1 (that is, under circumstances when it would no longer be called the solute). When the solute is present in small quantities, its fugacity deviates widely from Raoult's law. However, as N_2 approaches zero, the fugacity does approach a linear dependence on N_2. This limiting linear relationship is represented by the dotted line in Fig. 1—it is Henry's law in graphical representation.

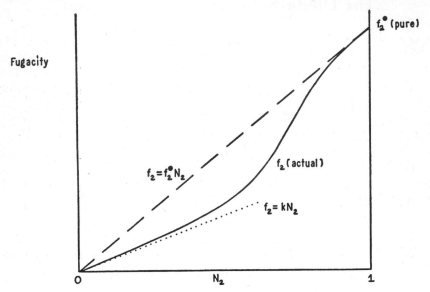

Fig. 1.   Distinction between Henry's law and Raoult's law.

If mole fraction is not a convenient unit of composition, one may transform the statement of the law into other units. Since the law need be applicable in only very dilute solutions,

$$N_2 = \frac{n_2}{n_1 + n_2} \simeq \frac{n_2}{n_1}, \tag{18.3}$$

where n_2 is the number of moles of solute and n_1 that of the solvent. Consequently, equation (18.1) may be revised to read

$$f_2 = k' \frac{n_2}{n_1} = k''m, \tag{18.4}$$

since m, the molality, is the number of moles of solute per 1000 grams of solvent. It is essential to note that even when Henry's law is valid over a wide range of composition, equations (18.3) and (18.4) are still approximate.

II. NERNST'S DISTRIBUTION LAW

If a small quantity of a solute, A, is distributed between two immiscible solvents, for example, I_2 between carbon tetrachloride and water, then at equilibrium the fugacity of the solute must be the same in both phases, since the partial molal free energies or escaping tendencies must be identical.

$$A \text{ (in solvent } a) = A \text{ (in solvent } b);$$

$$f_2 = f_2'. \tag{18.5}$$

If the solutions are sufficiently dilute so that Henry's law is valid for the solute in each phase, we may write

Phase a: $\qquad\qquad\qquad f_2 = k_a N_2;$ $\qquad\qquad\qquad\qquad$ (18.6)

Phase b: $\qquad\qquad\qquad f_2' = k_b N_2'.$ $\qquad\qquad\qquad\qquad$ (18.7)

In view of equation (18.5), it follows that

$$\frac{N_2'}{N_2} = \frac{k_a}{k_b} = K; \tag{18.8}$$

that is, the ratio of mole fractions at equilibrium in the two solvents is a constant, K. This formulation of Nernst's distribution law may be transformed readily into other concentration units, so that we may write also

$$\frac{m'}{m} = \kappa, \tag{18.9}$$

where κ is also a constant, but differs in magnitude from K. Again, it should be noted that equation (18.9) is valid only in a limited range, even for ideal solutions.

III. RAOULT'S LAW

It can be shown readily that if the solute obeys Henry's law in very dilute solutions, the solvent follows Raoult's law in the same series of solutions.

It is evident that such a proof should involve some expression correlating thermodynamic properties of solvent and solute. We may start from the differential equation for the partial molal free energies, since these are closely related to the fugacities, in which we are directly interested. From the general expression (13.44) it follows that at constant pressure and temperature

$$N_1 \, d\bar{F}_1 + N_2 \, d\bar{F}_2 = 0. \tag{18.10}$$

Since $\qquad\qquad\qquad \bar{F}_i = RT \ln f_i + B_i,$ $\qquad\qquad\qquad$ (18.11)

$$d\bar{F}_i = RT \, d \ln f_i$$

at constant temperature and pressure. Hence equation (18.10) may be transformed into

$$N_1 \, d \ln f_1 + N_2 \, d \ln f_2 = 0, \tag{18.12}$$

or

$$N_1 \frac{\partial \ln f_1}{\partial N_1} + N_2 \frac{\partial \ln f_2}{\partial N_1} = 0. \tag{18.13}$$

In any two-component solution

$$N_1 + N_2 = 1. \tag{18.14}$$

Therefore

$$dN_1 = -dN_2, \tag{18.15}$$

and hence equation (18.13) may be revised to read

$$N_1 \frac{\partial \ln f_1}{\partial N_1} - N_2 \frac{\partial \ln f_2}{\partial N_2} = 0, \tag{18.16}$$

or

$$\left(\frac{\partial \ln f_1}{\partial \ln N_1} \right)_{P,T} = \left(\frac{\partial \ln f_2}{\partial \ln N_2} \right)_{P,T}. \tag{18.17}$$

So far we have a perfectly general analysis for any nondissociating two-component solution at constant temperature and pressure. Since our immediate problem is the dilute solution, let us introduce the defining characteristic—Henry's law—and see how it can be used to reduce equation (18.17) further. Since

$$f_2 = kN_2 \tag{18.18}$$

and

$$\ln f_2 = \ln k + \ln N_2, \tag{18.19}$$

$$\left(\frac{\partial \ln f_2}{\partial \ln N_2} \right)_{P,T} = 1. \tag{18.20}$$

Equation (18.17) therefore becomes

$$\left(\frac{\partial \ln f_1}{\partial \ln N_1} \right)_{P,T} = 1. \tag{18.21}$$

Integration at constant pressure and temperature leads to

$$\int d \ln f_1 = \int d \ln N_1, \tag{18.22}$$

$$\ln f_1 = \ln N_1 + \ln \kappa_1, \tag{18.23}$$

where κ_1 is an integration constant. Removing logarithms, we obtain

$$f_1 = \kappa_1 N_1. \tag{18.24}$$

Evidently, then, if the solute obeys Henry's law, the fugacity of the solvent must also be a linear function of its mole fraction. If we consider boundary conditions, that is, the known value of f_1 at some known N_1, we can evaluate the constant.   Thus, when N_1 is 1, that is, when the solvent is pure, f_1 is $f_1^\bullet$.   Consequently, if equation (18.24) is to be valid,

$$\kappa_1 = f_1^\bullet. \tag{18.25}$$

Thus we have, in general,

$$f_1 = f_1^\bullet N_1; \tag{18.26}$$

that is, Raoult's law is obeyed by the solvent for solutions containing very little solute.

IV.   VAN'T HOFF'S LAW OF OSMOTIC PRESSURE

If we set up an apparatus such as is illustrated schematically in Fig. 2, in which two portions of pure solvent are separated by a membrane, M, which the solvent can permeate, the liquid levels will be equal and the

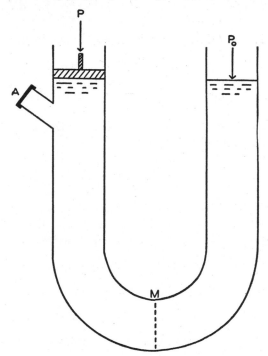

Fig. 2.   Schematic diagram of apparatus for measurement of osmotic pressure.

pressures P and P_0 will be equal at equilibrium.   Such a system will remain in equilibrium, so that we may write

$$f_1^\bullet \text{ (left)} = f_1^\bullet \text{ (right)}. \tag{18.27}$$

If now we add some solute to which the membrane is impermeable through the sidearm, A, then with adequate mixing the solute will be distributed uniformly throughout the left-hand chamber but will be absent from the right-hand one. Solvent will then be observed to move from the right to the left side. Apparently

$$f_1 \text{ (left) } < f_1^* \text{ (right)}. \tag{18.28}$$

This movement can be prevented and equilibrium restored if the pressure P is made sufficiently greater than P_0. With adequate pressure compensation, then,

$$f_1 \text{ (left) } = f_1^* \text{ (right)}. \tag{18.29}$$

Obviously, in this isothermal situation f_1 depends upon the pressure and the quantity of added solute. Hence we may write

$$d \ln f_1 = \left(\frac{\partial \ln f_1}{\partial P} \right)_{T,N_2} dP + \left(\frac{\partial \ln f_1}{\partial N_2} \right)_{P,T} dN_2. \tag{18.30}$$

If equilibrium is maintained, there is no *net* change in f_1, and hence

$$d \ln f_1 = 0.$$

Therefore $$\left(\frac{\partial \ln f_1}{\partial P} \right)_{T,N_2} dP = - \left(\frac{\partial \ln f_1}{\partial N_2} \right)_{P,T} dN_2. \tag{18.31}$$

In view of equation (15.41), the partial derivative in the left-hand side of equation (18.31) must be given by

$$\left(\frac{\partial \ln f_1}{\partial P} \right)_{T,N_2} = \frac{\bar{v}_1}{RT}. \tag{18.32}$$

To evaluate the derivative on the right-hand side, we make explicit recognition of the validity of Raoult's law for the solvent. Thus for the two-component system under consideration we may write

$$f_1 = f_1^* N_1 = f_1^*(1 - N_2), \tag{18.33}$$

which may be converted to the logarithmic form

$$\ln f_1 = \ln f_1^* + \ln (1 - N_2) \tag{18.34}$$

and differentiated, at constant pressure and temperature, to give

$$d \ln f_1 = d \ln (1 - N_2) = - \frac{dN_2}{1 - N_2}. \tag{18.35}$$

Since in very dilute solutions $1 - N_2$ does not differ substantially from unity, we may transform equation (18.35) into

$$\left(\frac{\partial \ln f_1}{\partial N_2} \right)_{P,T} = -1. \tag{18.36}$$

Substitution from equations (18.32) and (18.36) into equation (18.31) leads to

$$\frac{\bar{v}_1}{RT} dP = dN_2. \tag{18.37}$$

In very dilute solutions, $\bar{v}_1$, the partial molal volume of the solvent, is substantially the same as that of the pure solvent, $v_1^{\bullet}$. Hence equation (18.37) may be integrated from the infinitely dilute region to some finite but small mole fraction, N_2, to give

$$P - P_0 = \frac{RT}{v_1^{\bullet}} N_2. \tag{18.38}$$

The difference between the pressure on the solution and that on the solvent, $P - P_0$, which is necessary to maintain equilibrium, is defined as the osmotic pressure, and may be represented by the symbol π. Equation (18.38) may be written, therefore, as

$$\pi = \frac{RT}{v_1^{\bullet}} N_2 \tag{18.39}$$

and also as
$$\pi = \frac{RT}{v_1^{\bullet}} \frac{n_2}{n_1} = \frac{n_2 RT}{V}, \tag{18.40}$$

where V is the total volume of the solvent.

V. VAN'T HOFF'S LAW OF FREEZING-POINT LOWERING

Let us consider a pure solid phase, such as ice, in equilibrium with a pure liquid phase, such as water, at some specified temperature and pressure. If the two phases are in equilibrium,

$$f_s = f_{1}^{\bullet}, \tag{18.41}$$

where f_s represents the fugacity of the pure solid. If solute is added to the system, and if it dissolves only in the liquid phase and not in the solid phase, then the fugacity of the solvent will be lowered:

$$f_1 < f_1^{\bullet}. \tag{18.42}$$

To re-establish equilibrium, f_s must be lowered also. This can be accomplished by decreasing the temperature. In that case, however, it is necessary to recognize that the fugacity of the liquid solvent will also be lowered by the drop in temperature, as well as by the addition of solute. Nevertheless, equilibrium can be re-established if

$$df_s = df_1, \tag{18.43}$$

or
$$d \ln f_s = d \ln f_1. \tag{18.44}$$

Since the fugacity of the solid phase depends only upon the temperature, whereas that of the solvent depends upon both temperature and concentration of added solute, the total differentials of equation (18.44) may be expressed in terms of the appropriate partial derivatives as follows:

$$\left(\frac{\partial \ln f_s}{\partial T}\right)_P dT = \left(\frac{\partial \ln f_1}{\partial T}\right)_{P,N_2} dT + \left(\frac{\partial \ln f_1}{\partial N_2}\right)_{P,T} dN_2. \quad (18.45)$$

The two temperature derivatives may be replaced by the appropriate terms with partial molal enthalpies [equation (15.53)], and the remaining term by its equivalent from equation (18.36). Thus we obtain

$$\frac{\bar{H}^* - H_s^\bullet}{RT^2} dT = \frac{\bar{H}^* - \bar{H}_1}{RT^2} dT - dN_2, \quad (18.46)$$

which may be rearranged to

$$dT = -\frac{RT^2}{\bar{H}_1 - H_s^\bullet} dN_2. \quad (18.47)$$

$\bar{H}_1$ is the partial molal heat content of the liquid solvent; $H_s^\bullet$, the molal heat content of the pure solid form of the solvent. In sufficiently dilute solution $\bar{H}_1$ may be approximated by the heat content of one mole of pure liquid solvent. In such a case

$$\bar{H}_1 - H_s^\bullet \simeq \bar{H}_1^\bullet - H_s^\bullet = \text{heat of fusion of one mole of solvent}$$

$$= \Delta H_{\text{fusion solvent}}. \quad (18.48)$$

Hence equation (18.47) may be written

$$dT = -\frac{RT^2}{\Delta H_{\text{fusion solvent}}} dN_2. \quad (18.49)$$

For very dilute solutions the change in freezing point is small, so that T^2 and ΔH may be considered constant. Under these circumstances, equation (18.49) may be integrated to give

$$\Delta T = -\frac{RT^2}{\Delta H_{\text{fusion solvent}}} N_2. \quad (18.50)$$

VI. VAN'T HOFF'S LAW OF ELEVATION OF BOILING POINT

By methods analogous to those just outlined for the treatment of the lowering of the freezing point, it is possible to show that the boiling-point elevation for a dilute solution containing a nonvolatile solute is given by the expression

$$dT = \frac{RT^2}{\Delta H_{\text{vaporization solvent}}} dN_2. \quad (18.51)$$

It should be emphasized once again that all these laws of the dilute solution become more and more accurate as the solution approaches infinite dilution. For any finite concentration the expressions derived may or may not be good approximations to the observed behavior. The laws of the dilute solution are thus limiting laws. For a more satisfactory treatment of solutions of finite concentrations where deviations from the limiting laws become appreciable, the use of a new function becomes desirable. This new function is described in the following chapters.

Exercises

1. Derive equation (18.51), the van't Hoff expression for the elevation of the boiling point.

2. If a molecule of solute dissociates into two particles in dilute solution, Henry's law may be expressed by the relation

$$f_2 = K(N_2)^2, \tag{18.52}$$

where N_2 is the mole fraction of solute without regard to dissociation.

(a) Derive the Nernst law for the distribution of this solute between two solvents, in one of which it dissociates and in the other of which it does not.

(b) Derive the other laws of the dilute solution for such a dissociating solute.

3. Derive the laws of the dilute solution for a solute, one molecule of which dissociates into ν particles.

CHAPTER 19

Activity and Activity Coefficients of Non-Electrolytes: Standard States

IN THE preceding chapter we considered the laws which may be used to describe the behavior of dilute solutions of non-electrolytes. As has been emphasized repeatedly, these laws are strictly valid only in the limit of infinite dilution. For finite concentrations, a more exact description is possible when precise measurements are available.

In principle one should be able to use the fugacity function to describe deviations from ideality or from limiting laws even for liquids and solids, because, as was pointed out in Chapter 15, fugacities for these condensed states can be evaluated from information on the fugacities of gases and vapors in equilibrium with the solid or liquid. In practice, however, these fugacities frequently are difficult to determine, particularly for salt-like solids which are practically nonvolatile under ordinary conditions. Nevertheless, since generally we are interested only in *changes* in partial molal free energy, and therefore only in *ratios* of fugacities, our inability to establish absolute fugacities is no handicap. We can calculate changes in free energy in terms of relative values of the fugacity. Hence a new technique has been developed to provide a convenient method of dealing with relative fugacities.

I. DEFINITIONS

A. Activity. The *relative fugacity* or *activity*, a, may be defined as the ratio of the fugacity, f, in any given state to the fugacity, $f°$, in some standard state, generally taken at the same temperature:

$$a = \frac{f}{f°}. \tag{19.1}$$

Reference to our initial definition of fugacity [equation (15.34)] will show at once that the activity is a direct measure of the difference in partial molal free energies of the chosen and the standard states, because

$$\bar{F} - \bar{F}° = RT \ln f + B(T) - [RT' \ln f° + B(T')]. \tag{19.2}$$

Since both states are at the same temperature,

$$\bar{F} - \bar{F}° = RT \ln \frac{f}{f°} = RT \ln a. \tag{19.3}$$

B. Activity coefficient.   For many purposes it is convenient to define an additional function, the *activity coefficient*, as the ratio of the activity to the concentration of the component in the given state.   When the composition is expressed in terms of mole fraction, an activity coefficient is written as

$$\gamma_r = \frac{a_r}{N};$$ (19.4)[1]

it is sometimes referred to as the *rational activity coefficient*.   (The term "rational" arises from the fact that in an ideal solution γ_r is always unity, as will be shown later.)   On the other hand, when molality is the composition variable, another activity coefficient is defined by the equation

$$\gamma = \frac{a}{m}.$$ (19.5)

The coefficient given by equation (19.5) is called the *practical activity coefficient*, since molality is the most common method of expressing concentrations, particularly of dilute aqueous solutions.   An activity coefficient based on molarity can also be defined, but such a coefficient is not very convenient because it is not related in a simple fashion to other thermodynamic quantities (such as the partial molal enthalpy).

II. CHOICE OF STANDARD STATES

From the nature of the definition, it is evident that the activity of a given component may have any numerical value, depending upon the state chosen for reference, because if $f°$ is altered, so is a.   There is little reason besides convenience for one state to be chosen as the standard in preference to any other.   Indeed, it will frequently be convenient to change standard states as we proceed from one type of problem to another.   Nevertheless, the experience of several decades has illustrated the convenience of certain choices, which have now been generally adopted.   Unless a clear statement is made to the contrary, then, we shall assume the following conventional standard states in all of our subsequent discussions.

A. Gases.   Since absolute values of the fugacities of gases can be measured readily, it is convenient to define the standard state at any fixed temperature as that in which the gas has a fugacity of one atmosphere.   Hence the activity and fugacity of a gas are numerically identical:

$$a = \frac{f}{f°} = \frac{f}{1} = |f|.$$ (19.6)[2]

---

[1] The subscript r has been placed on a, as well as on γ, to emphasize that the value of a depends upon the choice of standard state.   This choice, in turn, depends upon the type of composition variable which is used, that is, mole fraction or molality, as will be shown in Section II.

[2] The symbol $|\ |$ is used to indicate absolute value.   In equation (19.6), a and f are equal in absolute value though they differ dimensionally.

The activity of an ideal gas is numerically equal to its pressure, since the fugacity and pressure are identical. Thus

$$a = |f| = |P|. \tag{19.7}$$

B. Liquids and solids.

1. *Pure substances.* In most problems involving pure substances, it is convenient to chose the pure solid or the pure liquid, at each and every temperature and at a pressure of one atmosphere, as the standard state. According to this convention, the activity of a pure solid or pure liquid at a pressure of one atmosphere is taken as 1.

In problems involving solutions, however, the choice of the pure component as the standard state is not always convenient, particularly for the solute. It will be necessary, therefore, to consider in some detail the distinctions usually made between solvent and solute in defining reference states.

2. *Solvent.* The reference state generally used for the component of a solution labeled the solvent is easily visualized. It is the pure substance at the same temperature and pressure as the solution:

$$f_1^\bullet(N_1 = 1) \equiv f_{1(\text{standard state})} = f_1^\circ. \tag{19.8}$$

It should be emphasized, perhaps, that in contrast to the case for gases, in general

$$f_1^\circ \neq 1 \text{ atm.}$$

On the other hand, the activity of the pure solvent is unity, since the standard state is the pure solvent; that is,

$$a_{1(\text{standard state})} = a_1^\circ = \frac{f_1^\circ}{f_1^\circ} = 1. \tag{19.9}$$

In any solution of finite concentration, the activity of the solvent must be less than unity.[3] If the fugacity of the solvent may be represented by the solid line in Fig. 1, then the activity curve, Fig. 2, would have the same form, since if each value of the fugacity is divided by the same constant, f_1°, the relative values remain the same.

It should be evident also that the broken line in Fig. 2 has a slope of unity, even though that in Fig. 1 does not; for in Fig. 2 when N_1, the abscissa, is unity, then so is a_1, the ordinate, equal to 1. Furthermore, the rational activity coefficient, a_1/N_1, at any concentration of solvent, N_t, is given by the ratio Y/X, because Y is a_1, and, since the slope of the broken line is 1, X is equal to N_1. In the example illustrated by Fig. 2, the ac-

---

[3] The activity of the solvent in solution cannot be greater than unity (except in a solution supersaturated with respect to solvent). If it were, it would be possible to separate spontaneously a portion of pure solvent from the solution, for the fugacity of the solvent in the solution, and hence its escaping tendency, would be greater than that of the pure substance.

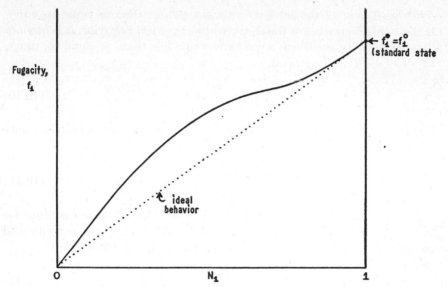

Fig. 1.   Fugacity and standard state for the solvent.

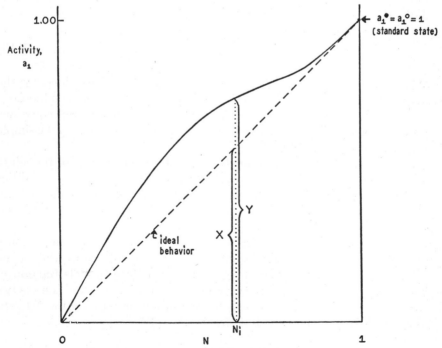

Fig. 2.   Activity and standard state for the solvent.

tivity coefficient of the solvent is always greater than or equal to unity. If the solid line lay below the (broken) line for ideal behavior, it is obvious that the activity coefficient would always be less than, or equal to, unity. In a real solution, in general,

$$\gamma_1 = \frac{a_1}{N_1} \neq 1. \tag{19.10}$$

For pure solvent, or, in other words, in the limit of the infinitely dilute solution, however,

$$\gamma_1 = \frac{a_1}{N_1} \doteq \frac{a_1^\circ}{N_1} = 1. \tag{19.11}$$

It is of interest at this point to consider the values of these functions for the solvent in an ideal solution. From the definition of the activity and the definition of an ideal solution [equation (17.1)], it follows that

$$a_1 = \frac{f_1}{f_1^\circ} = \frac{N_1 f_1^\bullet}{f_1^\circ} = \frac{N_1 f_1^\circ}{f_1^\circ} = N_1. \tag{19.12}$$

In other words, the activity of the solvent in an ideal solution equals its mole fraction. Furthermore, the activity coefficient of the solvent is unity at all concentrations in an ideal solution, for

$$\gamma_1 = \frac{a_1}{N_1} = \frac{N_1}{N_1} = 1. \tag{19.13}$$

Obviously these are desirable properties for these functions, since, for the ideal solution, changes in free energy can be calculated from mole fractions (Chapter 17); hence it is convenient to have our generalized equations for the calculation of free energies in terms of activities reduce to the simple formula for the ideal solution.

3. *Solute.* The procedure for choosing the standard state for the solute differs from that used for the solvent. Although we shall use molalities rather than mole fractions in almost all of our work, it is convenient, as an aid in grasping the significance of the reference state for the solute, to consider the "rational" choice first.

a. *Rational basis.* If we are using mole fraction, N_2, as the measure of solute concentrations, it is desirable to choose our standard state for the solute in such a fashion that the activity will reduce to mole fraction in very dilute solutions where Henry's law is applicable, because free-energy changes may be calculated in very dilute solutions in terms of the fugacity as given by Henry's law. It is desirable, then, that

$$\frac{a_2}{N_2} \to 1 \text{ as } N_2 \to 0. \tag{19.14}$$

Let us consider, as an example, a solute whose fugacity relationship is known throughout the entire concentration range and may be represented by the solid curve in Fig. 3. For the very dilute solution, the actual fugacity curve approaches the line representing Henry's law. If the activity is to satisfy the requirement specified by equation (19.14), then

$$\lim_{N_2 \to 0} \left(\frac{a_2}{N_2} \right) = \lim_{N_2 \to 0} \left(\frac{f_2}{f^{\circ}_2 N_2} \right) = 1. \tag{19.15}$$

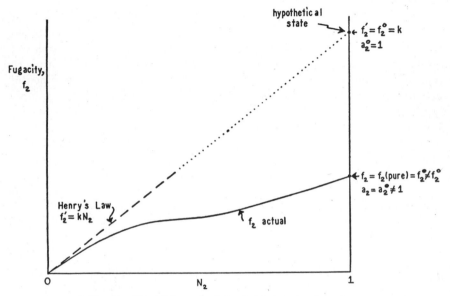

Fig. 3. Establishment of standard state for the solute.

But in the limit of the infinitely dilute solution, where Henry's law is valid (Fig. 3), the following condition must be obeyed also:

$$\lim_{N_2 \to 0} \left(\frac{f_2}{N_2} \right) = \lim_{N_2 \to 0} \left(\frac{f'_2}{N_2} \right) = k, \tag{19.16}$$

or

$$\lim_{N_2 \to 0} \left(\frac{f_2}{k N_2} \right) = 1. \tag{19.17}$$

It is apparent, then, that if equations (19.15) and (19.17) are to be valid simultaneously, the standard state should be that state in which f°_2 is numerically equal to k, the constant in Henry's law. In terms of the graph shown in Fig. 3, that state can be found by extrapolation of the broken line representing Henry's law to a concentration of $N_2 = 1$, since

$$f'_2 = k \quad \text{when} \quad N_2 = 1. \tag{19.18}$$

This fugacity, shown in the upper right-hand corner of Fig. 3, is conveniently selected as the standard fugacity for the solute.

It is apparent from the graph that

$$f_{2(\text{standard state})} \equiv f_2^\circ \neq f_{2(\text{pure solute})}. \tag{19.19}$$

In other words, the actual fugacity of the pure solute is not, in general, equal to the fugacity of the solute in the state chosen as standard. In fact, there is, in the example in Fig. 3, *no actual state* with a fugacity corresponding to that of the standard state.

In an introduction to this question of standard states, the selection for the solute may seem unnecessarily devious and roundabout. A much simpler approach would be a method of selection similar to that employed for the solvent, with pure solute being taken as the reference state. In practice, however, the latter procedure would require either experimental information on the pure solute in the same physical state (for example liquid or solid) as the solution, or data for solutions of sufficiently high concentration of solute so that Raoult's law might be approached and might be used for the extrapolation to obtain $f_{2(\text{actual})}$ at $N_2 = 1$. Such information of course, generally is not available. In essence this means that the details, of the solid curve in Fig. 3 may be known only for small values of N_2, and not necessarily as illustrated in the graph, throughout the entire concentration range. Thus there are generally no means for the determination of the actual value which f_2 would approach for pure (supercooled) liquid solute. On the other hand, even with data available only at low concentrations of solute, it is feasible, generally, to find the limiting slope, or in other words, the constant in Henry's law and hence, in principle, to determine f_2° for the *hypothetical* state which is selected as the standard state.

When data are actually available for the solute over substantially the entire concentration range, from mole fraction zero to unity, one may use conveniently either of two states, the hypothetical mole-fraction-unity or the actual mole-fraction-unity (that is, pure solute) as the standard state. It should be evident from the preceding discussion that in either case the activity of the solute in any solution in the example in Fig. 3 is always less than unity, since with either definition of the standard state f_2 in that state is greater than in any of the solutions.

On the other hand, the numerical magnitudes of the activity coefficients for the example in Fig. 3 will show a striking dissimilarity. If the standard state is determined by extrapolation of Henry's law, the values of the activities will be given by a curve such as in Fig. 4, with all values less than unity. The slope of the line which corresponds to Henry's law must be unity, since $a_2^\circ = 1$ when $N_2 = 1$. Consequently, the activity coefficient, a_2/N_2, must be given by y/x, since y is a_2 and x must equal N_2 at the point N_4. From the graph it is apparent that this activity coefficient is less than one in the example under consideration.

On the other hand, the activity coefficient for the solute in the same solution would be greater than 1 if the pure (liquid or supercooled liquid) solute were chosen as the standard state. If this alternative standard state is adopted, the scale for the ordinates in Fig. 4 must be changed so that $a_2^\bullet$ is 1, as is shown in Fig. 5. The activity of the solute in the hypothetical state which was formerly the standard state is now much greater than 1, as is also indicated in Fig. 5. Similarly, since the scale of ordinates has been condensed by the new choice of standard state, each activity is numerically greater than that assigned to the solute in the same solution on the basis of the other choice of standard state. Finally, the activity coefficient,

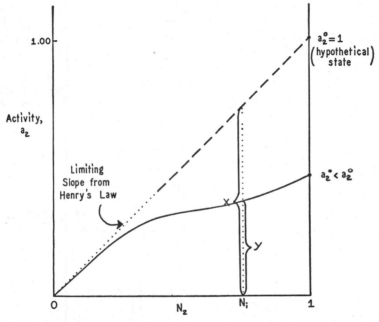

Fig. 4. Activity and activity coefficient when standard state is set by extrapolation of Henry's law.

a_2/N_2, for the solute of mole fraction N_i, now must be greater than unity, as can be seen by reference to Fig. 5: a_2/N_2 must be equal to y/x, since again y is a_2 and x is again equal to N_2.[4]

It should be emphasized that the choice of standard state makes no difference in the value obtained for the change in free energy accompanying the transfer of the solute from one solution to another of different concentration. It should be apparent that choice of reference state will not affect

---

[4] The equality of x and N_2 in Fig. 5 may not be apparent at first glance, because the axes of ordinates and abscissas are drawn to different scales. Nevertheless, since x lies on a line with a slope of unity (at $N_2 = 1$, $a_2^\circ = 1$), it should be apparent that x and N_2 are numerically identical.

differences in partial molal free energy;   that is,

$$\Delta F = \bar{\mathrm{F}}_2 - \bar{\mathrm{F}}_2' = \bar{\mathrm{F}}_2 - \bar{\mathrm{F}}_2^\circ - (\bar{\mathrm{F}}_2' - \bar{\mathrm{F}}_2^\circ).$$ (19.20)

No matter which state is chosen as the standard, it effectively drops out because $\bar{\mathrm{F}}^\circ$ is both added and subtracted in equation (19.20).   Or from another point of view, by extending equation (19.20) a little further by explicit introduction of equation (19.3), we can see that

$$\Delta F = RT \ln a_2 - RT \ln a_2'$$ (19.21)

$$= RT \ln \frac{f_2}{f_2^\circ} - RT \ln \frac{f_2'}{f_2^\circ}.$$ (19.22)

Therefore $$\Delta F = RT \ln \frac{f_2}{f_2'}.$$ (19.23)

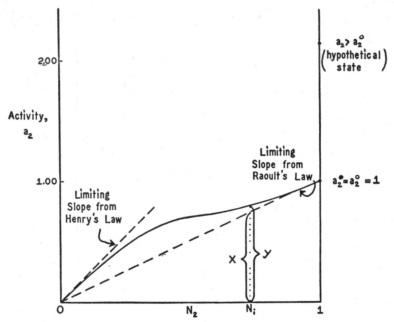

Fig. 5.   Activity and activity coefficient when pure solute is made the standard state.

Thus, once again it is obvious that f_2° can be canceled in calculations of changes in free energy.   Hence, although numerical values of a_2 and a_2' depend upon choice of standard state, their ratio, being a ratio of the corresponding fugacities, is independent of the standard state;   therefore ΔF is also independent of the standard state.

  b.   *Practical basis*.   With an adequate understanding of the "rational" basis for defining activities of a solute, we should encounter no difficulties in grasping the "practical" method.   We must recognize that, in general, we have information for solutions in concentration ranges which are numerically small on a mole fraction basis. Molality is used more commonly as the unit of concentration, particularly for aqueous solutions, toward which greatest interest has been directed.

Deviations from Henry's law may be represented, for a typical solute, by a graph such as is illustrated in Fig. 6. When molality is the composition variable, we may write as a substitute for Henry's law

$$f_2' = k'm. \tag{19.24}$$

As will be shown later, the constant, k', may be determined, in principle, if we obtain the limiting slope at infinite dilution for the solid curve in Fig. 6.

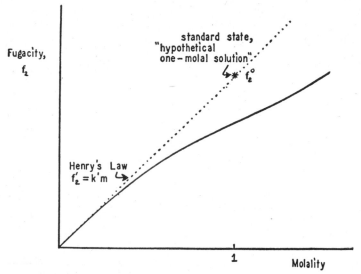

Fig. 6.   Establishment of standard state for solute.

Once again we recognize that the standard fugacity should be selected in such a fashion that

$$\frac{a_2}{m} \to 1 \quad \text{as} \quad m \to 0, \tag{19.25}$$

since free-energy changes for the solute in very dilute solutions may be calculated in terms of the fugacity as given by the substitute for Henry's law, in the form of equation (19.24).   In other words,

$$\lim_{m \to 0} \left(\frac{a_2}{m} \right) = \lim_{m \to 0} \left(\frac{f_2}{f_2^\circ m} \right) = 1. \tag{19.26}$$

But since the limiting law, equation (19.24), is also valid in the infinitely dilute solution, then the following condition must be obeyed also:

$$\lim_{m \to 0}\left(\frac{f_2}{k'm}\right) = 1. \qquad (19.27)$$

Since equations (19.26) and (19.27) must be valid simultaneously, the standard state should be chosen in such a manner that f_2° is numerically equal to k'. In terms of the graph shown in Fig. 6, such a state may be found by extrapolation of the (dotted) line of the limiting law to one-molal concentration, since on that straight line

$$f_2' = k' \qquad \text{when} \qquad m = 1. \qquad (19.28)$$

Therefore $\qquad\qquad f_2^\circ = k'. \qquad\qquad\qquad\qquad\qquad (19.28a)$

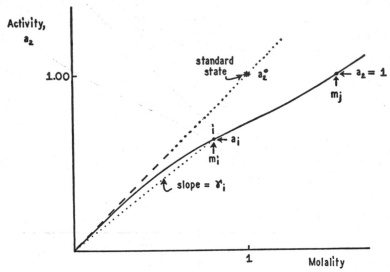

Fig. 7. Activity and activity coefficient when the standard state is established on a molality basis.

The standard state, chosen on this basis, is indicated by the asterisk in Fig. 6. Again, it should be evident that this reference state is a "hypothetical one-molal solution," that is, a state which has the fugacity that a one-molal solution would have *if it obeyed the limiting law*. It should be apparent also that it is *misleading* to say that the standard state of the solute is the infinitely dilute solution, because f_2 equals zero in an infinitely dilute solution. Thus if f_2° were zero, the activity at any finite concentration, being f_2/f_2°, would be infinite.

We can readily modify Fig. 6 into the corresponding *activity*-molality curve. We need only revise the axis of ordinates so that $a_2^\circ = 1$ for the hypothetical one-molal solution which we have chosen as the standard state (Fig. 7). It should be apparent from Fig. 7 that the activity of the solute may have values above unity. It is instructive to note also that the activity coefficient, γ_i, is given graphically by the slope of the dotted line from origin to the point i representing a particular state (molality m_i), since the slope of this dotted line is

$$\frac{a_i}{m_i} = \gamma_i. \tag{19.29}$$

It is also evident in Fig. 7 that for the particular system depicted there is a solution of some finite concentration, m_j, for which the activity is unity, in other words, with a fugacity equal to the standard fugacity. It would be misleading, nevertheless, to call this solution of molality m_j *the* standard state, since it would not have a partial molal enthalpy equal to that of the infinitely dilute solution (except perhaps by coincidence). As we shall demonstrate shortly, $\bar{H}_2^\circ$ is equal to $\bar{H}_2$ of an infinitely dilute solution. The so-called "standard state" is therefore a hypothetical one in which f° equals f at m_j and $\bar{H}^\circ$ equals $\bar{H}$ at infinite dilution.

III. GENERAL RELATION BETWEEN EQUILIBRIUM CONSTANT AND FREE ENERGY

With the establishment of the activity function, we are in a position to derive a general expression relating ΔF° of a reaction to the equilibrium constant and hence to eliminate the restrictions imposed upon previous relationships.

For simplicity we shall consider a reaction involving two substances, A and B, in states with activities a_A and a_B, respectively, being converted to two new substances, C and D, at activities a_C and a_D, respectively, with a free-energy change ΔF:

$$A \, (a_A) + B \, (a_B) = C \, (a_C) + D \, (a_D); \quad \Delta F. \tag{19.30}$$

To equation (19.30) let us add a series of equations which will yield the equation for the corresponding reaction with each substance at *unit* activity. For each of these additional reactions let us indicate the value of ΔF:

$$A \, (a = 1) = A \, (a_A) \quad ; \quad \Delta \bar{F} = RT \ln \frac{f_A}{f_A^\circ} = RT \ln a_A, \tag{19.31}$$

$$B \, (a = 1) = B \, (a_B) \quad ; \quad \Delta \bar{F} = RT \ln \frac{f_B}{f_B^\circ} = RT \ln a_B, \tag{19.32}$$

$$C \, (a_C) = C \, (a = 1) \quad ; \quad \Delta \bar{F} = RT \ln \frac{f_C^\circ}{f_C} = RT \ln \frac{1}{a_C}, \tag{19.33}$$

$$D \, (a_D) = D \, (a = 1) \quad ; \quad \Delta \bar{F} = RT \ln \frac{f_D^\circ}{f_D} = RT \ln \frac{1}{a_D}. \tag{19.34}$$

Addition of equations (19.30) to (19.34) leads to

$$A \ (a = 1) + B \ (a = 1) = C \ (a = 1) + D \ (a = 1), \quad (19.35)$$

with a free energy ΔF° (since each substance is at unit activity) given by the equation

$$\Delta F^\circ = \Delta F + RT \ln \frac{a_A a_B}{a_C a_D}, \quad (19.36)$$

which can be rearranged, with inversion of the activity ratio, to give

$$\Delta F = \Delta F^\circ + RT \ln \frac{a_C a_D}{a_A a_B}. \quad (19.37)$$

This is a general relation for the calculation of ΔF for any reaction in which the components are not at unit activity from available values of ΔF° and the actual activities, a_A, a_B, a_C, and a_D.

A further important expression is obtained from equation (19.37) if the activities considered in reaction (19.30) represent the values obtained under equilibrium conditions. In such a situation

$$\Delta F_{P,T} = 0 \quad \text{for reaction (19.30).} \quad (19.38)$$

Therefore
$$-\ln \left(\frac{a_C a_D}{a_A a_B} \right)_{\text{equilibrium}} = \frac{\Delta F^\circ}{RT} = \text{constant} \quad (19.39)$$

and thus
$$\left(\frac{a_C a_D}{a_A a_B} \right)_{\text{equilibrium}} = \text{constant} = K. \quad (19.40)$$

Hence we obtain a general relation between the thermodynamic equilibrium constant, K, and the standard change in free energy, ΔF°, for a reaction:

$$\Delta F^\circ = -RT \ln K. \quad (19.41)$$

IV. DEPENDENCE OF ACTIVITY ON PRESSURE

The standard fugacity for any pure substance or for any component of a solution is defined, generally, as that for the system at some fixed, specified pressure. In most problems concerning solutions, this specified pressure is one atmosphere. In view of our definition it follows that at constant temperature

$$\left(\frac{\partial \ln f^\circ}{\partial P} \right)_T = 0. \quad (19.42)$$

Consequently, the pressure coefficient of the activity must be the same as that for the fugacity:

$$\left(\frac{\partial \ln a}{\partial P} \right)_T = \left(\frac{\partial \ln (f/f^\circ)}{\partial P} \right)_T = \left(\frac{\partial \ln f}{\partial P} \right)_T = \frac{\bar{v}}{RT}. \quad (19.43)$$

The term with $\bar{v}$ in equation (19.43) is obtained from equation (15.41). Hence changes in activity with variations in pressure may be calculated if the partial molal volume of the component is known.

V. DEPENDENCE OF ACTIVITY ON TEMPERATURE

By analogy with previous discussions of temperature coefficients of the free-energy functions, we should expect the expression for the temperature dependence of the activity to involve some enthalpy function. It is necessary, therefore, to obtain some clear statements on the enthalpies of the reference or standard states.

A. Standard partial molal enthalpies. So long as we deal with pure solids and liquids, little needs to be said concerning the meaning of standard enthalpy, for by convention it is the enthalpy of the substance at the specified temperature and at one atmosphere pressure. With solutions, however, confusion arises occasionally. If we maintain clear recognition of our definitions of standard states for the activity, it will not be difficult to determine the states to which standard enthalpies must refer.

1. *Solvent.* As was pointed out previously, pure solvent at the same temperature and pressure (generally one atmosphere) as the solution is usually chosen as the reference state for the solvent. It is clear, then, without further discussion that

$$\bar{H}_{1(\text{standard state})} = \bar{H}_1^{\bullet} = \bar{H}_1^{\circ}. \tag{19.44}$$

2. *Solute.* For the solute, the standard state is the *hypothetical* one-molal solution with the fugacity which the solute would have *if it obeyed the limiting law at one-molal concentration.* Thus the partial molal enthalpy of the solute in the standard state is *not* necessarily that of the *actual* one-molal solution.

To determine $\bar{H}_2$ for the standard state of the solute, we start with the definition of f_2° given by equation (19.28a). In view of this correspondence of f_2° and k', it follows that the limiting law may be expressed as

$$f_2 = k'm = f_2^{\circ}m, \tag{19.45}$$

or in logarithmic form

$$\ln f_2 = \ln f_2^{\circ} + \ln m. \tag{19.46}$$

From equation (15.53) we can obtain the temperature coefficients of the two fugacities, each at constant molality:[5]

$$\left(\frac{\partial \ln f_2}{\partial T}\right)_{P,m} = \left(\frac{\partial \ln f_2^{\circ}}{\partial T}\right)_{P,m'=1}; \tag{19.47}$$

$$\frac{\bar{H}^* - \bar{H}_2}{RT^2} = \frac{\bar{H}^* - \bar{H}_{2(\text{standard state})}}{RT^2} \tag{19.48}$$

---

[5] At each and every temperature, f_2° is chosen along the corresponding line for the limiting law at a molality, m', of 1. Thus m' is fixed at unit molality.

Equation (19.45) is valid for the solute in the *real* solution only at the limit of infinite dilution. It follows then that the $\bar{H}_2$ in equation (19.48) is applicable to the solute in the real solution only at the limit of infinite dilution. Putting in this restriction explicitly, we obtain

$$\frac{\bar{H}^* - \bar{H}_{2\text{(infinite dilution)}}}{RT^2} = \frac{\bar{H}^* - \bar{H}_{2\text{(standard state)}}}{RT^2}. \tag{19.49}$$

Hence
$$\bar{H}_{2\text{(standard state)}} = \bar{H}_{2\text{(infinite dilution)}} = \bar{H}_2^{\,\circ}. \tag{19.50}$$

In other words, the partial molal heat content of the solute in the standard state (the hypothetical one-molal solution) is the same as the partial molal heat content of the solute in an infinitely dilute solution. For this reason, the infinitely dilute solution is frequently called the reference state for enthalpies of components in solution. In contrast to $\bar{F}_2$, $\bar{H}_2$ does not approach negative infinity as the molality goes to zero.

B. Equation for temperature coefficient. Having established that the enthalpies of an infinitely dilute solution may be used as standard heat contents for either solute or solvent, we are in a position to derive the equation for the temperature coefficient of the activity of either component in a real solution. Starting with the basic definition in equation (19.1), we may write

$$\left(\frac{\partial \ln a}{\partial T}\right)_{P,m} = \left(\frac{\partial \ln f}{\partial T}\right)_{P,m} - \left(\frac{\partial \ln f^{\circ}}{\partial T}\right)_{P,m'=1} \tag{19.51}$$

$$= \frac{\bar{H}^* - \bar{H}}{RT^2} - \frac{\bar{H}^* - \bar{H}^{\circ}}{RT^2}. \tag{19.52}$$

Therefore
$$\left(\frac{\partial \ln a}{\partial T}\right)_{P,m} = -\frac{\bar{H} - \bar{H}^{\circ}}{RT^2}. \tag{19.53}$$

Equation (19.53) is valid for either solute or solvent, since each step in its derivation is valid for either component. In conformity with previous definitions of relative enthalpies [equation (14.6)], we may modify equation (19.53) to read

$$\left(\frac{\partial \ln a}{\partial T}\right)_{P,m} = -\frac{\bar{L}}{RT^2}. \tag{19.54}$$

Thus the determination of the temperature dependence of the activity requires a knowledge of the relative partial molal enthalpy of the component under consideration.

If we start with equation (19.54), it is a simple matter to show that the temperature dependence of the *activity coefficient* may be expressed by the same relationship. It is obvious from equation (19.5) that

$$a = m\gamma \tag{19.55}$$

and, hence, for the solute

$$\left(\frac{\partial \ln a}{\partial T}\right)_{P,m} = \left(\frac{\partial \ln \gamma}{\partial T}\right)_{P,m} = -\frac{\bar{L}}{RT^2}. \tag{19.56}$$

This relation must hold also for the solvent, because if we start with the definition given by equation (19.4), we shall still obtain the relationship indicated in (19.56). We need only recall that if the molality is constant, so is the mole fraction.

VI. STANDARD ENTROPY

In the preceding discussions it has been pointed out that there may be a concentration m_j of the solute in the real solution which has an activity of unity, that is, an activity equal to that of the "hypothetical one-molal" standard state. It has been mentioned also that $\bar{H}_2$ of the solute in the standard state equals that of the solute at infinite dilution. One might naturally inquire, then, whether or not the *partial molal entropy* of the solute in the standard state, $\bar{s}_2^{\circ}$, corresponds to that in either of these two solutions.

We may proceed to answer this question by considering the thermodynamic properties of some state of the solute with $\bar{s}_2$ numerically equal to the partial molal entropy of the solute in the standard state, $\bar{s}_2^{\circ}$. At the molality of this state there will also be corresponding values of the partial molal enthalpy, $\bar{H}_2$, and the partial molal free energy, $\bar{F}_2$. For the latter, we may write

$$\bar{F}_2 = RT \ln f_2 + B(T). \tag{19.57}$$

Since there is no reason to assume that $\bar{F}_2$ of this solution is equal to $\bar{F}_2^{\circ}$, we shall also write the equation for the latter:

$$\bar{F}_2^{\circ} = RT \ln f_2^{\circ} + B(T). \tag{19.58}$$

Hence $\qquad \bar{F}_2 - \bar{F}_2^{\circ} = RT \ln f_2 - RT \ln f_2^{\circ}. \tag{19.59}$

Taking temperature derivatives, we obtain

$$\left(\frac{\partial \bar{F}_2}{\partial T}\right)_{P,m} - \left(\frac{\partial \bar{F}_2^{\circ}}{\partial T}\right)_{P,m'=1} = -(\bar{s}_2 - \bar{s}_2) = RT\left(\frac{\partial \ln f_2}{\partial T}\right)_{P,m}$$
$$+ R \ln f_2 - RT\left(\frac{\partial \ln f_2^{\circ}}{\partial T}\right)_{P,m'=1} - R \ln f_2^{\circ}. \tag{19.60}$$

Equation (19.60) can be modified to

$$-T(\bar{s}_2 - \bar{s}_2^{\circ}) = RT^2\left(\frac{\partial \ln f_2}{\partial T}\right)_{P,m} - RT^2\left(\frac{\partial \ln f_2^{\circ}}{\partial T}\right)_{P,m'=1}$$
$$+ RT \ln f_2 - RT \ln f_2^{\circ}. \tag{19.61}$$

TABLE 1
STANDARD STATES* FOR THERMODYNAMIC CALCULATIONS
(For every case it is assumed that the temperature has been specified)

Physical State	Free Energy	Enthalpy†	Entropy
Pure gas	Real gas at unit fugacity. Generally at pressure near one atmosphere	Real gas at zero pressure. (One atmosphere pressure is used in some tables. To be thermodynamically consistent with the reference state of unit fugacity specified for the free energy, the standard state for enthalpy values ought to be zero pressure. See Exercise 1)	Hypothetical ideal gas at one atmosphere pressure. See Chapter 11 on third law of thermodynamics
Pure liquid or pure solid	One atmosphere pressure	One atmosphere pressure	One atmosphere pressure
Solvent in a solution	Pure solvent at one atmosphere pressure	Pure solvent at one atmosphere pressure	Pure solvent at one atmosphere pressure
Solute in a solution	Hypothetical one-molal solution obeying limiting law corresponding to Henry's law.‡ There may also be a finite concentration, not equal to zero, at which the activity of the solute is unity. This may be considered as the standard state *only in free-energy calculations*	Hypothetical one-molal solution obeying limiting law corresponding to Henry's law. The value of the partial molal enthalpy in the standard state, $\bar{H}_2$, is *always* equal to that at infinite dilution. Hence the *infinitely dilute solution* may be used as *reference state* in all *enthalpy calculations*, but not in free-energy or entropy calculations	Hypothetical one-molal solution obeying limiting law corresponding to Henry's law. There may also be a solution of finite concentration, not equal to zero but also not having an activity of unity, with a partial molal entropy of solute equal to that in the reference state

* No mention is made in this table of a standard state for volumes. At first thought, this may be surprising. It should be realized, however, that this omission is a consequence of our method of defining standard states. Although we have devoted a great deal of attention to the variation of f_2° with temperature, we have avoided any such consideration of the dependence of f_2° on pressure by the simple expedient of defining the standard state as one at a *fixed pressure* of one atmosphere. Consequently, the partial derivative $(\partial \bar{F}^\circ / \partial P)_T$ is zero, since $\bar{F}^\circ$ does not vary with pressure.

If we had permitted f_2° to vary with pressure, as well as with temperature, then it can be shown, by considerations analogous to those described in Section V of this chapter, that $\bar{v}_2^\circ$, the partial molal volume of the solute in its standard state, is equal to that of the solute at infinite dilution.

† The standard state for the heat capacity is the same as that for the enthalpy. For a proof of this statement for the solute in a solution, see Exercise 2.

Taking account of the relation between enthalpies and the temperature coefficient of $\ln f$, and of equations (19.57) and (19.58), we obtain

$$-T(\bar{s}_2 - \bar{s}_2{}^\circ) = (\bar{H}_2{}^\circ - \bar{H}_2) + (\bar{F}_2 - \bar{F}_2{}^\circ). \qquad (19.62)$$

It is evident that at infinite dilution, that is, when $m = 0$,

$$\bar{s}_2 \neq \bar{s}_2{}^\circ \qquad \text{(at } m = 0\text{),} \quad (19.63)$$

since $\qquad\qquad \bar{F}_2 \neq \bar{F}_2{}^\circ \qquad \text{(at } m = 0\text{),} \quad (19.64)$

even though $\qquad \bar{H}_2 = \bar{H}_2{}^\circ \qquad \text{(at } m = 0\text{).} \quad (19.65)$

Hence the partial molal entropy of the solute in the standard state is not that of the solute at infinite dilution. Similarly, it is apparent that at the molality m_j (Fig. 7), where a_2 is unity,

$$\bar{s}_2 \neq \bar{s}_2{}^\circ \qquad \text{(at molality where } a_2 = 1\text{),} \quad (19.66)$$

for $\qquad\qquad \bar{H}_2 \neq \bar{H}_2{}^\circ \qquad \text{(at molality where } a_2 = 1\text{),} \quad (19.67)$

even though $\qquad \bar{F}_2 = \bar{F}_2{}^\circ \qquad \text{(at molality where } a_2 = 1\text{).} \quad (19.68)$

Thus $\bar{s}_2$ can be equal to $\bar{s}_2{}^\circ$ only for a solution with some molality, m_k, at which

$$(\bar{H}_2{}^\circ - \bar{H}_2) = (\bar{F}_2{}^\circ - \bar{F}_2). \qquad (19.69)$$

The particular value of the molality, m_k, at which this occurs will differ from solute to solute.

Surveying the thermodynamic properties of the hypothetical one-molal solution which has been chosen as the standard state for the solute, we can arrive at the following conclusions. It is characterized by values of the thermodynamic functions which may be represented by $\bar{F}_2{}^\circ$, $\bar{H}_2{}^\circ$, and $\bar{s}_2{}^\circ$. There frequently is also a *real* solution at some molality, m_j (Fig. 7), for which $\bar{F}_2 = \bar{F}_2{}^\circ$, that is, for which the activity has a value of unity. Values of $\Delta\bar{F}^\circ$ can be applied, therefore, to reactions involving this solute at this specific molality, m_j. On the other hand, $\Delta\bar{H}^\circ$, obtained for example

---

‡ A more elegant (though more difficult to visualize) formulation of the procedure for the selection of the standard state for a solute may be made as follows. Since

$$\bar{F}_2 = \bar{F}_2{}^\circ + RT \ln a \qquad (19.74)$$

and $\qquad\qquad\qquad a = m\gamma, \qquad\qquad\qquad\qquad (19.75)$

it follows that

$$\bar{F}_2{}^\circ = \bar{F}_2 - RT \ln m - RT \ln \gamma. \qquad (19.76)$$

Therefore the state in which the solute has a partial molal free energy of $\bar{F}_2{}^\circ$ may be found from the following limit, since γ approaches unity as m approaches zero:

$$\lim_{m \to 0} (\bar{F}_2 - RT \ln m) = \bar{F}_2{}^\circ. \qquad (19.77)$$

With this method of formulation it is also possible to show that there frequently is a real solution at some molality m_j for which $\bar{F}_2 = \bar{F}_2{}^\circ$, that $\bar{H}_2{}^\circ$ corresponds to $\bar{H}_2$ for a real solution at infinite dilution, and that $\bar{s}_2{}^\circ$ equals $\bar{s}_2$ for a real solution at a molality m_k, which is neither zero nor m_j.

from the appropriate temperature derivative of $\Delta\bar{F}^\circ$, does not apply to the molality m_j, since $\bar{H}_2^{\,\circ}$ corresponds to $\bar{H}_2$ for a real solution only at infinite dilution. Furthermore, $\bar{S}_2^{\,\circ}$ has a value equal to $\bar{S}_2$ for some *real* solution only at a molality m_k, which is neither zero nor m_j. Consequently, $\Delta\bar{S}^\circ$, obtained perhaps from a combination of $\Delta\bar{F}^\circ$ and $\Delta\bar{H}^\circ$, refers to a solution which differs in composition from that to which $\Delta\bar{F}^\circ$ refers as well as from that to which $\Delta\bar{H}^\circ$ can be applied. Thus the values of these three thermodynamic quantities, $\bar{F}_2^{\,\circ}$, $\bar{H}_2^{\,\circ}$, and $\bar{S}_2^{\,\circ}$, in the hypothetical standard state of the solute correspond, in general, to *three different real solutions*.

So far no attention has been paid to the solvent. The problem in this case is quite simple: if we derive an equation corresponding to (19.62), which is for the solute, we obtain

$$-T(\bar{S}_1 - \bar{S}_1^{\,\circ}) = (\bar{H}_1^{\,\circ} - \bar{H}_1) + (\bar{F}_1 - \bar{F}_1^{\,\circ}). \tag{19.70}$$

For pure solvent

$$\bar{F}_1^{\,\bullet} = \bar{F}_1^{\,\circ} \tag{19.71}$$

and

$$\bar{H}^{\,\bullet} = \bar{H}_1^{\,\circ}. \tag{19.72}$$

It is apparent immediately that

$$\bar{S}_1^{\,\bullet} = \bar{S}_1^{\,\circ} \tag{19.73}$$

and hence that the pure solvent is the standard state for the partial molal entropy as well as for $\bar{F}_1$ and $\bar{H}_1$.

Thus there are some marked differences in the conventions established for the standard states of solvent and solute. To assist in keeping these distinctions clear, Table 1 has been prepared. It contains a summary of the characteristics of the standard states for various thermodynamic functions not only for solutions, as described in this chapter, but also for pure substances, as discussed earlier in this text.

Exercises

1. For a pure real gas, the fugacity is defined so that

$$f = P \qquad \text{as} \qquad P \to 0.$$

Show that on this basis the enthalpy of the gas in the standard state must be equal to that at zero pressure. (*Hint:* if the gas were ideal,

$$f = kP = 1 \times P$$

at all pressures. For the real gas, $f^\circ = 1 = k$. Proceed by analogy with the discussion for the solute in solutions.)

2. (a) Rearrange equation (19.53) so that RT^2 occurs as a factor on the left-hand side.

(b) Differentiate both sides of the equation obtained in Part (a) with respect to T.

(c) Prove that $\bar{c}_{p(2)}^{\,\bullet}$ is equal to $\bar{c}_{p(2)}$ for the infinitely dilute solution. Keep in mind that the temperature coefficient of $\ln \gamma_2$ is zero in an infinitely dilute solution.

3.   It is necessary frequently to convert solute activity coefficients based on mole fraction to a molality basis, or vice versa.   The equation for making this conversion may be derived in the following way.

(a)   Starting from the fundamental definitions of activity on each concentration basis, prove that for a solute

$$\frac{a_N}{a_m} = \frac{f_m^\circ}{f_N^\circ},$$

where the subscript N refers to a mole fraction basis and m to a molality basis.

(b)   Show further that

$$\frac{\gamma_N}{\gamma_m} = \frac{m}{N} \frac{k'}{k},$$

where k and k' are defined by equations (19.16) and (19.24).

(c)   In very dilute solutions, Henry's law is valid in the form of either equation (19.16) or (19.24).   Prove that

$$k' = k \frac{M_1}{1000},$$

where M_1 is the molecular weight of the solvent.

(d)   Show that for any concentration

$$\frac{\gamma_N}{\gamma_m} = 1 + \frac{mM_1}{1000}.$$

CHAPTER 20

Determination of Activity and Activity Coefficients of Non-Electrolytes

HAVING established the definitions and conventions used in connection with the activity function, we are in a position to understand the experimental methods which have been used to determine numerical values of this quantity.

I. MEASUREMENTS OF VAPOR PRESSURE

If the vapor pressure of a substance is sufficiently great to be determined experimentally, the procedure for calculating activities is straightforward. Since in practice the method of choice of standard state differs for the solvent and solute, it is necessary to consider each component separately.

A. Solvent. The activity of the solvent is related in a very simple fashion to the vapor pressure, when the latter may be taken as a reliable measure of the fugacity. Since the pure solvent is the reference state, the activity a_1 of the solvent in any chosen solution is given by the expression

$$a_1 = \frac{f_1}{f_1^\circ} = \frac{f_1}{f_1^\bullet} = \frac{f \text{ (vapor of solvent over solution)}}{f \text{ (vapor of pure solvent)}}. \qquad (20.1)$$

In most cases, vapor pressures are sufficiently low so that their behavior is described accurately by the ideal gas law. In such a situation,

$$a_1 = \frac{p_1}{p_1^\circ}, \qquad (20.2)$$

where p represents the partial pressure of the vapor of the solvent.

B. Solute. If the solute is sufficiently volatile to allow a determination of its vapor pressure over the solution, then the calculation of its activity is simple also. Once again, we may assume that the vapor behaves essentially like an ideal gas at the pressures normally encountered. Thus

$$a_2 = \frac{f_2}{f_2^\circ} = \frac{p_2}{p_2^\circ}. \qquad (20.3)$$

In this equation, however, p_2° is not the vapor pressure of pure solute, but rather the vapor pressure that the solute would have in a one-molal solution *if it obeyed Henry's law*. It is necessary, then, to determine p_2° from the limiting behavior of the solute in an infinitely dilute solution.

In such a solution, where Henry's law is obeyed,

$$f_2 = k'm. \tag{20.4}$$

But if the vapor behaves as an ideal gas, we may write also

$$p_2 = k'm. \tag{20.5}$$

Obviously, if Henry's law were obeyed up to one-molal concentration,

$$p_2 \text{ (hypothetical one-molal solution)} = p_2^\circ; \tag{20.6}$$

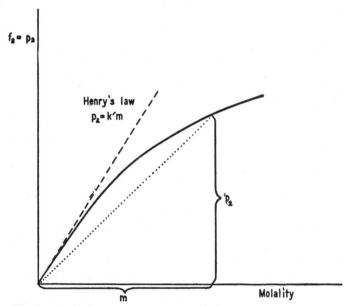

Fig. 1.   Partial pressures of solute as a function of concentration.

from equation (20.5) it follows that when m is unity,

$$p_2^\circ = |k'|. \tag{20.7}[1]$$

Thus we must calculate k', the Henry's law constant, in order to determine p_2°.

In practice, k' may be obtained easily from an extrapolation of experimental values of the ratio p_2/m. If the partial pressure of the solute is a nonlinear function of the molality, such as is illustrated in Fig. 1, the ratio p_2/m (which is the slope of the dotted line) will have a concentration

---

[1] Again the symbol | ⌉ is used to indicate absolute value.

dependence such as is shown in Fig. 2, since it is apparent that the slope of the dotted line decreases with increasing molality. Nevertheless, it is evident that the slope of the dotted line in Fig. 1 approaches that of Henry's law, as the molality approaches zero. Hence the intercept on the axis of ordinates in Fig. 2 should be the same for the solid curve as for the broken line representing Henry's law; or, analytically speaking,

$$\lim_{m \to 0} \left(\frac{p_2}{m} \right) = k' = |p_2^\circ|. \tag{20.8}$$

Thus if the vapor behaves as an ideal gas, the activity of the solute may be expressed as

$$a_2 = \frac{p_2}{p_2^\circ}, \tag{20.9}$$

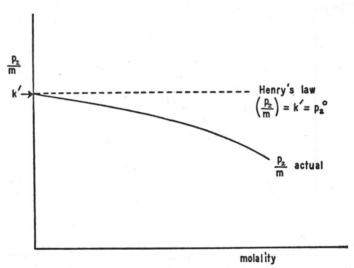

Fig. 2. Determination of partial pressure of solute in its standard state.

an equation analogous to (20.2). It should be emphasized, however, that p_2° is the vapor pressure in the hypothetical one-molal solution, not that of pure solute, solid or supercooled.

II. DISTRIBUTION OF SOLUTE BETWEEN TWO IMMISCIBLE SOLVENTS

If the activity of a solute is known in one solvent, then its activity in another solvent immiscible with the first may be determined from concentration-distribution measurements. As an example, we might consider a rather extreme situation, such as that illustrated in Fig. 3, where the shapes of the fugacity curves are different in two different solvents. The limiting behavior at infinite dilution, Henry's law, is indicated for each solution. It is evident from the graphs that the standard states are different in the two solvents, since the hypothetical one-molal solutions have different fugacities.

If the solute in solution A is in equilibrium with that in solution B, its escaping tendency must be the same in both solvents. Consequently, its fugacity, f_2, at equilibrium must also be identical in both solvents. Nevertheless, the solute will have different activities in solutions A and B, since

$$a_2 = \frac{f_2}{f_2^\circ}, \tag{20.10}$$

and $(f_2^\circ)_A$ differs from $(f_2^\circ)_B$ even though the phases are in equilibrium and f_2 is identical for both solutions.

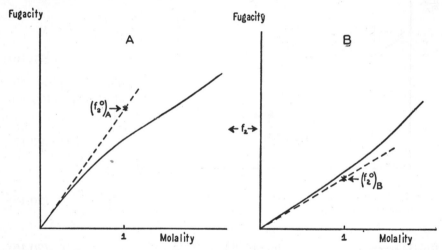

Fig. 3. Comparison of fugacity-molality curves for solute in two immiscible solvents.

If a solute is in equilibrium between two immiscible solvents, and if the activity of the solute is known in one of the solvents, then the activity in the other solvent may be calculated readily. To make this calculation, we recognize that at equilibrium

$$(f_2)_A = (f_2)_B. \tag{20.11}$$

Furthermore, since

$$(a_2)_A = \frac{(f_2)_A}{(f_2^\circ)_A} \tag{20.12}$$

and

$$(a_2)_B = \frac{(f_2)_B}{(f_2^\circ)_B}, \tag{20.13}$$

it follows from equation (20.11), that

$$(a_2)_A (f_2^\circ)_A = (a_2)_B (f_2^\circ)_B, \tag{20.14}$$

or
$$(a_2)_B = (a_2)_A \frac{(f_2^\circ)_A}{(f_2^\circ)_B}. \tag{20.15}$$

To calculate $(a_2)_B$ from $(a_2)_A$, we must find, therefore, the ratio of the fugacities in the respective standard states; but as we have pointed out repeatedly in this and in the preceding chapters, the f_2°'s are numerically equal to the corresponding constants in Henry's law for each solution. Hence our problem is to find the ratio of these k's.

Since for these solutions

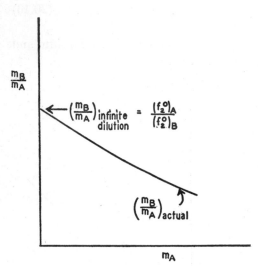

$$\left.\begin{array}{l}(f_2)_A = k_A m_A \\ (f_2)_B = k_B m_B\end{array}\right\} \text{ at } m = 0, \tag{20.16}$$

then for infinitely dilute solutions A and B in equilibrium with each other

$$k_A m_A = k_B m_B. \tag{20.17}$$

Consequently,

Fig. 4. Extrapolation of distribution data to obtain conversion factor for activities.

$$\lim_{m \to 0} \left(\frac{m_B}{m_A}\right) = \frac{k_A}{k_B}. \tag{20.18}$$

Furthermore,
$$\lim_{m \to 0} \left(\frac{m_B}{m_A}\right) = \frac{(f_2^\circ)_A}{(f_2^\circ)_B}. \tag{20.19}$$

Thus, equation (20.15) may be modified to read

$$(a_2)_B = (a_2)_A \left(\frac{m_B}{m_A}\right)_{\text{infinite dilution}}. \tag{20.20}$$

In practice, then, it is necessary to calculate the molality ratio of the solute between the two solvents, plot this ratio as a function of concentration (Fig. 4), and extrapolate to infinite dilution. The limiting value of this ratio may be used then to calculate $(a_2)_B$ for any concentration in solvent B, if $(a_2)_A$ is known for a solution with solvent A, in equilibrium with a solution with solvent B.

III. MEASUREMENTS OF ELECTROMOTIVE FORCE

Potential measurements are primarily of use in the determination of activities of electrolytes, a type of solute which we have yet to consider. Nevertheless, there are several situations in which e.m.f. measurements can be used to obtain information on activities of non-electrolytes.   In particular, the activities of components of alloys can be calculated from cells such as the following for lead amalgam:

$$\text{Pb (amalgam, } N_2'); \quad \text{Pb(CH}_3\text{COO)}_2, \text{ CH}_3\text{COOH}; \quad \text{Pb (amalgam, } N_2).$$
$$(20.21)$$

In this cell, two amalgams with different mole fractions of lead are dipped into a common electrolyte solution containing a lead salt.   The activities of lead in these amalgams can be calculated from e.m.f. measurements with this cell.

If we adopt the usual convention[2] of writing the chemical reaction in the cell as occurring so that electrons will move in the outside conductor from left to right, we find at the left-hand electrode

$$\text{Pb (amalgam, } N_2') = \text{Pb}^{++} + 2e \qquad (20.22)$$

and at the right-hand electrode

$$\text{Pb}^{++} + 2e = \text{Pb (amalgam, } N_2). \qquad (20.23)$$

Since the electrolyte containing Pb^{++} is common to both electrodes, the net reaction is the sum of (20.22) and (20.23):

$$\text{Pb (amalgam, } N_2') = \text{Pb (amalgam, } N_2). \qquad (20.24)$$

The change in free energy accompanying this reaction, at a fixed temperature and pressure, is given by the expression

$$\Delta F = RT \ln f_2 + B(T) - RT \ln f_2' - B(T) \qquad (20.25)$$
$$= RT \ln \frac{f_2}{f_2'},$$

or, in terms of activities,

$$\Delta F = RT \ln \frac{f_2/f_2^\circ}{f_2'/f_2^\circ} = RT \ln \frac{a_2}{a_2'}. \qquad (20.26)$$

---

[2] The reaction which occurs in the cell indicated in (20.21) may be written in either of two ways: Pb (amalgam, N_2') → Pb (amalgam, N_2) or Pb (amalgam, N_2) → Pb (amalgam, N_2').   Thus the cell may represent either of these two transformations, the one, of course, being the reverse of the other.   In quoting values of $\mathcal{E}$, and hence of ΔF, in connection with this cell, it is obviously necessary to specify the reaction to which one is referring.   We shall adopt the convention that the reaction represented by the notation of (20.21) is that which takes place when the left-hand electrode gives electrons to the outer conductor and hence when the right-hand electrode is positive and absorbs electrons from the outer conductor.

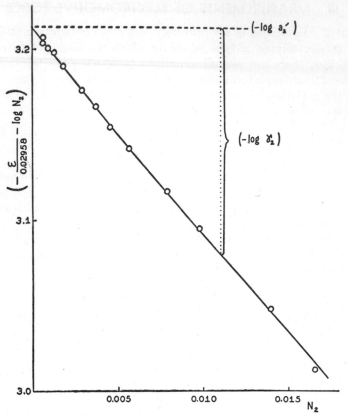

Fig. 5.   Extrapolation of e.m.f. data to obtain constant for use in the calculation of activities in lead amalgams.   Not all of the experimental points have been plotted on this graph.

If concentrations in the alloy are expressed in mole fraction units, then

$$a_2 = N_2 (\gamma_r)_2,$$ (20.27)

and if we introduce the relation between ΔF and the potential, $\mathcal{E}$ [equation (8.126)],

$$\Delta F = -n\mathfrak{F}\mathcal{E},$$ (20.28)

equation (20.26) may be transformed into

$$\mathcal{E} = -\frac{RT}{n\mathfrak{F}} \ln N_2 - \frac{RT}{n\mathfrak{F}} \ln (\gamma_r)_2 + \frac{RT}{n\mathfrak{F}} \ln a_2'.$$ (20.29)

In practice, the composition of one amalgam, for example, that of mole

TABLE 1

E.M.F. OF Pb (AMALGAM, $N_2' = 6.253 \times 10^{-4}$); Pb(CH$_3$COO)$_2$, CH$_3$COOH; Pb (AMALGAM, N_2) at 25 °C*

N_2	$-\varepsilon$, volt	a_2	$(\gamma_r)_2$
0.0006253	0.000000	0.0006123	0.979
.0006302	.000204	.0006222	
.0009036	.004636	.0008784	
.001268	.008911	.001225	0.966
.001349	.009659	.001298	
.001792	.013114	.001699	
.002055	.014711	.001921	0.935
.002744	.018205	.002526	
.002900	.018886	.002666	
.003086	.019656	.002829	0.917
.003203	.020068	.002920	
.003729	.021827	.003350	
.003824	.022111	.003424	0.895
.004056	.022802	.003612	
.004516	.023954	.003951	
.005006	.025160	.004341	0.867
.005259	.025692	.004525	
.005670	.026497	.004817	
.006085	.027256	.005111	0.840
.006719	.028340	.005560	
.007858	.029951	.006302	
.007903	.030010	.006333	0.801
.008510	.030771	.006717	
.009737	.032062	.007428	
.01125	.033437	.008274	0.735
.01201	.033974	.008616	
.01388	.035226	.009508	
.01406	.035323	.009574	0.681
.01456	.035609	.009793	
.01615	.036375	.01040	
.01650 (satd)	.036394	.01060	

*M. M. Haring, M. R. Hatfield, and P. P. Zapponi, *Trans. Electrochem. Soc.*, **75**, 473 (1939).

fraction N_2', may be kept constant and the potentials measured for a series of cells in which N_2 is varied. A typical series of data for lead amalgam is shown in Table 1.

To obtain values of a_2 or of $(\gamma_r)_2$ from these data, we rearrange equation (20.29) to read

$$\varepsilon + \frac{RT}{n\mathfrak{F}} \ln N_2 = + \frac{RT}{n\mathfrak{F}} \ln a_2' - \frac{RT}{n\mathfrak{F}} \ln (\gamma_r)_2. \qquad (20.30)$$

From the data in Table 1 we can evaluate the left-hand side of equation (20.30). A graph is made then of this quantity (or a related one) vs. N_2 (Fig. 5). In view of our definition of activity on a mole fraction basis, it follows that as

$$N_2 \to 0, \tag{20.31}$$

$$(\gamma_r)_2 \to 1. \tag{20.32}$$

Consequently, we may write

$$\lim_{N_2 \to 0} \left[\frac{\varepsilon}{2.303 \ (RT/n\mathfrak{F})} + \log N_2 \right] = \log a_2'. \tag{20.33}$$

This limit is obtained readily by an extrapolation such as is shown in Fig. 5.

Once a_2' is known, it is a simple matter to calculate a_2 at various mole fractions from the equation

$$\varepsilon - \frac{RT}{n\mathfrak{F}} \ln a_2' = - \frac{RT}{n\mathfrak{F}} \ln a_2. \tag{20.34}$$

The results obtained for the lead amalgams of the cell in (20.21) are assembled in Table 1. A few typical activity coefficients have been calculated also. In connection with the variation of $(\gamma_r)_2$ with concentration, it may be of interest to note, in addition, the graphical significance of $\log (\gamma_r)_2$ (Fig. 5).

IV. DETERMINATION OF THE ACTIVITY OF ONE COMPONENT FROM KNOWN VALUES OF THE ACTIVITY OF THE OTHER

A. Fundamental relations. Since we are attempting to obtain the properties of one component of a binary solution from those of the other, we must start from an equation expressing the interdependence of the behaviors of the two components. In the general discussion of partial molal quantities in Chapter 13, we arrived at such a relationship, for constant P and T:

$$n_1 \, d\bar{G}_1 + n_2 \, d\bar{G}_2 = 0. \tag{20.35}$$

Since the activity is derived fundamentally from the partial molal free energy, let us convert equation (20.35) into the specific form it acquires for $\bar{F}$:

$$n_1 \, d\bar{F}_1 + n_2 \, d\bar{F}_2 = 0. \tag{20.36}$$

In view of the relation between $\bar{F}$ and a given by equation (19.3), we may write, at constant temperature,

$$d\bar{F} = RT \, d \ln a. \tag{20.37}$$

Consequently, equation (20.36) becomes

$$n_1 \, d \ln a_1 + n_2 \, d \ln a_2 = 0. \tag{20.38}$$

Similarly, if we had started with the equation corresponding to (20.35) but in terms of mole fractions,

$$N_1 \, d\bar{G}_1 + N_2 \, d\bar{G}_2 = 0, \tag{20.39}$$

we would have obtained the expression

$$N_1 \, d \ln a_1 + N_2 \, d \ln a_2 = 0. \tag{20.40}$$

B. Calculation of activity of solvent from that of solute. If adequate data are available for the activity of the solute, the activity of the solvent may be obtained by rearrangement and integration of equation (20.40).

$$d \ln a_1 = -\frac{N_2}{N_1} \, d \ln a_2. \tag{20.41}$$

Before this equation is integrated, however, it is desirable, in the interests of greater precision,[3] to convert it into a corresponding one for activity coefficients.

In a binary solution, in which $N_1 + N_2 = 1$,

$$dN_1 = -dN_2 \tag{20.42}$$

and, consequently,

$$\frac{N_1}{N_1} \, dN_1 = -\frac{N_2}{N_2} \, dN_2, \tag{20.43}$$

or

$$d \ln N_1 = -\frac{N_2}{N_1} \, d \ln N_2. \tag{20.44}$$

Subtraction of equation (20.44) from (20.41) leads to the expression

$$d \ln \frac{a_1}{N_1} = -\frac{N_2}{N_1} \, d \ln \frac{a_2}{N_2}, \tag{20.45}$$

which upon integration from a mole fraction of zero to some value N_2 gives

$$\log \frac{a_1}{N_1} = -\int_{N_2=0}^{N_2} \frac{N_2}{N_1} \, d \log \frac{a_2}{N_2}. \tag{20.46}$$

Thus, in practice, it is merely necessary to plot N_2/N_1 vs. $\log \gamma_2$ and to integrate from infinite dilution to any concentration N_2. Since both or-

---

[3] Greater precision can be obtained if a function is used which has a minimum of variation in the important region of a graph.

dinates and abscissas approach zero at infinite dilution, no difficulty is encountered in carrying out the integration required in equation (20.46).

C. Calculation of activity of solute from that of solvent. By analogy with the preceding problem, we may rearrange equation (20.45) to read

$$d \log \frac{a_2}{N_2} = - \frac{N_1}{N_2} d \log \frac{a_1}{N_1}, \tag{20.47}$$

which upon integration gives

$$\log \frac{a_2}{N_2} = - \int_{N_2=0}^{N_2} \frac{N_1}{N_2} d \log \frac{a_1}{N_1}. \tag{20.48}$$

In practice, however, the required integration is difficult to carry out, particularly by graphical methods. The basic reason for the difficulty, as can be seen readily in equation (20.48), lies in the fact that N_1/N_2 approaches infinity in the infinitely dilute solution.

One method of overcoming this difficulty in computation utilizes the following procedure. Instead of setting the lower limit in the integration of equation (20.47) at infinite dilution, let us use a temporary lower limit of a finite concentration, N_2'. Thus we obtain, in place of equation (20.48),

$$\log \left(\frac{a_2/N_2}{a_2'/N_2'} \right) = - \int_{N_2'}^{N_2} \frac{N_1}{N_2} d \log \frac{a_1}{N_1}. \tag{20.49}$$

The integration required in equation (20.49) offers no computational difficulties, in contrast to that required in (20.48).

Using equation (20.49), we can now obtain precise values of the ratio $\left(\frac{a_2/N_2}{a_2'/N_2'} \right)$ for a series of values of N_2. To obtain a_2/N_2, we need to determine a_2'/N_2'.

This determination can be carried out readily as follows. It is evident that

$$\left(\frac{a_2}{N_2} \right) = \frac{(a_2/N_2)}{(a_2/N_2)*}, \tag{20.50}$$

since $(a_2/N_2)*$, representing this ratio at infinite dilution, must be 1 in view of our method of defining the standard state for the solute. It follows, then, that

$$\left(\frac{a_2}{N_2} \right) = \frac{(a_2/N_2)}{(a_2'/N_2')} \frac{(a_2'/N_2')}{(a_2/N_2)*}, \tag{20.51}$$

or $$\left(\frac{a_2}{N_2} \right) = \frac{(a_2/N_2)}{(a_2'/N_2')} \Big/ \frac{(a_2/N_2)*}{(a_2'/N_2')} = \frac{\gamma_2}{\gamma_2'} \Big/ \frac{\gamma_2*}{\gamma_2'}. \tag{20.52}$$

In view of equation (20.52), we can see that absolute values of (a_2/N_2) can be obtained if the relative activity coefficients, γ_2/γ_2', are extrapolated to infinite dilution, so that γ_2^*/γ_2' can be evaluated. Such an extrapolation can be made, as is indicated in Fig. 6, with a high degree of precision. In this manner, the uncertainties inherent in a direct integration of equation (20.48) are avoided.

A typical example of this type of problem is considered in Exercise 2.

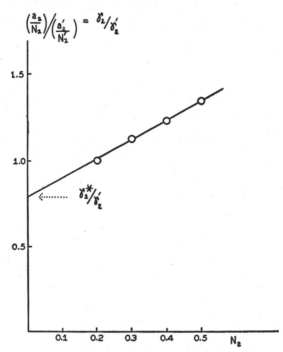

Fig. 6. Extrapolation of relative activity coefficients to obtain γ_2^*/γ_2', for calculation of absolute activity coefficients. In this particular problem, the reference state has been taken at $N_2 = 0.2$.

V. MEASUREMENTS OF FREEZING POINTS

Perhaps the method of most general applicability for determining activities of non-electrolytes in solutions is that based on measurements of the lowering of the freezing point of a solution. Since measurements are made of the properties of the solvent, activities of the solute must be calculated by methods based on principles described in the preceding section.

Elaborate procedures have been developed for obtaining activity coefficients from freezing-point and thermochemical data. To avoid duplication, however, the details will not be outlined at this point, since a completely general discussion, applicable to solutions of electrolytes as well as non-electrolytes, will be presented in the following chapter.

Exercises

1. The following data on the partial pressures (in mm Hg) of toluene and of acetic acid at 69.94°C have been taken from the International Critical Tables, Volume III, pages 217, 223, and 288. For the purposes of this exercise, assume that the partial pressure of each component is identical with its fugacity.

N_1 (toluene)	N_2 (acetic acid)	p_1 (toluene)	p_2 (acetic acid)
0.0000	1.0000	0	136
.1250	.8750	54.8	120.5
.2310	.7690	84.8	110.8
.3121	.6879	101.9	103.0
.4019	.5981	117.8	95.7
.4860	.5140	130.7	88.2
5349	.4651	137.6	83.7
.5912	.4088	154.2*	78.2
.6620	.3380	155.7	69.3
.7597	.2403	167.3	57.8
.8289	.1711	176.2	46.5
.9058	.0942	186.1	30.5
.9565	.0435	193.5	17.2
1.0000	.0000	202	0

*This is evidently a typographical error. The correct figure is near 145.2.

(a) Draw a graph of f_1 vs. N_1. Indicate Raoult's law by a dotted line.

(b) Draw a graph of f_2 vs. N_2. Indicate Raoult's law and Henry's law, each by a dotted line.

(c) Find the constant in Henry's law for acetic acid in toluene solutions by extrapolating a graph of f_2/N_2 vs. N_2 to infinite dilution.

(d) Calculate the activities and activity coefficients of acetic acid on the basis of an f_2° established from Henry's law. Plot these values vs. N_2.

(e) Calculate the activities and activity coefficients of acetic acid when the pure liquid is taken as the standard state. Plot these values on the same graph as in Part (d).

(f) Calculate the activities and activity coefficients of toluene, the solvent in these solutions.

2. The following partial pressures (in mm Hg) have been measured for the acetone component of an acetone-chloroform solution [J. von Zawidzki, Z. *physik. Chem.*, **35**, 129 (1900)]:

N_1 (acetone)	p_1 (acetone)
1.000	344.5
.9405	322.9
.8783	299.7
.8165	275.8
.7103	230.7
.6387	200.3
.5750	173.7

(a) Calculate the activity coefficients, a_1/N_1, for the acetone.

(b) Make a graph of N_1/N_2 vs. log a_1/N_1.

(c) Find $\left(\dfrac{a_2/N_2}{a_2'/N_2'}\right)$ at values of N_2 of 0.2, 0.3, 0.4, and 0.5. Let the solution of $N_2 = 0.2$ be the temporary reference solution.

(d) Find values of γ_2 at the same mole fractions, N_2, as in Part (c).

3. Potentials[4] of lithium amalgam electrodes in the cell

Li (amalgam, N_2), LiCl in acetonitrile, Li (amalgam, $N' = 0.0003239$) at 25°C are given below.

N_2 (lithium)	ε
0.0003239	0.0000
.001846	.0458
.002345	.0517
.004218	.0684
.008779	.0894
.01300	.1006
.02265 (satd)	.1189

(a) Plot ε vs. N_2. Note the "limit" as N_2 approaches zero. $\longrightarrow \infty$

(b) Calculate the activities and activity coefficients of lithium in the amalgams.

4. Calculate the e.m.f. of the following cells at 25°C. In each case consider the activity coefficient as unity.

(a) H_2 (76 cm of Hg), HCl (Xm), H_2 (70 cm of Hg).

(b) H_2 (70 cm of Hg), HCl (Xm), H_2 (76 cm of Hg).

(c) Tl (in amalgam, $N_{Tl} = 0.0001$), TlCl (Xm), Tl (in amalgam, $N_{Tl} = 0.001$).

(d) Tl (in amalgam, $N_{Tl} = 0.001$), Tl_2SO_4 (Xm), Tl (in amalgam, $N_{Tl} = 0.0001$).

(e) Cd (in amalgam, $N_{Cd} = 0.001$), $CdCl_2$ (Xm), Cd (in amalgam, $N_{Cd} = 0.0001$).

(f) Cd (in amalgam, $N_{Cd} = 0.0001$), $CdSO_4$ (Xm), Cd (in amalgam, $N_{Cd} = 0.001$).

---

[4] G. Spiegel and H. Ulich, Z. physik. Chem., **A178**, 187 (1937).

CHAPTER 21

Activity and Activity Coefficients: Strong Electrolytes

I. DEFINITIONS AND STANDARD STATES FOR DISSOLVED ELECTROLYTES

Unless one has had some introduction to the subject, it may be somewhat of a surprise to learn that it is necessary to consider electrolytes in a special category in the discussion of activity coefficients. From the considerations of standard states outlined in Chapter 19, it would seem to make no difference in the selection of standard states whether one is dealing with dissociable or nondissociable solutes. In principle, all one does to establish the standard state is the following: make a graph of the fugacity as a function of the molality; determine the limiting slope as m approaches zero; and extrapolate this limiting slope, Henry's law, back to unit molality. This point of intersection on the line for Henry's law is the standard fugacity for the solute. Offhand, it would seem simple enough to carry out the same type of extrapolation, in principle, for electrolytes as well as for non-electrolytes.

In actual practice a difficulty arises because of the fact that a graph of the relative fugacity of a strong electrolyte vs. the molality looks like the example illustrated in Fig. 1 for aqueous solutions of HCl. The experimental details which permit one to obtain the data for such a graph will be discussed later in this chapter. For the present, it will suffice to emphasize the shape of the graph, in particular the fact that as the molality approaches zero, the fugacity approaches zero with zero slope; that is (if K represents the proportionality constant),

$$\lim_{m \to 0} \frac{d(Kf)}{dm} = 0. \tag{21.1}$$

Clearly, in this situation it is impossible to extrapolate the limiting slope back to $m = 1$ to obtain the fugacity (or relative fugacity) of the "hypothetical 1-molal solution" which represents the standard state; for the only value one would obtain is zero, and hence the activity in any real solution, being f/f°, would always be infinite. Evidently, then, the method of choosing standard states for solutes which are non-electrolytes is not possible for electrolytic solutions.

A. Uni-univalent electrolytes. Since a plot of the relative fugacity of an electrolyte, such as aqueous HCl, vs. the first power of the molality does not give a limiting slope different from zero, we might examine graphs in which the axes are other functions of these variables. In such attempts, we should soon observe that a plot of the relative fugacity vs. the *square* of the molality, for a uni-univalent electrolyte, does approach a finite limiting slope (*not zero*) as the molality approaches zero. Thus, in Fig. 2 the same data are plotted as in Fig. 1, but the square of the molality has been taken as the abscissa. It is apparent immediately that the slope approaches a finite limiting value not zero.[1] Apparently, then, for a uni-univalent electrolyte, in the limit of infinite dilution

$$f_2 \propto m^2. \tag{21.2}$$

We may say, therefore, that the limiting law (substituting for Henry's law) for such a solute should take the form

$$f_2 = km^2. \tag{21.3}[2]$$

Having found a graph which does have a finite limiting slope, we can proceed to define the standard state for this uni-univalent electrolyte in a fashion analogous to that used for non-electrolytes. Once again, we recognize that if the limiting law, in the form of equation (21.3), is approached in very dilute solutions, then all free-energy changes involving the solute can be calculated in these dilute solutions from fugacities as expressed by this law. Thus,

$$\Delta \bar{F}_2 = RT \ln f_2' - RT \ln f_2 \tag{21.4}$$

$$= RT \ln \frac{(m')^2}{m^2}. \tag{21.5}$$

Hence the activity for the uni-univalent electrolyte, defined again as a ratio of fugacities,

$$a_2 = \frac{f_2}{f_{2(\text{standard state})}} = \frac{f_2}{f_2^{\circ}}, \tag{21.6}$$

---

[1] A precise estimate of this limiting slope, however, requires a much more expanded graph in the region of small concentrations.

[2] A similar conclusion may be reached from elementary considerations of the application of the law of mass action to the equilibrium

$$HCl\ (g) = H^+\ (aq) + Cl^-\ (aq)$$

in very dilute solutions. The equilibrium constant is

$$\frac{(H^+)\ (Cl^-)}{(HCl)} = \frac{(m)(m)}{p_{HCl}} = K.$$

Hence
$$f \cong p = \frac{1}{K} m^2 = K'm^2.$$

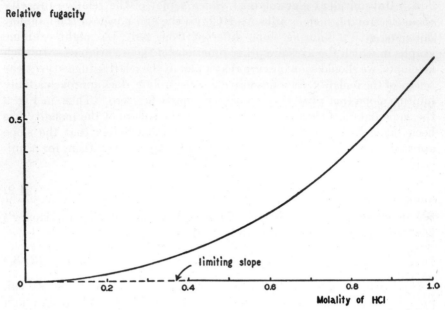

Fig. 1.   Relative fugacity versus molality for aqueous hydrochloric acid.   [Based on data from G. N. Lewis and M. Randall, *Thermodynamics*, page 336.]

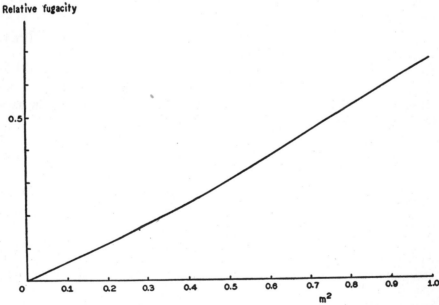

Fig. 2.   Relative fugacity versus the square of the molality, for aqueous hydrochloric acid.   [Data same as for Fig. 1.]

should also give an equation of the form of (21.5) for the free-energy change in very dilute solutions. Therefore, for very dilute solutions it is convenient to have

$$\frac{a_2}{m^2} \to 1 \text{ as } m \to 0. \tag{21.7}$$

Nevertheless, the activity is still defined by equation (21.6). Consequently,

$$\lim_{m \to 0} \left(\frac{a_2}{f_2/f_2^\circ} \right) = \lim_{m \to 0} \left(\frac{a_2}{km^2/f_2^\circ} \right) = 1. \tag{21.8}$$

Equations (21.7) and (21.8) can be satisfied simultaneously if the standard state is chosen so that its fugacity, f_2°, shall be numerically equal to k, the constant in the limiting law for the uni-univalent electrolyte. The value of k can be determined by extrapolation of the limiting law back to $m^2 = 1$. Thus the standard state for a uni-univalent electrolyte is the *hypothetical* one which would have a fugacity, f_2°, equal to k at unit molality if the solute obeyed the limiting law, as expressed by equation (21.3), up to this concentration.

Although we have spoken of the experimental behavior of a uni-univalent electrolyte in the preceding discussion, we have not introduced any assumptions as to dissociation in order to derive any of the preceding relations. As far as thermodynamics is concerned, no such details need be considered. We can take the limiting law in the form of equation (21.3) merely as an expression of the experimental facts and derive the thermodynamic relations which we desire. Nevertheless, in view of the general applicability of the ionic theory, it is desirable to express some of our definitions and results in a form which takes cognizance of this theory.

Thus, we can associate the occurrence of the molality as a square term in equation (21.3) with the presence of two particles, and we may even write

$$f_2 = k(m_+)(m_-) = km^2, \tag{21.9}$$

where (m_+) represents the molality of the cation and (m_-) that of the anion of a completely dissociated uni-univalent electrolyte. The second part of equation (21.9) follows from the fact that

$$m = m_+ = m_-. \tag{21.10}$$

Similarly, it is frequently helpful to speak of individual ion activities, a_+ of the cation, a_- of the anion. By analogy with preceding systems, these ion activities should be defined, for convenience, in such a fashion that they approach the molality of the ion at infinite dilution:

$$a_+/m_+ \rightarrow 1 \text{ as } m \rightarrow 0, \tag{21.11}$$

$$a_-/m_- \rightarrow 1 \text{ as } m \rightarrow 0. \tag{21.12}$$

By substitution in equation (21.9), it becomes evident that in very dilute solutions of a uni-univalent electrolyte

$$f_2 = k(a_+)(a_-). \tag{21.13}$$

Since

$$\frac{f_2}{k} = \frac{f_2}{f_2^{\circ}} = a_2, \tag{21.14}$$

we obtain the following expression:

$$a_2 = (a_+)(a_-). \tag{21.15}$$

Although, so far, equation (21.15) has been demonstrated to be valid only for infinitely dilute solutions, we can make it valid for all concentrations *by definition;* because if we define a_+ and a_- so that the product $(a_+)(a_-)$ gives a_2, the individual ion activities will have the properties we desire, namely, their product is the activity of the solute, and each one approaches the molality as m approaches zero.

No thermodynamic method has yet been devised to enable one to determine individual ion activities, neither a_+ nor a_-. Since each of these approaches m, for a uni-univalent electrolyte, at infinite dilution, a_+ must also equal a_- at infinite dilution. At any finite concentration, however, the deviation between a_+ and a_- is unknown, although it may be negligibly small in dilute solutions. Nevertheless, in a solution of any concentration, the *mean* activity of the ions can be determined. By the mean activity, $a_{\pm}$, we refer to the geometric mean, which for a uni-univalent electrolyte is defined by the equation

$$a_{\pm} = (a_+a_-)^{\frac{1}{2}} = a_2^{\frac{1}{2}}. \tag{21.16}$$

In correspondence to the treatment of non-electrolytes, we may also define an *activity coefficient,* γ_i, for each ion in an electrolyte solution. For each ion of a uni-univalent electrolyte we may make the following definitions:

$$\gamma_+ = \frac{a_+}{m_+}, \tag{21.17}$$

$$\gamma_- = \frac{a_-}{m_-}. \tag{21.18}$$

These individual-ion activity coefficients have the desired property of approaching unity at infinite dilution, since each ratio a_i/m_i approaches unity. However, individual-ion activity coefficients, like the corresponding activities, cannot be determined. It is customary, therefore, to deal with the mean activity coefficient, $\gamma_\pm$, which can be related to measurable quantities as follows, for a uni-univalent electrolyte:

$$\gamma_\pm = (\gamma_+\gamma_-)^{1/2} = \left(\frac{a_+}{m_+}\frac{a_-}{m_-}\right)^{1/2},\qquad(21.19)$$

$$\gamma_\pm = \frac{a_\pm}{m} = \frac{a_2^{1/2}}{m}.\qquad(21.20)$$

B. Multivalent electrolytes.

1. *Symmetrical salts.* For salts in which anion and cation have the same valence, activities and related quantities are defined in exactly the same way as for uni-univalent electrolytes. For $MgSO_4$, for example, a finite limiting slope would be obtained if the relative fugacity were plotted against m^2. Furthermore, m_+ equals m_-. Consequently, the treatment of symmetrical salts does not differ from that just described for uni-univalent electrolytes.

2. *Unsymmetrical salts.* For salts which do not contain equal numbers of positive and negative ions, some further discussion is necessary to remove possible ambiguities in the meanings of various terms associated with activities. As an example, let us consider a salt, such as $BaCl_2$, which dissociates into one cation and two anions. By analogy with the case of a binary electrolyte, we may define the ion activities in terms of the expression

$$a_2 = (a_+)(a_-)(a_-) = a_+(a_-)^2.\qquad(21.21)$$

The mean ionic activity, $a_\pm$, in this case also is taken to be the geometric mean of the individual ion activities:

$$a_\pm = [a_+(a_-)^2]^{1/3} = a_2^{1/3}.\qquad(21.22)$$

So far, the definitions parallel those for uni-univalent salts. In turning to mean activity coefficients, $\gamma_\pm$, however, a factor enters in the treatment of $BaCl_2$ which did not occur previously. We should like to have individual ion activities obey the following relations in a very dilute solution of molality m:

$$a_+ = m_+ = m,\qquad(21.23)$$

$$a_- = m_- = 2m.\qquad(21.24)$$

Since equation (21.22) is to be valid at all concentrations, including the very dilute ones, the insertion of equations (21.23) and (21.24) into (21.22) leads to the following expression for very dilute solutions:

$$a_\pm = [(m)(2m)^2]^{1/3} = (4)^{1/3}m.\qquad(21.25)$$

Obviously, then, if we defined $\gamma_\pm$ as the ratio $a_\pm/m$, where m is the molality of the salt, this activity coefficient would not approach unity at infinite dilution, but another constant, $4^{1/3}$. Although there may be no fundamental objection to such a situation, it would not be as convenient as one in which $\gamma_{\pm}$ approached unity for every type of salt.

To fulfill this latter requirement, therefore, we shall define γ_+ and γ_- in such a way that $\gamma_{\pm}$ for a ternary electrolyte, such as $BaCl_2$, is given by the equation

$$\gamma_{\pm} = \frac{a_{\pm}}{4^{1/3}m}. \tag{21.26}$$

With this definition, $\gamma_{\pm}$ will approach unity in infinitely dilute solutions.

There is another way of viewing equation (21.26) which will make it look more nearly like that for the uni-univalent electrolyte. The basic reason for the difference in the case of the ternary $BaCl_2$ is the fact that the molality of the cation is not equal to that of the anion, but rather $m_- = 2m_+$. Thus, a definition of the mean activity coefficient, $\gamma_{\pm}$, as the ratio $a_{\pm}/m$, where m is the molality of the salt, is one-sided, since $m = m_+$ but m is *not* equal to m_-. This difficulty can be resolved, however, by defining a *mean molality* for the ions in a manner analogous to that used for the activity. Thus, for a salt like $BaCl_2$ we may write

$$m_{\pm} = [(m_+)(m_-)^2]^{1/3} \tag{21.27}$$

$$= [(m)(2m)^2]^{1/3} = 4^{1/3}m. \tag{21.28}$$

If we now define the mean activity coefficient in terms of the mean ion activity and the mean ion molality, respectively, we shall have the expression

$$\gamma_{\pm} = \frac{a_{\pm}}{m_{\pm}} = \frac{a_2^{1/3}}{m_{\pm}}, \tag{21.29}$$

which looks formally equivalent to equation (21.20) and yet has the same properties as equation (21.26).

One more comparison of the ternary electrolyte with the binary may be instructive. For the uni-univalent electrolyte, the standard fugacity was determined from a calculation of the limiting slope in a plot of relative fugacity vs. m^2 (Fig. 2) and the utilization of this limiting slope to extrapolate back to $m = 1$. The point on the line for the limiting law corresponding to $m = 1$ is taken as the standard fugacity. For a ternary electrolyte, it is necessary to plot the relative fugacity against m^3 in order to obtain a limiting slope, at infinite dilution, not equal to zero. In other words, for very dilute solutions of a ternary electrolyte, the empirical situation can be represented by an equation of the form

$$f_2 = km^3. \tag{21.30}$$

Again k may be thought of as the limiting law constant for such a dissociating solute (Fig. 3).   Again, also, k may be evaluated by extrapolation of the line of the limiting slope back to unit molality (Fig. 3).

In this case, however, we shall not take this point on the line for the limiting law as the standard fugacity.   If we did, a_2 in very dilute solutions would be given by the expression

$$a_2 = \frac{f_2}{f_2^{\circ}} = \frac{f_2}{k} = m^3.$$
(21.31)

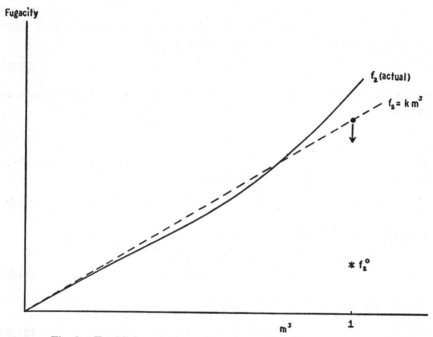

Fig. 3.   Establishment of standard state for a ternary electrolyte.

Such a consequence would conflict with our previous definitions of a_+ and a_-, which led to equation (21.25) for $a_{\pm}$ and hence to the following expression for a_2 in very dilute solutions:

$$a_2 = (a_{\pm})^3 = 4m^3.$$
(21.32)

Since it is more convenient in practice to retain the properties of $a_{\pm}$ specified in equations (21.22)–(21.24), we must choose our standard fugacity so that equation (21.32), rather than (21.31), should be valid.   This choice can be made very readily if we take a hypothetical point, at unit molality, for which the fugacity is one-fourth of the limiting law constant,

k (see Fig. 3). Thus, in place of equation (21.31), we should write

$$a_2 = \frac{f_2}{f_2^\circ} = \frac{f_2}{k/4}. \tag{21.33}$$

This equation is applicable at all concentrations, and in view of equation (21.30) it reduces to equation (21.32) in the infinitely dilute solution.

Once again, we can see that it may be convenient to choose our standard state in different ways when we face different problems.

3. *General case.* Without going into any detailed discussion, we shall find that extension of the preceding considerations to the general electrolyte can be made readily as follows. If an electrolyte $A_{\nu_+} B_{\nu_-}$ dissociates into ν_+ positive ions and ν_- negative ions, its fugacity, f_2, in very dilute solutions is given by the following expression:

$$f_2 = k m^{(\nu_+ + \nu_-)} = k m^\nu. \tag{21.34}$$

The individual ion activities are related to the activity of the solute, a_2, by the equation

$$a_2 = (a_A)^{\nu_+} (a_B)^{\nu_-}, \tag{21.35}$$

where a_A represents the activity of the cation and a_B that of the anion. The mean ionic activity, $a_\pm$, is defined as

$$a_\pm = a_2^{1/(\nu_+ + \nu_-)} = [(a_A)^{\nu_+} (a_B)^{\nu_-}]^{1/\nu}; \tag{21.36}$$

the mean ionic molality, $m_\pm$, as

$$\begin{aligned} m_\pm &= [(\nu_+ m)^{\nu_+} (\nu_- m)^{\nu_-}]^{1/\nu} \\ &= m[(\nu_+)^{\nu_+} (\nu_-)^{\nu_-}]^{1/\nu}; \end{aligned} \tag{21.37}$$

and the mean activity coefficient, $\gamma_\pm$, as

$$\gamma_\pm = \frac{a_\pm}{m_\pm} = [(\gamma_+)^{\nu_+} (\gamma_-)^{\nu_-}]^{1/\nu}. \tag{21.38}$$

As an aid in grasping the significance of these equations, Table 1 has been prepared to illustrate their application to several special cases. The student should convince himself of the self-consistency of these expressions.

C. Mixed electrolytes. So long as attention is focused on a single electrolyte in a mixture, no further modifications are necessary in the definitions and concepts described in the preceding sections. The significance of mean molality remains to be illustrated.

1. *Mean ionic molality.* In a solution of mixed electrolytes common ions may be present, which must be recognized in calculations of the mean ionic molality of any given component. For example, in a solution of 0.1 m NaCl and 0.2 m MgCl$_2$, the mean ionic molality, $m_\pm$, for NaCl is

TABLE 1

THERMODYNAMIC FUNCTIONS FOR DISSOLVED SOLUTES

	Sucrose	NaCl	Na_2SO_4	$AlCl_3$	$MgSO_4$	$A_{\nu_+}B_{\nu_-}$
Limiting law: $f_2 =$	km	km^2	km^3	km^4	km^2	$km^{(\nu_+ +\nu_-)}$
$a_2 =$	a	$(a_+)(a_-)$	$(a_+)^2(a_-)$	$(a_+)(a_-)^3$	$(a_+)(a_-)$	$(a_+)^{\nu_+}(a_-)^{\nu_-}$
$a_\pm =$	$\cdots\cdots$	$[(a_+)(a_-)]^{\frac{1}{2}}$	$[(a_+)^2(a_-)]^{\frac{1}{3}}$	$[(a_+)(a_-)^3]^{\frac{1}{4}}$	$[(a_+)(a_-)]^{\frac{1}{2}}$	$[(a_+)^{\nu_+}(a_-)^{\nu_-}]^{1/(\nu_+ +\nu_-)}$
$m_\pm =$	$\cdots\cdots$	m	$4^{\frac{1}{3}}m$	$3^{\frac{3}{4}}m$	m	$[(\nu_+)^{\nu_+}(\nu_-)^{\nu_-}]^{1/(\nu_+ +\nu_-)}\,m$
$\gamma_\pm =$	$\cdots\cdots$	$\dfrac{a_\pm}{m_\pm}$	$\dfrac{a_\pm}{m_\pm}$	$\dfrac{a_\pm}{m_\pm}$	$\dfrac{a_\pm}{m_\pm}$	$\dfrac{a_\pm}{m_\pm}$

$$m_{\pm}(\text{NaCl}) = [(m_{\text{Na}^+})(m_{\text{Cl}^-})]^{\frac{1}{2}}$$
$$= [(0.1)(0.5)]^{\frac{1}{2}} = 0.224m. \qquad (21.39)$$

For the MgCl_2, in turn,

$$m_{\pm}(\text{MgCl}_2) = [(m_{\text{Mg}^{++}})(m_{\text{Cl}^-})^2]^{\frac{1}{3}}$$
$$= [(0.2)(0.5)^2]^{\frac{1}{3}} = 0.368m. \qquad (21.40)$$

Other problems can be solved readily from these examples. It is merely necessary to keep in mind in dealing with mixed electrolytes that more than one component may contribute to the concentration of a single species of ion.

2. *Ionic strength.* We shall be interested also in the combined effects of several electrolytes on the activity of a single one. For this purpose, it has been found convenient, from both empirical experience and electrostatic theory of electrolyte solutions, to introduce a quantity known as the *ionic strength.* This quantity allows one to compare the effects of salts of different valence type.

The contribution of each ion to the ionic strength, μ, is obtained by multiplication of the molality of the ion by the square of its valence. One-half of the sum of these contributions for all the ions present is defined as the ionic strength. Thus

$$\mu = \tfrac{1}{2} \sum_i m_i z_i^2, \qquad (21.41)[3]$$

where m_i is the molality and z_i the valence or charge of the ion. The factor $\tfrac{1}{2}$ has been included so that the ionic strength will correspond to the molality for a uni-univalent salt. Thus, for NaCl

$$\mu = \tfrac{1}{2}[m(1)^2 + m(1)^2] = m. \qquad (21.42)$$

On the other hand, for BaCl_2

$$\mu = \tfrac{1}{2}[m(2)^2 + 2m(1)^2] = 3m, \qquad (21.43)$$

since $m_+ = m$ and $m_- = 2m$, where m is the molality of the salt. It should be a simple matter for the student to verify the corresponding relations for other salts in Table 2.

II. DETERMINATION OF ACTIVITIES OF STRONG ELECTROLYTES

All of the methods used in the study of non-electrolytes can be applied *in principle* to the determination of activities of electrolyte solutes also. In practice, however, several methods are very difficult to apply to electrolytes because of the impracticability of obtaining data for solutions sufficiently dilute to allow the necessary extrapolations to infinite dilution.

---

[3] In many publications the ional concentration, Γ, is used in place of the ionic strength, μ. These two quantities are related in a very simple fashion: $\Gamma = 2\mu$.

Thus, some data are available for the vapor pressures of the halogen acids in their aqueous solutions; but even in these relatively favorable examples, these measurements by themselves do not permit one to determine the activities of the solutes because significant data cannot be accumulated at concentrations below 4-molal.

TABLE 2

RELATIONS BETWEEN IONIC STRENGTH AND MOLALITY

Salt	NaCl	Na$_2$SO$_4$	AlCl$_3$	MgSO$_4$	$A_{\nu_+}B_{\nu_-}$ *
μ =	m	$3m$	$6m$	$4m$	$\frac{1}{2}\nu_-\left(\dfrac{\nu_-}{\nu_+}+1\right)z_-^2 m$

* z_- refers to the charge on the anion, B, of the salt $A_{\nu_+}B_{\nu_-}$.

Activity data for electrolytes are usually obtained by one of three independent experimental methods: (1) electromotive-force measurements; (2) solubility determinations; and (3) freezing-point determinations. We shall consider in some detail the procedure involved in each of these cases.

Much information on activities of electrolytes has also been obtained recently by a technique known as the *isopiestic method*, in which a comparison is made of the concentrations of two solutions with equal vapor pressures. However, we shall not discuss this approach since its principle can be understood readily once the basis of the freezing-point method has been mastered.

A. Electromotive-force measurements. The electromotive force (e.m.f.) of a cell is a quantity whose magnitude depends upon the free-energy change for the reaction which occurs in the cell. In general, then, the e.m.f. of the cell depends upon the activity of the electrolyte in the cell. For example, in the cell composed of a hydrogen electrode and a silver-silver chloride electrode, represented by the notation

$$H_2, \; HCl \; (m), \; AgCl, \; Ag, \qquad (21.44)$$

the e.m.f. depends upon the free-energy change in the reaction[4]

$$\tfrac{1}{2}H_2 + AgCl = HCl \; (m) + Ag. \qquad (21.45)$$

This free-energy change, in turn, depends upon the activity of the aqueous hydrochloric acid, as well as upon the activities of the other chemical reactants and products.

---

[4] The reaction which occurs in the cell in (21.44) may be written in either of two ways: $\tfrac{1}{2}H_2 + AgCl \rightarrow HCl + Ag$ or $HCl + Ag \rightarrow \tfrac{1}{2}H_2 + AgCl$. Thus, the cell may represent either of these two transformations, the one, of course, being the reverse of the other. In quoting values of ε, and hence of ΔF, in connection with cell (21.44), it is obviously necessary to specify the reaction to which one is referring. We shall adopt the convention that the reaction represented by the notation of (21.44) is that which occurs when the left-hand electrode gives electrons to the outer conductor, and hence when the right-hand electrode is positive and absorbs electrons from the outer conductor.

To obtain an explicit relation between these quantities, we shall start with the general expression relating free-energy changes and activities [equation (19.37)]. For reaction (21.45)

$$\Delta F = \Delta F^\circ + RT \ln \frac{a_{HCl} a_{Ag}}{a_{H_2}^{\frac{1}{2}} a_{AgCl}}. \tag{21.46}$$

Since the e.m.f. is measured under conditions such that the reaction is opposed by the potentiometer and hence occurs reversibly,

$$\Delta F = -n\mathfrak{F}\mathcal{E}. \tag{21.47}$$

Therefore, equation (21.46) may be transformed to read

$$-n\mathfrak{F}\mathcal{E} = -n\mathfrak{F}\mathcal{E}^\circ + RT \ln \frac{a_{HCl} a_{Ag}}{a_{H_2}^{\frac{1}{2}} a_{AgCl}}, \tag{21.48}$$

or

$$\mathcal{E} = \mathcal{E}^\circ - \frac{RT}{n\mathfrak{F}} \ln \frac{a_{HCl} a_{Ag}}{a_{H_2}^{\frac{1}{2}} a_{AgCl}}. \tag{21.49}$$

With the conditions under which these e.m.f. measurements can be made, the partial pressure of the hydrogen gas may be maintained at one atmosphere, or a correction may be made for small deviations from it. Hence, in view of our convention that the activity of a gas is equal to its fugacity, and since the fugacity of hydrogen does not differ significantly from its partial pressure near one atmosphere, a_{H_2} may be set equal to unity in equation (21.49). Similarly, in view of our convention with respect to the activity of pure solids, a_{AgCl} and a_{Ag} may each be set equal to unity also. Consequently, equation (21.49) may be reduced to

$$\mathcal{E} = \mathcal{E}^\circ - \frac{RT}{n\mathfrak{F}} \ln a_{HCl}. \tag{21.50}$$

Thus, the cell indicated by (21.44) is a very convenient one for determinations of the activity of dissolved hydrochloric acid.

So far, however, we have no value of $\mathcal{E}^\circ$, the potential of the cell when the HCl is also at unit activity. The value measured for $\mathcal{E}$ would depend only upon the state of the system, since ΔF for a reaction is independent of the choice we make of standard states. Hence, it is evident that the value we calculate for $\mathcal{E}^\circ$ depends upon the value we assign to a_{HCl} (that is, upon how we define the standard state), since the right-hand side of equation (21.50) contains only two terms, the sum of which must be a fixed number, $\mathcal{E}$, for any specified solution of HCl (for example, 0.75955 volt at 0.005376-molal). To evaluate $\mathcal{E}^\circ$, then, we must introduce our conventions with respect to standard states for electrolytic solutes.

For dissolved HCl we may write

$$a_{HCl} = a_{H^+} a_{Cl^-} = (m_{H^+} \gamma_{H^+})(m_{Cl^-} \gamma_{Cl^-}) \qquad (21.51)$$

as a consequence of equations (21.15), (21.17), and (21.18). The introduction of equation (21.51) into (21.50), with slight rearrangement, leads to

$$\mathcal{E} = \mathcal{E}° - \frac{RT}{n\mathcal{F}} \ln \left[(m_{\pm})^2 (\gamma_{\pm})^2 \right], \qquad (21.52)$$

which is equivalent to

$$\mathcal{E} = \mathcal{E}° - 2 \frac{RT}{n\mathcal{F}} \ln (m_{\pm} \gamma_{\pm}). \qquad (21.53)$$

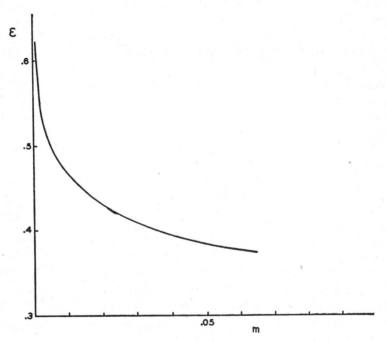

Fig. 4. Potentials at 25 °C of the cell: H_2, HCl (m), AgCl, Ag.

Since we have, as yet, no way of determining $\gamma_{\pm}$, we still cannot evaluate $\mathcal{E}°$. At this point, however, we can make use of one of the consequences of our specification of the standard state for the solute, namely, that γ must approach unity as the molality of the solute approaches zero. As m approaches zero, however, $\mathcal{E}$ rises rapidly toward infinity (Fig. 4) and hence no extrapolation of equation (21.53), as it stands, can be made to obtain $\mathcal{E}°$.

The equation can be rearranged, however, to overcome this indeterminacy. For this purpose we first break the logarithmic term into two parts, with the m and γ factors separated:

$$\varepsilon = \varepsilon° - 2\frac{RT}{n\mathcal{F}}\ln m_\pm - 2\frac{RT}{n\mathcal{F}}\ln \gamma_\pm. \tag{21.54}$$

Then, by grouping the two terms which tend to become infinite at very low molalities, we obtain

$$\varepsilon + 2\frac{RT}{n\mathcal{F}}\ln m_\pm = \varepsilon° - 2\frac{RT}{n\mathcal{F}}\ln \gamma_\pm. \tag{21.55}$$

We may simplify our notation by dropping the subscripts of the m and γ terms and by denoting the left-hand side of equation (21.55) by a single symbol, $\varepsilon°{}'$:

$$\varepsilon°{}' = \varepsilon + 2\frac{RT}{n\mathcal{F}}\ln m = \varepsilon° - 2\frac{RT}{n\mathcal{F}}\ln \gamma. \tag{21.56}$$

Values of $\varepsilon°{}'$ are not difficult to calculate from measurements of ε and m. It is merely necessary, then, to plot $\varepsilon°{}'$ against some function of the molality, to extrapolate and thus obtain $\varepsilon°$, since it is clear from equation (21.56) that as

$$m \to 0,$$
$$\gamma \to 1,$$
$$\ln \gamma \to 0,$$

and
$$\varepsilon°{}' \to \varepsilon°. \tag{21.57}$$

As a first trial, we might be inclined to plot $\varepsilon°{}'$ against m. In practice it turns out, however, that the values of $\varepsilon°{}'$ (Fig. 5) approach the ordinate axis with a very steep slope which makes extrapolation impossible. On the other hand, both experience and theory indicate that $\sqrt{m}$ as abscissa permits a very simple extrapolation,[5] as is indicated in Fig. 6. With $\varepsilon°{}'$, then, as ordinate and $\sqrt{m}$ as abscissa we can plot the experimental data and obtain $\varepsilon°$ with a high order of precision.

---

[5] It should be pointed out that if a finite limiting slope is obtained in a plot of $\varepsilon°{}'$ vs. $\sqrt{m}$, an infinite slope *must* be obtained in a plot of $\varepsilon°{}'$ vs. m. If in very dilute solutions

$$\varepsilon°{}' = a + b\sqrt{m}$$

(where a and b are constants), then

$$\frac{\partial \varepsilon°{}'}{\partial \sqrt{m}} = b.$$

But
$$\frac{\partial \varepsilon°{}'}{\partial m} = \frac{b}{2\sqrt{m}}$$

which approaches infinity as $m \to 0$. In fact, if $m^{1/2}$ as abscissa gives a finite limiting slope, any other power of m will give either a zero or an infinite limiting slope.

Alternative methods of extrapolation have been tried[6] to improve the precision of the extrapolation to an even greater degree.

Once having obtained $\mathcal{E}^\circ$, we can calculate mean activity coefficients readily, because as is evident from equation (21.56), the expression

$$\ln \gamma = \frac{n\mathcal{F}}{2RT} (\mathcal{E}^\circ - \mathcal{E}^{\circ\prime}) \tag{21.58}$$

is valid at all concentrations.

The mean activity coefficient, $\gamma_\pm$, for HCl is plotted vs. m in Fig. 7 as an

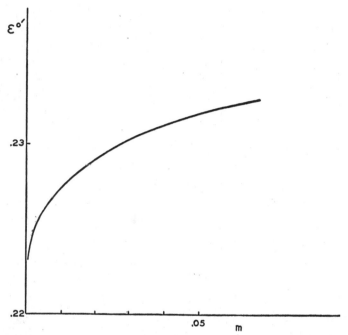

Fig. 5. $\mathcal{E}^{\circ\prime}$ at 25 °C as a function of the molality, for the cell: H_2, HCl (m), AgCl, Ag.

example of the behavior of a uni-univalent electrolyte in aqueous solution. From these data the activity, a_2, has been calculated and is illustrated as a function of the square of the molality in Fig. 8, the dashed line indicating the limiting slope. The point on the dashed line corresponding to $m^2 = 1$ is the "hypothetical one-molal solution" which is the standard state. It should be noted that the activity of HCl is unity at a concentration slightly below 1.2-molal $(m^2 = 1.43)$ and therefore $f_2 = f_2^\circ$ and $\bar{F}_2 = \bar{F}_2^\circ$ for this solution, but that this is *not the standard state* because it does not have the

[6] H. S. Harned and B. B. Owen, *The Physical Chemistry of Electrolytic Solutions*, Reinhold Publishing Corporation, New York, 1943, pages 328–331.

partial molal enthalpy and other thermodynamic properties characteristic of the infinitely dilute solution. On the other hand, $\bar{L}_2$ for the "hypothetical one-molal solution" is that of the infinitely dilute solution.

It may be worth while also to mention that at a concentration near 2-molal, a point beyond the range indicated in Fig. 8, the curve for a_2 crosses the dashed line. At this intersection a_2 equals m^2, and hence $\gamma_\pm$ must be unity. This phenomenon is a consequence of our choosing the standard state in such a manner that a_2/m^2 shall approach unity at infinite dilution; for as a result of this choice, the slope of the dashed line in Fig. 8 must be 1.

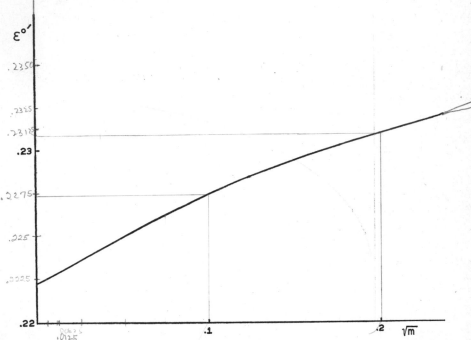

Fig. 6. Appropriate axes for the extrapolation of $\varepsilon^{\circ\prime}$ at 25 °C to determine ε° for the cell: H₂, HCl (m), AgCl, Ag.

Thus it should be apparent that e.m.f. measurements give activity values based on our initial definitions and conventions with regard to standard states.

B. Solubility measurements. As long as a pure solid, A, is in equilibrium with a dissolved solute, A', any modifications which occur in the solution, brought about, for example, by the addition of other electrolytes, and which change the quantity of dissolved solute (and hence its molality, m), must be due to variations in the activity coefficient, $\gamma_\pm$, since at a specified temperature and pressure the fugacity, and hence the activity, a_s, of the pure solid solute is constant. Therefore, at equilibrium, the equilibrium constant, K, must reduce to the following equation for a uni-univalent solute:

$$A \text{ (pure solid)} \rightleftharpoons A' \text{ (solute in solution)}; \quad K = \frac{a_2}{a_s} = \frac{a_2}{1} = a_+a_-$$

$$= a_\pm^2 = m_\pm^2 \gamma_\pm^2. \qquad (21.59)$$

To determine the mean activity coefficient, $\gamma_\pm$, it is convenient to convert equation (21.59) into logarithmic form:

$$\log \gamma_\pm = \log K^{\frac{1}{2}} - \log m_\pm. \qquad (21.60)$$

Activity coefficients are obtained as soon as the constant is evaluated.

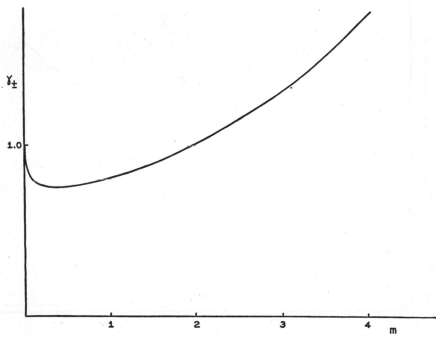

Fig. 7.   Mean activity coefficients of aqueous hydrochloric acid at 25 °C.   [Based on data from H. S. Harned and B. B. Owen, *The Physical Chemistry of Electrolytic Solutions*, page 547.]

The equilibrium constant or thermodynamic solubility product, K, may be evaluated readily from a plot of $\log m_\pm$ vs. some function of the total ionic strength, because as

$$\mu \to 0,$$

$$\gamma_\pm \to 1,$$

$$\log \gamma_\pm \to 0,$$

and
$$\log K^{\frac{1}{2}} \to \log m_\pm. \qquad (21.61)$$

Both theory and practice indicate that $\sqrt{\mu}$ is the proper abscissa. A typical extrapolation for some solutions of AgCl in various aqueous electrolytes is indicated in Fig. 9. From the extrapolated value for the constant K, the following equation may be written for the mean activity coefficient of AgCl (since $m_\pm$ equals m):

$$\log \gamma_\pm = -4.895 - \log m. \qquad (21.62)$$

C. Depression of the freezing point of the solvent. The temperature at which crystals of pure solvent are in equilibrium with a solution depends upon the activity of the solvent in the solution. Changes in activity of the

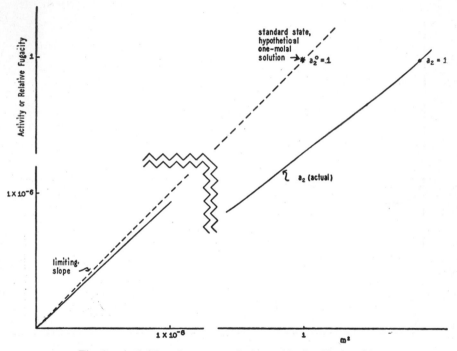

Fig. 8.   Activities of aqueous solutions of hydrochloric acid.

solvent can be determined, therefore, from measurements of the freezing-point depression due to the addition of solute. Since the activities of solvent and solute are related thermodynamically, it is possible, in turn, to calculate the activity of the solute once that for the solvent is known. The method may be used equally well with electrolytes as with non-electrolytes, and is capable of yielding activity coefficients of high precision.

Since the experimental data are used to calculate the properties of the solvent first, we shall begin by considering some thermodynamic relationships in connection with this component.

1. *Some activity relations for the solvent.* When pure solvent, such as

water, is in equilibrium with pure ice, the fugacity of H_2O must be the same in each phase:

$$f_{ice} = f_{pure\ water}.\qquad(21.63)$$

In this particular situation, each phase is also in its standard state, so that its activity is unity:

$$a_{ice} = a_{pure\ water} = 1.\qquad(21.64)$$

However, if we consider an aqueous solution, for which the temperature of equilibrium with pure ice is below 0°C, the situation is a little more com-

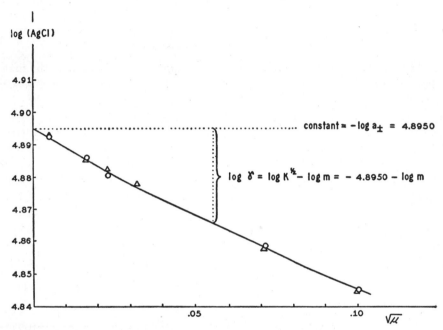

Fig. 9.   Activity coefficients of AgCl in aqueous solution, obtained from variation of solubility with ionic strength.   Added electrolyte: ◯, $NaNO_3$; △, KNO_3. [Based on data from S. Popoff and E. W. Neuman, *J. Phys. Chem.*, **34**, 1853 (1930).]

plex.   At equilibrium the fugacity of the ice must still equal that of the solvent water:

$$f_{ice} = f_{solvent}.\qquad(21.65)$$

However, *if we maintain our usual definitions of the standard states*

$$a_{ice} \neq a_{water};\qquad(21.66)$$

for the activity of the ice must be unity, since it is a pure solid, whereas that of the liquid solvent is not unity, since it is not the *pure* liquid phase. Thus, for the solvent the standard state would be the pure *supercooled* liquid, which cannot be present under equilibrium conditions below 0°C.

It would be convenient, nevertheless, if equation (21.66) were valid. It can be made so by choosing a different standard state for the ice. If we agree to keep a single standard state, pure supercooled liquid water at the specified temperature of the freezing point of the solution, for the ice as well as the solvent water, then

$$a'_{ice} = a_{water} \tag{21.67}$$

or

$$a_s = a_1, \tag{21.68}$$

where a'_{ice} or a_s refers to the activity of the ice on the basis of its new standard state. Thus

$$a_s = a'_{ice} = \frac{f_{ice}}{f_{pure\ supercooled\ water}}. \tag{21.69}$$

With this new standard we are in a position to turn attention to the first part of our problem—the establishment of a relation between a_s (and hence a_1) and the freezing-point depression of the solution.

2. *Activity of the solvent as a function of the freezing point of the solution.* In deriving an expression correlating the depression in freezing point with the activity of the solvent, we must recognize that we are dealing with a set of experiments in which the (equilibrium) temperature is being changed continuously. Hence it is perhaps reasonable to expect that we must start with some expression which involves a temperature coefficient of the activity. Such an expression can be obtained readily by a consideration of the following process (in which water is used merely as a typical example):

$$H_2O\ (pure\ water) \rightarrow H_2O\ (ice). \tag{21.70}$$

The free-energy change accompanying this transformation at any temperature is given by the relation

$$\Delta F = RT \ln \frac{f_{ice}}{f_{pure\ water}} = RT \ln a_s, \tag{21.71}$$

the latter part following from equation (21.69). The temperature coefficient of this free-energy change, and hence of a_s, follows immediately from equation (8.95):

$$\left(\frac{\partial(\Delta F/T)}{\partial T}\right)_P = R\left(\frac{\partial \ln a_s}{\partial T}\right)_P = -\frac{\Delta H}{T^2}. \tag{21.72}$$

Since the process in (21.70) is the solidification of H_2O, ΔH in equation (21.72) is the heat absorbed during solidification at the specified temperature. Hence we may write

$$\left(\frac{\partial \ln a_s}{\partial T}\right)_P = -\frac{\Delta H_{solidification}}{RT^2} = \frac{\Delta H_{fusion}}{RT^2}. \tag{21.73}$$

Before we can integrate this expression, we must give some thought to the properties of ΔH. The heat of solidification will also be temperature dependent, with a coefficient given by the expression

$$\left(\frac{\partial \Delta H}{\partial T}\right)_P = \Delta C_p, \tag{21.74}$$

where ΔC_p is the difference between the heat capacity of pure ice and pure (supercooled) water at the same temperature. By analogy with preceding equations [equations (4.38) and (4.41)] for heat capacities as a function of the temperature, we may express ΔC_p as a power series in terms of t, the *centigrade* temperature:

$$\Delta C_p = \Delta\Gamma_0 + \Delta\Gamma_1 t + \Delta\Gamma_2 t^2 + \dots, \tag{21.75}$$

where $\Delta\Gamma_0$, $\Delta\Gamma_1$, and $\Delta\Gamma_2$ are constants. An expression for the heat of solidification, ΔH_s, may then be obtained readily by integration of equation (21.74):

$$\Delta H_s = \Delta H_\theta + \Delta\Gamma_0 t + \Delta\Gamma_1 \frac{t^2}{2} + \Delta\Gamma_2 \frac{t^3}{3} + \dots. \tag{21.76}$$

The integration constant, ΔH_θ, being the heat of solidification at $t = 0$, obviously is the heat absorbed at the freezing point of the pure liquid. This temperature on the *absolute* temperature scale is represented by the symbol θ.

Having obtained an expression for ΔH_s as a function of the temperature, we are in a position to integrate equation (21.73). The explicit insertion of equation (21.76) into (21.73) leads to the relation

$$\frac{d \ln a_s}{dT} = -\frac{1}{R}\left[\frac{\Delta H_\theta}{T^2} + \frac{\Delta\Gamma_0 t}{T^2} + \frac{\Delta\Gamma_1 t^2}{2T^2} + \dots\right]. \tag{21.77}$$

Before integration is carried out, however, it is necessary to convert either t or T into a single, common temperature variable.

The conversion commonly used introduces the variable ϑ, the depression of the freezing point, which, in view of its definition, must be related for H_2O to t and T by the expressions

$$t = -\vartheta, \tag{21.78}$$

and
$$T = \theta - \vartheta. \tag{21.79}$$

Substitution of equations (21.78) and (21.79) into (21.77) leads to the relation

$$\frac{d \ln a_s}{dT} = -\frac{1}{R}\frac{1}{(\theta - \vartheta)^2}\left[\Delta H_\theta - \Delta\Gamma_0 \vartheta + \frac{\Delta\Gamma_1}{2} \vartheta^2 - \dots\right]. \tag{21.80}$$

The factor $1/(\theta - \vartheta)^2$ can be expanded into the series

$$\frac{1}{(\theta - \vartheta)^2} = \frac{1}{\theta^2[1 - (\vartheta/\theta)]^2} = \frac{1}{\theta^2}\left[1 + 2\frac{\vartheta}{\theta} + 3\frac{\vartheta^2}{\theta^2} + \ldots\right] \quad (21.81)$$

and multiplied into the bracket in equation (21.80). If coefficients of the same power of ϑ are grouped together, we obtain the equation[7]

$$-\frac{d \ln a_s}{dT} = A_1 + B_1\vartheta + C_1\vartheta^2 + \ldots, \quad (21.82)$$

where

$$A_1 = \frac{\Delta H_\theta}{R\theta^2}, \quad (21.83)$$

$$B_1 = \frac{2\Delta H_\theta}{R\theta^3} - \frac{\Delta\Gamma_0}{R\theta^2}, \quad (21.84)$$

$$C_1 = 3\frac{\Delta H_\theta}{R\theta^4} - 2\frac{\Delta\Gamma_0}{R\theta^3} + \frac{\Delta\Gamma_1}{2R\theta^2}. \quad (21.85)$$

Thus, for a given solvent at a fixed pressure, each of the coefficients A_1, B_1, and C_1 is a constant, since each contains terms which are specific properties of the pure solvent.

Equation (21.82) may now be integrated readily, since

$$dT = -d\vartheta, \quad (21.86)$$

as is apparent from equation (21.79), to give

$$\ln a_s = A_1\vartheta + \frac{B_1}{2}\vartheta^2 + \frac{C_1}{3}\vartheta^3 + \ldots. \quad (21.87)$$

For water as the solvent these constants may be evaluated easily to give the expression

$$\log a_s = -4.2091 \times 10^{-3}\vartheta - 0.2152 \times 10^{-5}\vartheta^2 + 0.359 \times 10^{-7}\vartheta^3$$
$$+ 0.212 \times 10^{-9}\vartheta^4 + 0.095 \times 10^{-11}\vartheta^5 + \ldots. \quad (21.88)$$

Thus we have in equation (21.87) a general expression for the activity of the solvent *at the temperature of its freezing point*. In many cases, particularly in solutions which are not concentrated, the activity of the solvent is practically temperature-independent, and hence the values obtained from equation (21.87) are useful at temperatures near 25°C also. For the highest precision, however, this temperature-independence may not be assumed. It is necessary, therefore, to turn our attention to the problem of calculating activities of the solvent at other temperatures from data at the freezing point.

3. *Activity of the solvent at temperatures other than the freezing point.*

[7] T. F. Young, *Chem. Rev.*, **13**, 103 (1933). Coefficients of higher powers for equation (21.82) are also given in this article.

Again it is reasonable to expect that we must start with an expression giving the temperature-dependence of the solvent activity [equation (19.54)]:

$$\left(\frac{\partial \ln a_1}{\partial T}\right)_{P,m} = -\frac{\bar{L}_1}{RT^2}. \tag{21.89}$$

Again, also, we recognize that calculations of the highest precision must take into account the temperature-dependence of $\bar{L}_1$, at any given molality. Thus

$$\left(\frac{\partial \bar{L}_1}{\partial T}\right)_{P,m} = \left(\frac{\partial(\bar{H}_1 - \bar{H}_1^\circ)}{\partial T}\right)_{P,m} = \left(\frac{\partial \bar{H}_1}{\partial T}\right)_{P,m} - \left(\frac{\partial \bar{H}_1^\circ}{\partial T}\right)_{P,m}$$

$$= \bar{c}_{p1} - \bar{c}_{p1}^\circ \equiv \Delta\bar{c}_{p1}. \tag{21.90}$$

For greater generality we shall not assume that $\Delta\bar{c}_{p1}$ is a constant, but rather that it varies linearly with temperature. Since calorimetric data are generally known best near 25°C, and since we are usually interested in calculating the activities at 25°C, we shall express the temperature-dependence of $\Delta\bar{c}_{p1}$ in terms of $\Delta\bar{c}_{p1}''$, the relative partial molal heat capacity of the solvent at 25°C. Thus at any temperature, T,

$$\Delta\bar{c}_{p1} = \Delta\bar{c}_{p1}'' + \Delta\Gamma_1'(T - 298.1), \tag{21.91}$$

where $\Delta\Gamma_1'$ is a constant. If we use this expression for the temperature coefficient of $\bar{L}_1$ in equation (21.90), we can integrate to obtain

$$\bar{L}_1 - \bar{L}_1'' = \int_{298.1}^{T} \Delta\bar{c}_{p1} \, dT$$

$$= \int_{298.1}^{T} [\Delta\bar{c}_{p1}'' + \Delta\Gamma_1'(T - 298.1)] \, dT, \tag{21.92}$$

where $\bar{L}_1''$ is the relative partial molal enthalpy of the solvent at 25°C. Putting in the limits, we obtain

$$\bar{L}_1 = \bar{L}_1'' + (\Delta\bar{c}_{p1}'' - 298.1\Delta\Gamma_1')(T - 298.1) + \frac{\Delta\Gamma_1'}{2}[T^2 - (298.1)^2]. \tag{21.93}$$

This equation can then be rearranged to give

$$\bar{L}_1 = \bar{L}_1'' - 298.1\Delta\bar{c}_{p1}'' + \frac{(298.1)^2\Delta\Gamma_1'}{2} + (\Delta\bar{c}_{p1}'' - 298.1\Delta\Gamma_1')T$$

$$+ \frac{\Delta\Gamma_1'}{2}T^2, \tag{21.94}$$

which may now be used for the integration of equation (21.89) over definite limits to obtain a_1'', the activity of the solvent at 25°C. Thus we may write, for a specific molality,

$$-2.303R \ \log \frac{a_1''}{a_1} = \int_T^{298.1} \frac{\bar{L}_1'' - 298.1 \, \Delta \bar{c}_{p1}'' + [(298.1)^2 \Delta \Gamma_1'/2]}{T^2} \, dT$$

$$+ \int_T^{298.1} \frac{\Delta \bar{c}_{p1}'' - 298.1 \, \Delta \Gamma_1'}{T} \, dT$$

$$+ \int_T^{298.1} \frac{\Delta \Gamma_1'}{2} \, dT. \tag{21.95}$$

When this equation is integrated over the limits indicated, and the various factors and terms are grouped together as coefficients of $\bar{L}_1''$, $\Delta \bar{c}_{p1}''$, and $\Delta \Gamma_1'$, respectively, we obtain an equation [8] of the form

$$x \equiv \log \frac{a_1''}{a_1} = -\bar{L}_1'' y + \Delta \bar{c}_{p1}'' z - \Delta \Gamma_1' \Omega, \tag{21.96}$$

where the $''$ refers to the value of the quantity at 298.1°K and

$$y = \frac{298.1 - T}{2.303 R (298.1) T}, \tag{21.97}$$

$$z = 298.1 y - \frac{1}{R} \log \frac{298.1}{T}, \tag{21.98}$$

$$\Omega = 298.1 \left(z + \frac{T - 298.1}{2} y \right). \tag{21.99}$$

The coefficients y, z, and Ω are clearly functions only of the temperature of the freezing point, the reference temperature, 25°C, and the fundamental constant R. Thus once T'' has been chosen, it is possible to calculate tables of these coefficients in terms of the freezing-point temperature, T, which are applicable to any solvent.

In equation (21.96) we now have an expression from which we can calculate the activity of the solvent at 25°C, a_1'', from the freezing point of the solution at the specified molality, given adequate calorimetric data for the same solution. For purposes of further calculations, it will be convenient to rearrange equation (21.96) to the form

$$\log a_1'' = \log a_1 - \bar{L}_1'' y + \Delta \bar{c}_{p1}'' z - \Delta \Gamma_1' \Omega, \tag{21.100}$$

or $\qquad \log a_1'' = \log a_1 + x. \tag{21.101}$

To complete our discussion of the calculation of activities from freezing-point measurements, we need to consider finally the procedure by which the information for the solvent may be used to calculate activity coefficients for the solute in the same solutions.

4. *Activity of the solute.* To calculate the activity of the solute from

---

[8] G. N. Lewis and M. Randall, *Thermodynamics*, McGraw-Hill Book Company, Inc., New York, 1923, pages 348–349. T. F. Young, *Chem. Rev.*, **13**, 103 (1933).

that of the solvent, we should start, of course, from the general differential relation between the two [equation (20.38)]:

$$d \ln a_2 = - \frac{n_1}{n_2} d \ln a_1. \tag{21.102}$$

It will be convenient, however, to deal with the molality, m, rather than the mole ratio n_1/n_2.   Since for water as solvent

$$m = \frac{n_2}{n_1/55.508} = 55.508 \frac{n_2}{n_1}, \tag{21.103}$$

we obtain in place of equation (21.102) the relation

$$d \ln a_2 = - \frac{55.508}{m} d \ln a_1'', \tag{21.104}$$

where a_1'' is used to indicate that our calculations are to be carried out for solvent and solute at a constant temperature of 25°C.

We can obtain an explicit equation for $d \ln a_1''$ if we modify equation (21.101) to

$$\ln a_1'' = \ln a_1 + 2.303x, \tag{21.105}$$

differentiate to obtain

$$d \ln a_1'' = d \ln a_1 + 2.303 \, dx, \tag{21.106}$$

and then use equation (21.87) to obtain $d \ln a_1$ (since $a_1 = a_s$ at the freezing point).   In this way we obtain the equation

$$d \ln a_2 = - \frac{55.508}{m} [(A_1 + B_1 \vartheta + C_1 \vartheta^2 + \ldots)d\vartheta + 2.303 \, dx]. \tag{21.107}$$

The terms in equation (21.107) can be rearranged to give

$$d \ln a_2 = - \frac{55.508 A_1}{m} d\vartheta - \frac{55.508}{m} (B_1 \vartheta + C_1 \vartheta^2 + \ldots) \, d\vartheta$$

$$- \frac{55.508}{m} 2.303 \, dx. \tag{21.108}$$

The first term on the right-hand side of equation (21.108) can be simplified further by reference to equation (21.83) for an explicit definition of A_1. Thus,

$$55.508 A_1 = \frac{55.508 \, \Delta H_\theta}{R\theta^2} = \frac{55.508(-1436.7)}{1.9864(273.1)^2}$$

$$= - \frac{1}{1.858} \equiv - \frac{1}{\lambda}. \tag{21.109}$$

Consequently, equation (21.108) may be rewritten in the form

$$d \ln a_2 = \frac{d\vartheta}{\lambda m} - \frac{55.508}{m}(B_1\vartheta + C_1\vartheta^2 + \ldots)\, d\vartheta - \frac{55.508}{m}2.303\, dx. \quad (21.110)$$

At this point it is appropriate to introduce a provision for the possibility that the solute may dissociate into two or more particles. If ν is the total number of particles produced by a single molecule of solute, then in view of equation (21.36) we may write

$$d \ln a_\pm = d \ln (a_2)^{1/\nu} = \frac{1}{\nu}\, d \ln a_2. \quad (21.111)$$

Multiplying both sides of equation (21.110) by $1/\nu$, we obtain

$$d \ln a_\pm = \frac{d\vartheta}{\nu\lambda m} - \frac{55.508}{\nu m}(B_1\vartheta + C_1\vartheta^2 + \ldots)d\vartheta - \frac{55.508}{\nu m}2.303\, dx. \quad (21.112)$$

This equation is adequate for the determination (by integration over specified limits) of the ratio of two activities at two finite concentrations, m and m'. It cannot be used as it stands, however, for obtaining absolute values of $a_\pm$ because the first term becomes indeterminate at zero molality.

To overcome this difficulty, we may adopt the following stratagem. Starting with equation (21.37) for $m_\pm$ and converting it to a logarithmic form,

$$\ln m_\pm = \ln m + \frac{1}{\nu} \ln [(\nu_+)^{\nu_+} (\nu_-)^{\nu_-}], \quad (21.113)$$

we find that

$$d \ln m_\pm = d \ln m, \quad (21.114)$$

since ν_+ and ν_- are constants. Subtracting $(d \ln m)$, then, from both sides of equation (21.112), we obtain the expression

$$d \ln a_\pm - d \ln m_\pm = d \ln \gamma_\pm = \frac{d\vartheta}{\nu\lambda m} - d \ln m$$

$$- \frac{55.508}{\nu m} (B_1\vartheta + C_1\vartheta^2 + \ldots)\, d\vartheta$$

$$- \frac{55.508}{\nu m} 2.303\, dx. \quad (21.115)$$

The first two terms on the right-hand side of equation (21.115), neither of which can be integrated readily from a lower limit of zero molality, can now be converted into several other terms which permit precise integrations. This conversion can be accomplished by recognizing that

$$d\left(\frac{\vartheta}{m}\right) = \frac{1}{m}\,d\vartheta - \frac{\vartheta}{m^2}\,dm. \tag{21.116}$$

Consequently, the following equation,

$$\frac{d\vartheta}{m} = d\left(\frac{\vartheta}{m}\right) + \frac{\vartheta}{m^2}\,dm = d\left(\frac{\vartheta}{m}\right) + \frac{\vartheta}{m}\,d\ln m, \tag{21.117}$$

may be used to substitute into the first term of equation (21.115). Therefore the first two terms on the right-hand side of equation (21.115) may now be combined as follows:

$$\frac{1}{\nu\lambda}\frac{d\vartheta}{m} - d\ln m = \frac{1}{\nu\lambda}\,d\left(\frac{\vartheta}{m}\right) + \frac{\vartheta}{\nu\lambda m}\,d\ln m - d\ln m \tag{21.118}$$

$$= \frac{1}{\nu\lambda}\,d\left(\frac{\vartheta}{m}\right) + \left(\frac{\vartheta}{\nu\lambda m} - 1\right)d\ln m. \tag{21.119}$$

For convenience in manipulation, Lewis and Randall have proposed a further definition:

$$\left(\frac{\vartheta}{\nu\lambda m} - 1\right) \equiv -j. \tag{21.120}$$

From this definition it follows that

$$-dj = \frac{1}{\nu\lambda}\,d\left(\frac{\vartheta}{m}\right). \tag{21.121}$$

With the aid of these revisions in notation it is finally possible to write an equation which can be used in practice to calculate $\gamma_{\pm}$ at the desired temperature, 25°C. Making the appropriate substitutions from equations (21.118)–(21.121) into (21.115), we obtain

$$d\ln\gamma_{\pm} = -dj - j\,d\ln m - \frac{55.508}{\nu m}(B_1\vartheta + C_1\vartheta^2 + \ldots)\,d\vartheta$$

$$- \frac{55.508}{\nu m}\,2.303\,dx, \tag{21.122}$$

which can be integrated from infinite dilution to any specified molality to give[9]

---

[9] Note that

$$\lim_{m\to 0}\frac{\vartheta}{m} = \nu\lambda,$$

since λ is the freezing-point constant for a nondissociating solute in water. consequently,

$$\lim_{m\to 0}\frac{\vartheta}{\nu\lambda m} = 1,$$

and hence at $m = 0$, $j = 0$.

$$\log \gamma_\pm = -\frac{j}{2.303} - \int_0^m j(d \log m)$$

$$-\frac{55.508}{2.303\nu} \int_0^m \frac{(B_1\vartheta + C_1\vartheta^2 + \ldots)}{m} \, d\vartheta - \frac{55.508}{\nu} \int_0^m \frac{dx}{m}. \quad (21.123)$$

In practice, it is found that each of these terms can be integrated from zero molality, and hence absolute values of $\gamma_\pm$ can be determined.

In the actual treatment of experimental data, several investigators have worked out special methods for increasing the precision of some of the integrations required, particularly that of the second term in equation (21.123). The details of these methods are beyond the objectives of this discussion but may be obtained by reference to Harned and Owen.[10]

III. ACTIVITY COEFFICIENTS OF SOME STRONG ELECTROLYTES

A. Experimental values. With the experimental methods described, as well as with several others, the activities of numerous strong electrolytes of various valence type have been calculated. Many of these data have been assembled and examined critically by Harned and Owen.[11]

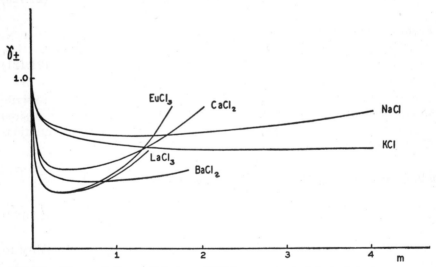

Fig. 10. Mean activity coefficients at 25 °C for some typical electrolytes in aqueous solution.

The behavior of a few typical electrolytes is illustrated in Fig. 10. By definition, $\gamma_\pm$ is unity at zero molality for all electrolytes. In every case, furthermore, $\gamma_\pm$ decreases rapidly with increasing molality at low values of

[10] H. S. Harned and B. B. Owen, *The Physical Chemistry of Electrolytic Solutions*, Reinhold Publishing Corporation, New York, 1943, pages 288–296.
[11] H. S. Harned and B. B. Owen, *op. cit.*

m. The steepness of this initial drop, however, varies with the valence type of the electrolyte. For a given valence type, $\gamma_{\pm}$ is substantially independent of the chemical nature of the constituent ions, so long as m is below about 0.01. At higher concentrations, curves for $\gamma_{\pm}$ begin to separate widely and to exhibit marked specific ion effects.

B. Theoretical correlation. No adequate theoretical model based on the atomic characteristics of the ions has yet been developed which is capable of accounting for the thermodynamic properties of aqueous solutions over wide ranges of concentration. For dilute solutions of completely ionized electrolytes, however, expressions have been derived[12] which predict exactly the limiting behavior of activity coefficients in an infinitely dilute solution, and which provide very useful equations for describing these quantities at small finite concentrations. Although it is beyond the objectives of this text to consider the development of the Debye-Hückel theory, it is desirable to present some of the final results because they are of such great value in the handling of experimental data.

In the limit of the infinitely dilute solution, according to the Debye-Hückel theory, individual ion activity coefficients are given by the equation

$$\log \gamma = -Az^2\sqrt{\mu}, \tag{21.124}$$

where A is a constant for a given solvent at a specified temperature. Values of A for aqueous solutions are listed in Table 3. In actual practice, equation (21.124) has been found useful up to $\sqrt{\mu}$ near 0.1, that is, for solutions with ionic strengths as high as 0.01. In practice, also, the mean activity coefficient, $\gamma_{\pm}$, would be required, but it is not difficult to show from equation (21.124) that *for an electrolyte with two kinds of ions*[13]

$$\log \gamma_{\pm} = -A|z_+z_-|\sqrt{\mu}, \tag{21.125}[14]$$

where z_+ and z_- are the number of charges on the cation and anion, respectively.

At ionic strengths near 0.01 it is convenient frequently to use the more complete form of the Debye-Hückel expression:

$$\log \gamma = \frac{-Az^2\sqrt{\mu}}{1 + Ba_i\sqrt{\mu}}, \tag{21.126}$$

where A has the same significance as in equation (21.124), B is a constant for a given solvent at a specified temperature (Table 3), and a_i may be thought of as the "effective diameter" of the ion *in the solution*. Since no

---

[12] P. Debye and E. Hückel, *Physik. Z.*, **24**, 185 (1923).

[13] For the general case of any number of ions, see Harned and Owen, *op. cit.*, page 35.

[14] The symbol | | is used to indicate absolute value, without regard to the sign of the charges.

truly independent method is available for evaluating a_i, this quantity is essentially an empirical parameter, but it is a fact that the a_i's are of a magnitude expected for ion sizes.

In connection with the use of equation (21.126), Kielland[15] has assembled a series of "effective diameters" for a large number of ions, grouped according to charge. These values of a_i are listed in Table 4. With these values of a_i, he has calculated individual ion activity coefficients from equation (21.126); these have been summarized in Table 5. Though the activity coefficient of an individual ion cannot be determined, nor perhaps even defined accurately,[16] these tables provide very useful semi-empirical methods for estimating γ's for use in connection with calculations of free-energy changes in solutions of ionic strength below 0.1.

TABLE 3

VALUES OF CONSTANTS IN DEBYE-HÜCKEL EQUATION FOR ACTIVITY COEFFICIENTS IN AQUEOUS SOLUTIONS*

Temperature, °C	A	$B \times 10^{-8}$
0	0.4883	0.3241
5	0.4921	0.3249
10	0.4960	0.3258
15	0.5000	0.3266
20	0.5042	0.3273
25	0.5085	0.3281.
30	0.5130	0.3290
35	0.5175	0.3297
40	0.5221	0.3305
45	0.5270	0.3314
50	0.5319	0.3321
55	0.5371	0.3329
60	0.5425	0.3338

* The constants listed are based on concentration measurements in terms of unit weight of solvent. Corresponding constants for concentrations in terms of unit volume of solution differ very slightly. Both sets may be found in an article by G. G. Manov, R. G. Bates, W. J. Hamer, and S. F. Acree, *J. Am. Chem. Soc.*, **65**, 1765 (1943). See also G. Scatchard, *ibid.*, **65**, 1249 (1943), and P. van Rysselberghe, *ibid.*, **65**, 1249 (1943).

For solutions above $\mu = 0.1$, various extensions of the Debye-Hückel theory have been proposed. Recently attempts have been made also to superimpose terms taking account of ion-solvent interactions. Some of these efforts are described in a recent article by Stokes and Robinson.[17]

[15] J. Kielland, *J. Am. Chem. Soc.*, **59**, 1675 (1937).
[16] E. A. Guggenheim, *J. Phys. Chem.*, **33**, 842 (1929); *ibid.*, **34**, 1541 (1930).
[17] R. H. Stokes and R. A. Robinson, *J. Am. Chem. Soc.*, **70**, 1870 (1948).

TABLE 4
Effective Diameters of Hydrated Ions in Aqueous Solution

$10^8 a_i$	Inorganic Ions: Charge 1
9	H^+
6	Li^+
4–4.5	Na^+, $CdCl^+$, ClO_2^-, IO_3^-, HCO_3^-, $H_2PO_4^-$, HSO_3^-, $H_2AsO_4^-$, $Co(NH_3)_4(NO_2)_2^+$
3.5	OH^-, F^-, NCS^-, NCO^-, HS^-, ClO_3^-, ClO_4^-, BrO_3^-, IO_4^-, MnO_4^-
3	K^+, Cl^-, Br^-, I^-, CN^-, NO_2^-, NO_3^-
2.5	Rb^+, Cs^+, NH_4^+, Tl^+, Ag^+

	Inorganic Ions: Charge 2
8	Mg^{++}, Be^{++}
6	Ca^{++}, Cu^{++}, Zn^{++}, Sn^{++}, Mn^{++}, Fe^{++}, Ni^{++}, Co^{++}
5	Sr^{++}, Ba^{++}, Ra^{++}, Cd^{++}, Hg^{++}, S^{--}, $S_2O_4^{--}$, WO_4^{--}
4.5	Pb^{++}, CO_3^{--}, SO_3^{--}, MoO_4^{--}, $Co(NH_3)_5Cl^{++}$, $Fe(CN)_5NO^{--}$
4	Hg_2^{++}, SO_4^{--}, $S_2O_3^{--}$, $S_2O_8^{--}$, SeO_4^{--}, CrO_4^{--}, HPO_4^{--}, $S_2O_6^{--}$

	Inorganic Ions: Charge 3
9	Al^{+++}, Fe^{+++}, Cr^{+++}, Sc^{+++}, Y^{+++}, La^{+++}, In^{+++}, Ce^{+++}, Pr^{+++}, Nd^{+++}, Sm^{+++}
6	$Co(ethylenediamine)_3^{+++}$
4	PO_4^{---}, $Fe(CN)_6^{---}$, $Cr(NH_3)_6^{+++}$, $Co(NH_3)_6^{+++}$, $Co(NH_3)_5H_2O^{+++}$

	Inorganic Ions: Charge 4
11	Th^{++++}, Zr^{++++}, Ce^{++++}, Sn^{++++}
6	$Co(S_2O_3)(CN)_5^{----}$
5	$Fe(CN)_6^{----}$

	Inorganic Ions: Charge 5
9	$Co(SO_3)_2(CN)_4^{-----}$

	Organic Ions: Charge 1
8	$(C_6H_5)_2CHCOO^-$, $(C_3H_7)_4N^+$
7	$[OC_6H_2(NO_2)_3]^-$, $(C_3H_7)_3NH^+$, $CH_3OC_6H_4COO^-$
6	$C_6H_5COO^-$, $C_6H_4OHCOO^-$, $C_6H_4ClCOO^-$, $C_6H_5CH_2COO^-$, $CH_2=CHCH_2COO^-$, $(CH_3)_2CHCH_2COO^-$, $(C_2H_5)_4N^+$, $(C_3H_7)_2NH_2^+$
5	$CHCl_2COO^-$, CCl_3COO^-, $(C_2H_5)_3NH^+$, $(C_3H_7)NH_3^+$
4.5	CH_3COO^-, CH_2ClCOO^-, $(CH_3)_4N^+$, $(C_2H_5)_2NH_2^+$, $NH_2CH_2COO^-$
4	$NH_3^+CH_2COOH$, $(CH_3)_3NH^+$, $C_2H_5NH_3^+$
3.5	$HCOO^-$, H_2 citrate$^-$, $CH_3NH_3^+$, $(CH_3)_2NH_2^+$

	Organic Ions: Charge 2
7	$OOC(CH_2)_5COO^{--}$, $OOC(CH_2)_6COO^{--}$, Congo red anion^{--}
6	$C_6H_4(COO)_2^{--}$, $H_2C(CH_2COO)_2^{--}$, $(CH_2CH_2COO)_2^-$
5	$H_2C(COO)_2^{--}$, $(CH_2COO)_2^{--}$, $(CHOHCOO)_2^{--}$
4.5	$(COO)_2^{--}$, H citrate^{--}

	Organic Ions: Charge 3
5	Citrate^{---}

TABLE 5

INDIVIDUAL ACTIVITY COEFFICIENTS OF IONS IN WATER AT 25°C

Param-eter	Volume Ionic Strength,* μ							
$10^8 a_i$	0.0005	0.001	0.0025	0.005	0.01	0.025	0.05	0.1
Ion Charge 1								
9	.975	.967	.950	.933	.914	.88	.86	.83
8	.975	.966	.949	.931	.912	.880	.85	.82
7	.975	.965	.948	.930	.909	.875	.845	.81
6	.975	.965	.948	.929	.907	.87	.835	.80
5	.975	.964	.947	.928	.904	.865	.83	.79
4.5	.975	.964	.947	.928	.902	.86	.82	.775
4	.975	.964	.947	.927	.901	.855	.815	.77
3.5	.975	.964	.946	.926	.900	.855	.81	.76
3	.975	.964	.945	.925	.899	.85	.805	.755
2.5	.975	.964	.945	.924	.898	.85	.80	.75
Ion Charge 2								
8	.906	.872	.813	.755	.69	.595	.52	.45
7	.906	.872	.812	.755	.685	.58	.50	.425
6	.905	.870	.809	.749	.675	.57	.485	.405
5	.903	.868	.805	.744	.67	.555	.465	.38
4.5	.903	.868	.805	.742	.665	.55	.455	.37
4	.903	.867	.803	.740	.660	.545	.445	.355
Ion Charge 3								
9	.802	.738	.632	.54	.445	.325	.245	.18
6	.798	.731	.620	.52	.415	.28	.195	.13
5	.796	.728	.616	.51	.405	.27	.18	.115
4	.796	.725	.612	.505	.395	.25	.16	.095
Ion Charge 4								
11	.678	.588	.455	.35	.255	.155	.10	.065
6	.670	.575	.43	.315	.21	.105	.055	.027
5	.668	.57	.425	.31	.20	.10	.048	.021
Ion Charge 5								
9	.542	.43	.28	.18	.105	.045	.020	.009

* Concentrations expressed in moles liter^{-1}.

Exercises

1. Prove that the following equation is valid for an electrolyte which dissociates into ν particles:

$$\left(\frac{\partial \ln \gamma_\pm}{\partial T}\right)_{P,m} = -\frac{\bar{L}_2}{\nu RT^2}. \tag{21.127}$$

2. For the equilibrium between a pure solute and its saturated solution,

$$\text{solute (pure)} \rightleftharpoons \text{solute (satd solution)},$$

the equilibrium constant, K, is given by

$$K = \frac{a_{2(\text{satd})}}{a_{2(\text{pure})}} = \frac{a_{2(\text{satd})}}{1}. \tag{21.128}$$

(a) Show that

$$\left(\frac{\partial \ln a_2}{\partial T}\right)_{\text{satd}} = -\frac{L_2^\bullet}{RT^2}. \tag{21.129}$$

(b) For the general case of an electrolyte which dissociates into ν particles, show that

$$\nu\left[\left(\frac{\partial \ln \gamma_\pm}{\partial T}\right)_{\text{satd}} + \left(\frac{\partial \ln m_\pm}{\partial T}\right)_{\text{satd}}\right] = -\frac{L_2^\bullet}{RT^2}. \tag{21.130}$$

(c) Considering $\ln \gamma_\pm$ as a function of temperature and molality (pressure being maintained constant), show that

$$\left(\frac{\partial \ln \gamma_\pm}{\partial T}\right)_{\text{satd}} = \left(\frac{\partial \ln \gamma_\pm}{\partial T}\right)_m + \left(\frac{\partial \ln \gamma_\pm}{\partial m}\right)_T\left(\frac{\partial m}{\partial T}\right)_{\text{satd}}. \tag{21.131}$$

(d) Derive the equation

$$\left(\frac{\partial \ln \gamma_\pm}{\partial T}\right)_m + \left(\frac{\partial m}{\partial T}\right)_{\text{satd}}\left[\left(\frac{\partial \ln \gamma_\pm}{\partial m}\right)_T + \left(\frac{1}{m}\right)_{\text{satd}}\right] = -\frac{L_2^\bullet}{\nu RT^2}. \tag{21.132}$$

(e) Making use of equation (21.127), derive the following expression:

$$\Delta H_{\text{soln}} = \nu RT^2 \left(\frac{\partial m}{\partial T}\right)_{\text{satd}}\left[\left(\frac{\partial \ln \gamma_\pm}{\partial m}\right)_T + \left(\frac{1}{m}\right)_{\text{satd}}\right], \tag{21.133}$$

where ΔH_{soln} is the heat absorbed per mole of solute dissolved in the (nearly) saturated solution.

(f) To what does equation (21.133) reduce if the solute is a non-electrolyte?

(g) To what does equation (21.133) reduce if the solution is ideal?

3. Calculate the electromotive force of the cell pair

$$H_2 \text{ (g), HCl (4}m\text{), AgCl, Ag—Ag, AgCl, HCl (10 } m\text{), H}_2 \text{ (g)},$$

from vapor-pressure data (Lewis and Randall, *Thermodynamics*, page 330).

4. The electromotive force at 25 °C of the cell

$$H_2 \text{ (g), HCl (0.002951}m\text{), AgCl, Ag}$$

is 0.52393 volt when the apparent barometric height (as read on a brass scale) is 75.10 cm of mercury at 23.8 °C, and when the hydrogen is bubbled to the atmosphere through a column of solution 0.68 cm high. Calculate the partial pressure of the hydrogen, and the electromotive force of the cell when the partial pressure of hydrogen is one atmosphere.

5. In the table below are given unpublished data of T. F. Young and N. Anderson on the potentials at 25°C of the cell

$$H_2, HCl\ (m),\ AgCl,\ Ag.$$

(a) Plot $\mathcal{E}^{\circ\prime}$ vs. an appropriate composition variable, draw a smooth curve, and determine $\mathcal{E}^{\circ}$ by extrapolation.

(b) On the graph in Part (a), show the significance of the activity coefficient.

(c) Draw the asymptote predicted by the Debye-Hückel limiting law for the curve in Part (a).

(d) Determine the (mean) activity coefficient of the ions of HCl at $m = 0.001$, 0.01, and 0.1, respectively. Compare these values with those computed from the Debye-Hückel limiting law.

(e) What error is introduced into the calculated activity coefficients by an error of 0.00010 volt in $\mathcal{E}$ or $\mathcal{E}^{\circ}$?

$\sqrt{m_\pm}$	$m_\pm$[18]	$\mathcal{E}$	$\log m_\pm$	$\mathcal{E}^{\circ\prime}$
.054322	.0029509	.52456	-2.53005	.22524
.044115	.0019461	.54541	-2.71083	.22471
.035168	.0012368	.56813	-2.90770	.22413
.029908	.0008945	.58464	-3.04842	.22399
.027024	.0007303	.59484	-3.13650	.22378
.020162	.0004065	.62451	-3.39094	.22334
.015033	.00022599	.65437	-3.64591	.22303
.011631	.00013528	.68065	-3.86877	.22296
.009704	.00009417	.69914	-4.02606	.22283
.007836	.00006140	.72096	-4.21185	.22267
.007343	.00005392	.72759	-4.26827	.22263
.005376	.000028901	.75955	-4.53909	.22255

[18] In the most dilute solutions studied, the molality of chloride ion was considerably greater than the molality of the hydrogen ion. Therefore the mean molality, $m_\pm$, is tabulated, rather than the molality of either ion.

6. In the following table are listed S. Popoff and E. W. Neuman's values [*J. Phys. Chem.*, **34**, 1853 (1930)] of the solubility of silver chloride in water containing "solvent" electrolytes at the concentrations indicated. According to the same authors, the solubility of silver chloride in pure water is 1.278×10^{-5} mole per liter.

(a) Using distinctive symbols to represent each of the four series of data, plot (with reference to a single pair of axes) the solubility of silver chloride vs. the concentration of each solvent electrolyte. Plot on the same graph the solubility of silver chloride against three times the concentration of barium nitrate.

(b) Show that these data are in accordance with the ionic-strength principle.

(c) Verify several of the tabulated values of the total ionic strength; then plot the logarithm of the reciprocal of the solubility vs. the square root of the ionic strength.

(d) Draw a line representing the Debye-Hückel limiting law, for comparison with the data.

(e) Determine the activity of silver chloride in a solution containing only silver chloride, and in solutions in which the ionic strength is 0.001 and 0.01, respectively.

(f) Calculate the solubility product of silver chloride. What is the activity of silver chloride in any of the saturated solutions?

Concentration of Solvent Electrolyte (mole liter^{-1})	Concentration of AgCl (mole liter^{-1})	μ	$\sqrt{\mu}$	$-\log(AgCl)$
KNO₃				
.000 012 80	1.280 $\times$ 10^{-5}	.000 025 6	.005 06	4.892 8
.000 260 9	1.301	.000 273 9	.016 55	4.885 7
.000 509 0	1.311	.000 522 1	.022 85	4.882 4
.001 005	1.325	.001 018	.031 91	4.877 8
.004 972	1.385	.004 986	.070 61	4.858 6
.009 931	1.427	.009 945	.099 72	4.845 6
NaNO₃				
.000 012 81	1.281	.000 025 6	.005 06	4.892 5
.000 264 3	1.300	.000 277 3	.016 65	4.886 1
.000 515 7	1.315	.000 528 9	.023 00	4.881 1
.005 039	1.384	.005 053	.071 08	4.858 9
.010 076	1.428	.010 090	.100 45	4.845 3
HNO₃				
.000 012 8	1.280	.000 025 6	.005 06	4.892 8
.000 723 3	1.318	.000 736 5	.027 14	4.880 1
.002 864	1.352	.002 877 5	.053 64	4.869 0
.005 695	1.387	.005 709	.075 56	4.857 9
.009 009	1.422	.009 023	.094 99	4.847 1
Ba(NO₃)₂				
.000 006 40	1.280	.000 032 0	.005 66	4.892 8
.000 036 15	1.291	.000 121 4	.011 02	4.889 1
.000 211 08	1.309	.000 646 3	.025 42	4.883 1
.000 706 4	1.339	.002 133	.046 18	4.873 2
.001 499	1.372	.004 511	.067 16	4.862 7
.002 192	1.394	.006 590	.081 18	4.855 7
.003 083	1.421	.009 263	.096 24	4.847 4

7. From the data on aqueous NaCl solutions reported by T. F. Young and J. S. Machin [*J. Am. Chem. Soc.*, **58**, 2254 (1936)] tabulated below, calculate:

(a)   the activity, a_1, of the solvent in each NaCl solution at the temperature at which that solution freezes;

(b)   the corresponding activity at 25°C, a_1''.

Values of y and z may be found in Lewis and Randall, *Thermodynamics*, page 613. Ω may be computed from equation (21.99).

m	ϑ	$\bar{L}_1$		
		0°C	12.5°C	25°C
1.0	3.388	7.35	4.75	2.97
2.0	6.965	21.83	14.96	9.96
2.8	10.099	33.97	23.63	15.74
4.0	15.376	48.71	33.51	20.97

8. From the vapor pressure data in the International Critical Tables, Volume III, page 297, calculate the activity of the water in 1.0, 2.0, 2.8, and 4.0 m NaCl at 25°C. On a graph, compare the activity data of this problem with the two series in the preceding one.

9. Calculate the e.m.f. of each of the following cells at 25 °C. Use approximate values of the activity coefficient, as calculated from the Debye-Hückel limiting law.

(a) H_2, HCl (0.0001m), Cl_2—Cl_2, HCl (0.001m), H_2.

(b) Mg, $MgSO_4$ (0.001m), Hg_2SO_4, Hg—Hg, Hg_2SO_4, $MgSO_4$ (0.0001m), Mg.

10. The following table contains data of H. S. Harned and L. F. Nims [*J. Am. Chem. Soc.*, **54**, 423 (1932)] for the cell

Ag, AgCl, NaCl (4m), Na (amalgam)—Na (amalgam), NaCl (0.1m), AgCl, Ag.

t, °C	ε	ε/T
15	0.18265	0.00063398
20	0.18663	0.00063675
25	0.19044	0.00063885
30	0.19407	0.00064028
35	0.19755	0.00064119

(a) Write the cell reaction.

(b) Find $\Delta \bar{H}_2$ or $\Delta \bar{L}_2$ for the reaction in Part (a) at 25 °C.

(c) Compare the result in Part (b) with that which you would obtain from the direct calorimetric data [Chapter 14, Exercise 6].

(d) How precise is the result in Part (b) if ε can be measured to ± 0.00010 volt?

(e) Obtain activity coefficients for the solutions of NaCl in this cell from Harned and Owen, *Physical Chemistry of Electrolytic Solutions*, page 557. Calculate the e.m.f. of the cell at 25 °C. Compare your result with the value listed in the table.

11. The solubility (moles per kilogram of H_2O) of cupric iodate, $Cu(IO_3)_2$, in aqueous solutions of KCl at 25°, as determined by R. M. Keefer [*J. Am. Chem. Soc.*, **70**, 476 (1948)], is given in the following table:

KCl m	$Cu(IO_3)_2$ m
0.00000	3.245×10^{-3}
0.00501	3.398
0.01002	3.517
0.02005	3.730
0.03511	3.975
0.05017	4.166
0.07529	4.453
0.1005	4.694

(a) Plot the logarithm of the reciprocal of the solubility vs. $\sqrt{\mu}$ and extrapolate to infinite dilution.

(b) Using the Debye-Hückel limiting law to evaluate activity coefficients, show that

$$\log (m_{Cu^{++}})(m_{IO_3^-})^2 = \log K + 3.051 \sqrt{\mu}. \qquad (21.134)$$

(c) Find $\log K$. Keefer reports a value of -7.1353.

(d) Draw a line representing the limiting law on the graph in Part (a).

(e) Keefer has found that the following equation,

$$\log (m_{Cu^{++}})(m_{IO_3^-})^2 = -7.1353 + \frac{3.036 \sqrt{\mu}}{1 + 1.08 \sqrt{\mu}}, \qquad (21.135)$$

represents the solubility data given in the table with great precision. (The constant 3.036 is slightly different from that of 3.051 used in equation (21.134) because it is based on older values of the electronic charge.) At the highest concentration of KCl, for example, the calculated solubility of $Cu(IO_3)_2$ is 4.697×10^{-3} mole (kg. $H_2O)^{-1}$. Verify this calculation.

CHAPTER 22

Free-Energy Changes for Processes Involving Solutions

THE EMPHASIS in the preceding few chapters on the calculation of activities of the components of a solution may tend to obscure our primary objective in chemical thermodynamics—the calculation of free-energy changes, and those of associated thermodynamic functions, for a specified reaction. The widespread efforts which have been made to treat deviations from ideality in solutions have had as their principal goal the establishment of an experimental and theoretical basis for precise calculations of changes in free energy in transformations involving solutions. It is appropriate, therefore, to conclude our discussions of the principles of chemical thermodynamics with a consideration of some typical calculations of free-energy changes in real solutions.

Many of these calculations require information on free-energy changes in ionic dissociations. For weak electrolytes these are obtained from measurements of equilibrium constants, which are then used to calculate ΔF°. Hence it is desirable to consider some of the methods available for the determination of K for weak electrolytes. First, however, it is necessary to consider in greater detail the significance of activity coefficients in solutions of weak electrolytes.

I. ACTIVITY COEFFICIENTS OF WEAK ELECTROLYTES

Let us consider a typical weak electrolyte such as acetic acid, whose ionization may be represented by the following equation:

$$HC_2H_3O_2 \rightleftharpoons H^+ + C_2H_3O_2^-. \tag{22.1}$$

In discussions in Chapter 21 of the activity coefficient,

$$\gamma = \frac{a}{m}, \tag{22.2}$$

for strong electrolytes, we always used the stoichiometric (or total) molality for m and disregarded the possibility of incomplete dissociation. Consequently, in the present example with acetic acid we might adopt a similar procedure and define

$$\gamma_+ = \frac{a_+}{m_s} \tag{22.3}$$

and
$$\gamma_- = \frac{a_-}{m_s},$$
(22.4)

where m_s represents the stoichiometric (or total) molality of acetic acid.

On the other hand, it is possible to measure (or closely approximate) the ionic concentrations of a weak electrolyte; hence it is more convenient to define ionic activity coefficients, f, based on the *actual* ionic concentrations, m_+ or m_-. Thus

$$f_+ = \frac{a_+}{m_+}$$
(22.5)

and
$$f_- = \frac{a_-}{m_-},$$
(22.6)

Similarly, for the undissociated species of molality m_u,

$$f_u = \frac{a_u}{m_u}.$$
(22.7)

Since the degree of dissociation, α, of a weak electrolyte such as acetic acid is given by the equation

$$\alpha = \frac{m_+}{m_s} = \frac{m_-}{m_s},$$
(22.8)

it follows that
$$\gamma_+ = \alpha f_+$$
(22.9)

and
$$\gamma_- = \alpha f_-.$$
(22.10)

With these two types of activity coefficient clearly differentiated, we are in a position to examine the principles of the experimental methods for obtaining dissociation constants of weak electrolytes.

II. DETERMINATION OF EQUILIBRIUM CONSTANTS FOR DISSOCIATION OF WEAK ELECTROLYTES

Within recent years, three experimental methods have been developed which are capable of determining dissociation constants with a precision of the order of tenths of 1 per cent. Each of these—the electromotive-force,[1] the conductance,[2] and the optical methods[3]— is based on classical techniques, but the methods of treatment of the data, as well as the experimental procedures, have been improved substantially. Since the optical method is rather limited, because the acid and conjugate base must show substantial differences in absorption of visible or ultraviolet light, we shall limit our discussion to the two electrical methods.

[1] H. S. Harned and R. W. Ehlers, *J. Am. Chem. Soc.*, **54**, 1350 (1932).
[2] D. A. MacInnes and T. Shedlovsky, *J. Am. Chem. Soc.*, **54**, 1429 (1932).
[3] H. von Halban and G. Kortüm, *Z. physik. Chem.*, **170**, 212, 351 (1934).

A. From electromotive-force measurements. If we are interested in the dissociation constant of a weak acid (or base), it is frequently possible to arrange a cell, without liquid junction, whose potential depends upon the ion concentrations in the solution and hence on the dissociation constant of the acid. As an example, we may consider acetic acid in a cell containing a hydrogen electrode and a silver-silver chloride electrode:

$$\text{H}_2; \quad \text{HC}_2\text{H}_3\text{O}_2 \ (m_1), \ \text{NaC}_2\text{H}_3\text{O}_2 \ (m_2), \ \text{NaCl} \ (m_3); \quad \text{AgCl, Ag.} \quad (22.11)$$

Since the reaction which is associated with this cell is

$$\tfrac{1}{2}\text{H}_2 \ (g) + \text{AgCl} \ (s) = \text{Ag} \ (s) + \text{HCl} \ (aq), \quad\quad (22.12)$$

the electromotive force must be given by the expression (Chapter 21, Section IIA)

$$\varepsilon = \varepsilon^\circ - \frac{RT}{\mathcal{F}} \ln a_{\text{HCl}} = \varepsilon_0 - \frac{RT}{\mathcal{F}} \ln (m_{\text{H}^+} m_{\text{Cl}^-} \gamma_{\text{H}^+} \gamma_{\text{Cl}^-}). \quad (22.13)$$

Since the molality m_{H^+} depends upon the acetic acid equilibrium, which we may indicate in a simplified notation by the equation

$$\text{HAc} = \text{H}^+ + \text{Ac}^-, \quad\quad (22.14)$$

we have a pathway by which the dissociation constant, K, for acetic acid may be introduced into the equation for the e.m.f. Thus, K is given by the expression

$$K = \frac{m_{\text{H}^+} m_{\text{Ac}^-}}{m_{\text{HAc}}} \frac{f_{\text{H}^+} f_{\text{Ac}^-}}{f_{\text{HAc}}}, \quad\quad (22.15)$$

which may be rearranged to give an equation for m_{H^+}, which then can be introduced into (22.13). Thus we obtain

$$\varepsilon = \varepsilon^\circ - \frac{RT}{\mathcal{F}} \ln \left(K \frac{m_{\text{HAc}}}{m_{\text{Ac}^-}} m_{\text{Cl}^-} \frac{\gamma_{\text{H}^+} \gamma_{\text{Cl}^-} f_{\text{HAc}}}{f_{\text{H}^+} f_{\text{Ac}^-}} \right). \quad\quad (22.16)$$

This equation can be broken up, for convenience, into several terms and transposed to read

$$\varepsilon - \varepsilon^\circ + \frac{RT}{\mathcal{F}} \ln \frac{m_{\text{HAc}} m_{\text{Cl}^-}}{m_{\text{Ac}^-}} = -\frac{RT}{\mathcal{F}} \ln K - \frac{RT}{\mathcal{F}} \ln \frac{\gamma_{\text{Cl}^-} f_{\text{HAc}}}{f_{\text{Ac}^-}}$$

$$\equiv -\frac{RT}{\mathcal{F}} \ln K'. \quad\quad (22.17)[4]$$

---

[4] Since f_{H^+} is defined in terms of m_{H^+}, the total molality of the hydrogen ion produced by the acetic acid, and γ_{H^+} is defined on the assumption that no molecules of undissociated HCl are produced, the m used is the same for each activity coefficient; hence $\gamma_{\text{H}^+} = f_{\text{H}^+}$. Therefore these activity coefficients may be cancelled out of equation (22.16) and hence they are absent from equation (22.17).

Since $\mathcal{E}^\circ$ has been evaluated already (Chapter 21, Section IIA), all of the terms on the left-hand side of equation (22.17) can be determined or estimated. Thus, $\mathcal{E}$ is the experimental value of the potential of the cell in (22.11); the molalities are given by the expressions

$$m_{Cl^-} = m_3, \tag{22.18}$$

$$m_{HAc} = m_1 - m_{H^+}, \tag{22.19}$$

and
$$m_{Ac^-} = m_2 + m_{H^+}. \tag{22.20}$$

In general, $m_{H^+} \ll m_1$ or m_2, so that it can be estimated from equation (22.15) by the insertion of an approximate value of K and the neglect of activity coefficients. Thus it is possible to obtain tentative values of $-(RT/\mathfrak{F}) \ln K'$ and hence of K' at various concentrations of acetic acid, sodium acetate, and sodium chloride, respectively. The ionic strength, μ, may be estimated as

$$\mu = m_2 + m_3 + m_{H^+}. \tag{22.21}$$

Then, if $-(RT/\mathfrak{F}) \ln K'$, or K', is plotted against the ionic strength and extrapolated to $\mu = 0$, it is evident from the right-hand side of equation (22.17) that as

$$\mu \to 0,$$

$$\gamma_{ratio} = \frac{\gamma_{Cl^-} f_{HAc}}{f_{Ac^-}} \to 1,$$

$$\log \gamma_{ratio} \to 0,$$

$$\frac{RT}{\mathfrak{F}} \ln K' \to \frac{RT}{\mathfrak{F}} \ln K,$$

and therefore
$$K' \to K. \tag{22.22}$$

A typical extrapolation of the data for acetic acid is illustrated in Fig. 1. At 25°C, the value of 1.754×10^{-5} has been found for K by this method.

If the equilibrium constant is not already known fairly well, the K determined by the procedure outlined may be looked upon as a first approximation. It may then be used to estimate m_{H^+} for insertion into equations (22.19), (22.20), and (22.21), and a second extrapolation may be carried out. In this way a second value of K may be obtained. The process may then be repeated until successive estimates of K agree within the precision of the experimental data.

B. From conductivity measurements. Like the electromotive-force method, conductance measurements have been used for many decades for the estimation of dissociation constants of weak electrolytes. If we use acetic acid as our example again, we find that the equivalent conductance, Λ, shows a strong dependence upon concentration, as is illustrated in Fig. 2. The rapid decline in Λ with increasing concentration is due largely to a decrease in the fraction of dissociated molecules.

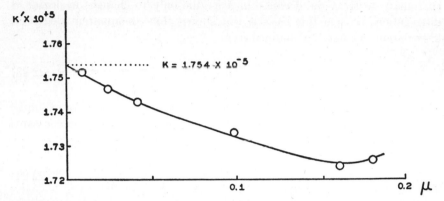

Fig. 1. Extrapolation of K' values in the determination of the ionization constant of acetic acid at 25°C. [Based on data from H. S. Harned and R. W. Ehlers, *J. Am. Chem. Soc.*, **54**, 1350 (1932).]

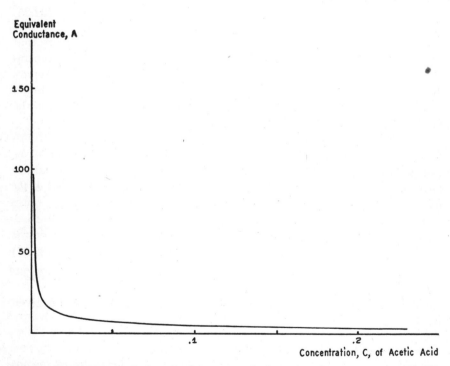

Fig. 2. Equivalent conductances of aqueous solutions of acetic acid at 25°C. [Based on data from D. A. MacInnes and T. Shedlovsky, *J. Am. Chem. Soc.*, **54**, 1429 (1932).]

In the classical treatment of the conductance of weak electrolytes it was customary to treat the decrease in Λ as due only to changes in degree of dissociation, α. On this basis it was shown that an apparent degree of dissociation, α', may be obtained from

$$\alpha' = \frac{\Lambda}{\Lambda_0},\tag{22.23}$$

where Λ_0 is the equivalent conductance of the weak electrolyte, for example, acetic acid, at infinite dilution. Hence the apparent dissociation constant, K', should be obtainable from the expression

$$K' = \frac{C'_{\mathrm{H^+}}C'_{\mathrm{Ac^-}}}{C'_{\mathrm{HAc}}} = \frac{(\alpha'C)(\alpha'C)}{(1 - \alpha')C} = \frac{(\Lambda/\Lambda_0)^2 C}{1 - (\Lambda/\Lambda_0)},\tag{22.24}$$

if C is the total (stoichiometric) concentration of acetic acid. Λ_0 generally was evaluated from data at infinite dilution for strong electrolytes. Thus, for acetic acid Λ_0 was obtained as follows:

$$\Lambda_0(\mathrm{H^+ + Ac^-}) = \Lambda_0(\mathrm{H^+ + Cl^-}) + \Lambda_0(\mathrm{Na^+ + Ac^-})$$
$$- \Lambda_0(\mathrm{Na^+ + Cl^-}).\tag{22.25}$$

For more precise calculations, however, it is necessary to recognize that the mobility (and hence the conductance) of ions changes with concentration, even when dissociation is complete, because of interionic forces. Thus, equation (22.23) must be oversimplified in its use of Λ_0 to evaluate α, since at any finite concentration the equivalent conductances of the $\mathrm{H^+}$ and $\mathrm{Ac^-}$ ions, even when dissociation is complete, do not equal Λ_0.

To allow for the change in mobility due to changes in ion concentrations, MacInnes and Shedlovsky[5] proposed the use of a quantity Λ_e in place of Λ_0. For acetic acid, for example, Λ_e is obtained from the equivalent conductances of HCl, NaAc, and NaCl at *a concentration, C_i, equal to that of the ions in the solution of acetic acid.* Thus, since

$$\Lambda_{\mathrm{HCl}} = 426.04 - 156.70\sqrt{C} + 165.5C\,(1 - 0.2274\sqrt{C}),\tag{22.26}$$

$$\Lambda_{\mathrm{NaAc}} = 90.97 - 80.48\sqrt{C} + 90C\,(1 - 0.2274\sqrt{C}),\tag{22.27}$$

and $\quad \Lambda_{\mathrm{NaCl}} = 126.42 - 88.53\sqrt{C} + 89.5C\,(1 - 0.2274\sqrt{C}),\tag{22.28}$

the effective conductance, Λ_e, of completely dissociated acetic acid is given by

$$\Lambda_e = \Lambda_{\mathrm{HCl}} + \Lambda_{\mathrm{NaAc}} - \Lambda_{\mathrm{NaCl}} = 390.59 - 148.61\sqrt{C_i} +$$
$$165.5C_i(1 - 0.2274\sqrt{C_i}).\tag{22.29}$$

---

[5] D. A. MacInnes and T. Shedlovsky, *J. Am. Chem. Soc.*, **54**, 1429 (1932).

Assuming now that the degree of dissociation at the stoichiometric molar concentration, C, is given by the expression

$$\alpha'' = \frac{\Lambda}{\Lambda_e}, \tag{22.30}$$

we obtain a further approximation for the dissociation constant:

$$K'' = \frac{C''_{H^+} C''_{Ac^-}}{C''_{HAc}} = \frac{(\alpha'' C)^2}{(1 - \alpha'') C} = \frac{(\Lambda/\Lambda_e)^2 C}{1 - (\Lambda/\Lambda_e)}. \tag{22.31}$$

If in addition we now insert appropriate activity coefficients we obtain a third approximation for the dissociation constant:

$$K''' = \frac{C''_{H^+} C''_{Ac^-}}{C''_{HAc}} \frac{f_{H^+} f_{Ac^-}}{f_{HAc}}. \tag{22.32}$$

This equation may be converted into logarithmic form to give

$$\log K''' = \log K'' + \log \left(\frac{f_{\pm}^2}{f_u} \right), \tag{22.33}$$

where
$$f_{\pm}^2 = f_{H^+} f_{Ac^-} \tag{22.34}$$
and
$$f_u = f_{HAc}. \tag{22.35}$$

To evaluate $\log K''$, it is necessary to know Λ_e and therefore C_i. Yet to know C_i we must have a value for α, which, in turn, depends upon a knowledge of Λ_e. In practice, this stalemate is overcome by a method of successive approximations. To begin, we take $\Lambda_e = \Lambda_0$ and make a first approximation for α from equation (22.30). With this value of α'' we can calculate a tentative C_i, which can be inserted in equation (22.29) to give a tentative value of Λ_e. From the latter and equation (22.30), a new value of α'' is obtained, which leads to a revised value for C_i and subsequently for Λ_e. This method is continued until successive calculations give substantially the same value of α''. Thus, for a 0.02000-molar solution of acetic acid, with an equivalent conductance of 11.563 ohms^{-1}, a first approximation to α is

$$\alpha' = \frac{11.563}{390.59} = 0.029604,$$

since $\Lambda_0 = 390.59$ ohms^{-1}. Therefore

$$C_i = \alpha' C = 0.029604\,(0.02000) = 0.00059208,$$
and
$$\sqrt{C_i} = 0.024333.$$

Insertion of this value of $\sqrt{C_i}$ into equation (22.29) leads to a value of

$$\Lambda_e = 387.07.$$

Coupling this value with 11.563 for Λ, we obtain

$$\alpha'' = \frac{11.563}{387.07} = 0.029873,$$

$$C_i = 0.00059746,$$

$$\sqrt{C_i} = 0.024443,$$

and
$$\Lambda_e = 387.06.$$

A third calculation of α'' gives 0.029874, which is substantially the same as the result of the second approximation and hence is available for use in equation (22.31).

Thus it is possible to obtain precise values of $\log K''$. The determination of K can then be carried out by an extrapolation procedure. As is evident from equation (22.33), $\log K'''$ differs from $\log K''$ because of the activity-coefficient term. Little can be said about the dependence of f_u on concentration,[6] but from the Debye-Hückel theory[7] we should expect $\log f_{\pm}^2$ to depend upon $\sqrt{\mu}$, the dependence approaching linearity with increasing dilution. Hence, if we plot $\log K''$ vs. $\sqrt{\mu}$, the extrapolated intercept should be $\log K$, where K is the true thermodynamic constant; because as

$$\sqrt{\mu} \to 0,$$

$$f_{\pm}^2 \to 1,$$

$$\frac{f_{\pm}^2}{f_u} \to 1,$$

$$\log \frac{f_{\pm}^2}{f_u} \to 0,$$

and therefore
$$\log K'' \to \log K. \tag{22.36}$$

The extrapolation of the data for acetic acid is illustrated in Fig. 3. The slope of the curve is not too large, and thus the intercept can be determined with confidence. MacInnes and Shedlovsky report a value for K of 1.753×10^{-5} at 25°C.

An alternative method of extrapolation, in which the slope is reduced almost to zero, may be carried out by the following modification of equation (22.33). Separating the activity coefficients, we obtain

$$\log K''' = \log K'' + \log f_{\pm}^2 - \log f_u. \tag{22.37}$$

There is little that we can say about $\log f_u$, except that, empirically speaking, it is a function of the first power of the concentration. However, we

[6] See M. Randall and C. F. Failey, *Chem. Rev.*, **4**, 291 (1927).

[7] The Debye-Hückel theory gives the *actual* ionic activity coefficient, f_i, rather than the stoichiometric activity coefficient. For strong electrolytes it is customarily assumed that dissociation is complete, and hence γ_i has been used in equations (21.124)–(21.126).

can be quite specific about the properties of $\log f_{\pm}^2$ in very dilute solution, because in that region the Debye-Hückel limiting law should be a good approximation. Thus we may introduce the expression

$$\log f_{\pm}^2 = 2 \log f_{\pm} = 2[-0.509 \, |z_+ z_-| \, \sqrt{\mu}] \qquad (22.38)$$
$$= -1.018\sqrt{\mu}$$

into equation (22.37) and rearrange the resultant equation into the relation

$$\log K''' + \log f_u = \log K'' - 1.018\sqrt{\mu} \equiv \log K''''. \qquad (22.39)$$

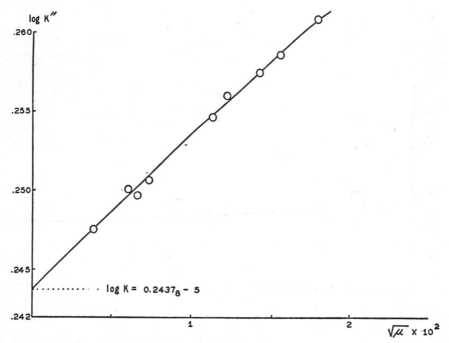

Fig. 3. Extrapolation of ionization constants of acetic acid.

Since both K'' and $\sqrt{\mu}$ can be calculated from experimental data, $\log K''''$ can be determined easily. A plot of $\log K''''$, against μ this time, gives a curve with small slope, such as is illustrated in Fig. 4. The determination of the intercept in this graph is probably easier than it is in Fig. 3, even for uni-univalent ions, and is certainly so when the dissociation process involves polyvalent ions.

III. SOME TYPICAL CALCULATIONS

A. Standard free energy of formation of aqueous solute: HCl. We have discussed in some detail the various methods which can be used to obtain the standard free energy of formation of a pure gaseous compound such as HCl (g). Since many of its reactions are carried out in water solution, it is desirable to know $\Delta F f°$ for HCl (aq) also.

Our problem can be reduced essentially to that of finding $\Delta F°$ for the reaction

$$HCl\ (g, a = f = 1) = HCl\ (aq, a_2 = 1), \qquad (22.40)$$

since to this $\Delta F°$ we can always add $\Delta F f°$ of HCl (g) to attain our objective. It is necessary to realize that although a_{HCl} in equation (22.40) is unity on both sides, the standard states are not the same for the gaseous and aqueous states, and hence the fugacities and partial molal free energies are not equal. To obtain $\Delta F°$ for reaction (22.40), we may express both activities in terms of a common standard state, or we may break up the reaction into a set of transformations for which we can find the fugacities or relative fugacities

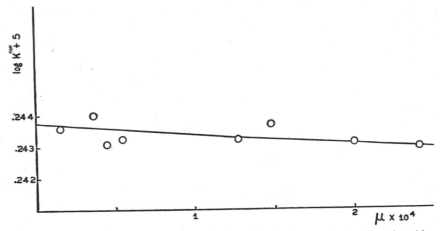

Fig. 4.   Alternative method of extrapolation of ionization constants of acetic acid.

of both reactant and product from available data. We shall adopt this latter course.

Since data for the activities of HCl in aqueous solution are available, and since in the gaseous phase at low pressures we may set the fugacity equal to the pressure, we could solve our problem if we could find some aqueous solution for which the equilibrium partial pressure of HCl is known. Such measurements have been carried out,[8] and the equilibrium partial pressures of HCl over several solutions are known. We shall use the information for one of these solutions, that of 4-molal concentration, for which the equilibrium partial pressure is 0.2395×10^{-4} atm.

[8] S. J. Bates and H. D. Kirschman, *J. Am. Chem. Soc.*, 41, 1991 (1919); cf. G. N. Lewis and M. Randall, *Thermodynamics*, McGraw-Hill Book Company, Inc., 1923, page 330.

Thus we can now write the following three equations, the sum of which is equivalent to reaction (22.40):

HCl (g, $a = f = p = 1$) = HCl (g, $p = 0.2395 \times 10^{-4}$ atm);
$$\Delta F_{298} = RT \ln (0.2395 \times 10^{-4}) = -6310 \text{ cal mole}^{-1}, \quad (22.41)$$

HCl (g, $p = 0.2395 \times 10^{-4}$ atm) = HCl (aq, $m = 4$);
$$\Delta F_{298} = 0 \text{ (equilibrium reaction)}, \quad (22.42)$$

HCl (aq, $m = 4$, $a_2' = 49.66$) = HCl (aq, $a_2 = 1$);

$$\Delta F_{298} = RT \ln \frac{a_2}{a_2'} = RT \ln \frac{1}{49.66} = -2313 \text{ cal mole}^{-1}. \quad (22.43)$$

The activity of HCl in a 4-molal solution, required for the ΔF in equation (22.43), was calculated from the mean activity coefficient, 1.762, taken from tables of Harned and Owen,[9] as follows:

$$a_2 = m_{\pm}^2 \gamma_{\pm}^2 = (4)^2(1.762)^2 = 49.66. \quad (22.44)$$

Thus we can now obtain the standard free-energy change associated with equation (22.40), since the sum of equations (22.41)–(22.43) leads to

HCl (g, $a = 1$) = HCl (aq, $a_2 = 1$); $\Delta F_{298}^{\circ} = -8623 \text{ cal mole}^{-1}$.
$$(22.45)$$

Having obtained the standard free-energy change accompanying the transfer of HCl from the gaseous to the aqueous state, we can add to it the standard free energy of formation of gaseous HCl,

½H₂ (g) + ½Cl₂ (g) = HCl (g); $\Delta Ff^{\circ} = -22{,}770 \text{ cal mole}^{-1}$,   (22.46)

and obtain the standard free energy of formation of aqueous HCl:

½H₂ (g) + ½Cl₂ (g) = HCl (aq); $\Delta Ff^{\circ} = -31{,}393 \text{ cal mole}^{-1}$.   (22.47)

B.   Standard free energy of formation of individual ions: HCl. No method has yet been developed for separating the free energy of formation of a strong electrolyte into the free energies for the formation of its constituent ions.   In fact, it has even been shown that the free energy of formation of an individual ion has no operational meaning.[10]   Nevertheless, for the purposes of tabulation and calculation it is possible to break up ΔFf° *arbitrarily* into two or more parts, corresponding to the number of ions formed, in a fashion analogous to that used in tables of standard electrode potentials.   In both cases the standard free energy of formation of aqueous H^+ is taken as zero at every temperature:

$$\tfrac{1}{2}H_2 \text{ (g)} = H^+ \text{ (aq)} + e; \quad \Delta Ff^{\circ} = 0. \quad (22.48)$$

[9] H. S. Harned and B. B. Owen, *The Physical Chemistry of Electrolytic Solutions,* Reinhold Publishing Corporation, New York, 1943, page 547.

[10] E. A. Guggenheim, *J. Phys. Chem.,* **33,** 842 (1929);   *ibid.,* **34,** 1541 (1930).

With this assumption it is possible to calculate the standard free energies of formation of other ions. For example, for Cl^- ion we proceed by adding appropriate equations to reaction (22.47). First we shall need the change associated with the reaction

$$HCl \ (aq, a_2 = 1) = H^+ \ (aq, a_+ = 1) + Cl^- \ (aq, a- = 1). \quad (22.49)$$

It can be shown readily, however, that we have already arbitrarily taken $\Delta F°$ for this reaction as zero by defining the individual ion activities [equation (21.15)] in such a fashion that

$$a_2 = (a_+)(a_-). \quad (22.50)$$

The equilibrium constant, K, for reaction (22.49) thus is unity:

$$K = \frac{a_{H^+}a_{Cl^-}}{a_{HCl}} = \frac{(a_+)(a_-)}{(a_2)} = 1. \quad (22.51)$$

Hence for this reaction

$$\Delta F° = 0. \quad (22.52)$$

Therefore, if to equation (22.47) we add (22.49) and subtract (22.48),

$$\tfrac{1}{2}H_2 \ (g) + \tfrac{1}{2}Cl_2 \ (g) = HCl \ (aq); \quad \Delta Ff°_{298} = -31,393 \ \text{cal mole}^{-1}, \quad (22.47)$$

$$HCl \ (aq) = H^+ \ (aq) + Cl^- \ (aq); \quad \Delta F° = 0, \quad (22.49)$$

$$H^+ \ (aq) + e = \tfrac{1}{2}H_2 \ (g); \quad \Delta F° = 0, \quad (22.48)$$

we obtain the standard free energy of formation of Cl^- ion:

$$\tfrac{1}{2}Cl_2 \ (g) + e = Cl^- \ (aq); \quad \Delta F°_{298} = -31,393 \ \text{cal mole}^{-1}. \quad (22.53)$$

This $\Delta F°$ corresponds, of course, to the value which may be calculated from the standard electrode potential.

C. Standard free energy of formation of solid solute in aqueous solution.

1. *Solute very soluble: NaCl.* The solution of this problem is analogous to that for HCl (aq) except that the step of bringing the pure solute into solution under equilibrium conditions will differ, since NaCl is a solid. Assuming that the standard free energy of formation of NaCl (s) is available,[11] the $\Delta Ff°_{298}$ for NaCl (aq) can be obtained by a summation of the following processes:

---

[11] W. M. Latimer, *Oxidation Potentials*, Prentice-Hall, Inc., New York, 1938.

Na (s) + ½Cl$_2$ (g) = NaCl (s); ΔF° = $-91{,}770$ cal mole^{-1}, (22.54)

NaCl (s) = NaCl (aq, satd, m = 6.12); ΔF = 0 (equilibrium), (22.55)

NaCl (m = 6.12, a_2' = 38.42) = NaCl (a_2 = 1);

$$\Delta F = RT \ln \frac{1}{a_2'} = -2163 \text{ cal mole}^{-1}, \qquad (22.56)$$

NaCl (a_2 = 1) = Na$^+$ (a_+ = 1) + Cl$^-$ (a_- = 1); ΔF = 0. (22.57)

Na (s) + ½Cl$_2$ (g) = Na$^+$ (a_+ = 1) + Cl$^-$ (u_- = 1);
$$\Delta F^{\,\circ}_{298} = -93{,}933 \text{ cal mole}^{-1}. \quad (22.58)$$

The value of a_2' in equation (22.56) is obtained as follows:

$$a_2' = (a_+)(a_-) = (m_\pm)^2(\gamma_\pm)^2 = (6.12)^2(1.013)^2 = 38.42. \quad (22.59)$$

Having the standard free energy of formation of the aqueous ions, we can also obtain that for the Na$^+$ ion alone, by subtracting equation (22.53) from (22.58). Thus we obtain

$$\text{Na (s) = Na}^+ \text{ (aq) } + e; \quad \Delta F^\circ_{298} = -62{,}540 \text{ cal mole}^{-1}. \quad (22.60)$$

2. *Slightly soluble solute: AgCl.* The procedure in this case is a little different from that for NaCl because of the use of the solubility-product method of expressing the solubility of a slightly soluble substance such as AgCl. Nevertheless, the calculation can be carried out readily, following the same general principle of adding individual reactions for which ΔF can be evaluated. Thus, for AgCl we can add the following equations:

Ag (s) + ½Cl$_2$ (g) = AgCl (s); ΔF° = $-26{,}220$ cal mole^{-1}, (22.61)

AgCl (s) = AgCl (aq, satd); ΔF = 0 (equilibrium), (22.62)

AgCl (aq, satd, a_2' = $a_+'a_-'$) = Ag$^+$ (a_+ = 1) + Cl$^-$ (a_- = 1);

$$\Delta F = RT \ln \frac{(a_+a_-)}{(a_+'a_-')_{\text{satd soln}}} = -RT \ln K_{\text{S.P.}}$$
$$= 13{,}356 \text{ cal mole}^{-1}. \quad (22.63)$$

Ag (s) + ½Cl$_2$ (g) = Ag$^+$ (a_+ = 1) + Cl$^-$ (a_- = 1);
$$\Delta F^\circ_{298} = -12{,}864 \text{ cal mole}^{-1}. \quad (22.64)$$

Once again, if we take the value in equation (22.53) for the free energy of formation of the Cl$^-$ ion, we can calculate a value for the Ag$^+$ ion by subtracting (22.53) from (22.64). Thus we obtain

$$\text{Ag (s) = Ag}^+ \text{ (a_+ = 1) } + e; \quad \Delta F^\circ_{298} = 18{,}529 \text{ cal mole}^{-1}. \quad (22.65)$$

D. Standard free energy of formation of ion of weak electrolyte. An illustrative example of this type of problem may be taken from data of Borsook and Schott[12] for the formation of the first anion of succinic acid,

---

[12] H. Borsook and H. F. Schott, *J. Biol. Chem.*, 92, 535 (1931).

$C_4H_5O_4^-$. The solubility of succinic acid in water at 25° is 0.715 mole $(1000 \text{ g. } H_2O)^{-1}$. In such a solution the acid is 1.12 per cent ionized ($\alpha = 0.0112$) and the undissociated portion has an activity coefficient of 0.87. Knowing the first dissociation constant of succinic acid, 6.4×10^{-5}, we can calculate $\Delta F°$ of formation of the $C_4H_5O_4^-$ ion by the addition of the following equations:

$$3H_2 \text{ (g)} + 4C \text{ (graphite)} + 2O_2 \text{ (g)} = C_4H_6O_4 \text{ (s)};$$
$$\Delta F°_{298} = -178{,}800 \text{ cal mole}^{-1}, \quad (22.66)$$

$$C_4H_6O_4 \text{ (s)} = C_4H_6O_4 \text{ (aq, satd)}; \quad \Delta F = 0 \text{ (equilibrium)}, \quad (22.67)$$

$$C_4H_6O_4 \text{ (aq, satd, } a_2') = C_4H_6O_4 \text{ (}a_2 = 1);$$
$$\Delta F = RT \ln (a_2/a_2') = 290 \text{ cal mole}^{-1}, \quad (22.68)$$

$$C_4H_6O_4 \text{ (}a_2 = 1) = H^+ \text{ (}a_+ = 1) + C_4H_5O_4^- \text{ (}a_- = 1);$$
$$\Delta F° = -RT \ln K = 5700 \text{ cal mole}^{-1}. \quad (22.69)$$

$$3H_2 \text{ (g)} + 4C \text{ (graphite)} + 2O_2 \text{ (g)} = H^+ \text{ (}a_+ = 1) + C_4H_5O_4^- \text{ (}a_- = 1);$$
$$\Delta F° = -172{,}810 \text{ cal mole}^{-1}. \quad (22.70)$$

Of these steps, only equation (22.68) needs any comment. For the free-energy change in this step, a_2' of the *undissociated* species of succinic acid in the saturated solution is obtained as follows:

$$a_2' = m_u f_u = m_{\text{stoichiometric}} (1 - \alpha) f_u = (0.715)(1 - 0.0112)(0.87). \quad (22.71)$$

E. Standard free energy of formation of moderately strong electrolyte. Moderately strong electrolytes, such as aqueous HNO_3, have generally been treated thermodynamically as completely dissociated substances. Thus, for HNO_3 (aq) the value for $\Delta Ff°$ of $-26{,}250$ cal mole^{-1} listed by Latimer[13] refers to the reaction

$$\tfrac{1}{2}H_2 \text{ (g)} + \tfrac{1}{2}N_2 \text{ (g)} + \tfrac{3}{2}O_2 \text{ (g)} = H^+ \text{ (aq, } a_+ = 1) + NO_3^- \text{ (aq, } a_- = 1); \quad (22.72)$$

the activity of the nitric acid is defined by the equation

$$a_{HNO_3} = a_{H^+} a_{NO_3^-} = m_s^2 \gamma_{\pm}^2, \quad (22.73)$$

where m_s is the stoichiometric (or total) molality of the acid.

Within recent years optical methods applicable to moderately strong electrolytes have been made increasingly precise;[14] it has proved feasible to determine by these methods dissociation constants, and hence concentrations of the undissociated species. Thus, for HNO_3 in aqueous solution at 25°C, K is 21 ± 4.[15] In defining this equilibrium constant, however,

---

[13] W. M. Latimer, *Oxidation Potentials*, Prentice-Hall, Inc., New York, 1938, page 83.
[14] For a recent review, see T. F. Young and L. A. Blatz, *Chem. Rev.*, **44**, 93 (1949).
[15] O. Redlich and J. Bigeleisen, *J. Am. Chem. Soc.*, **65**, 1883 (1943).

we have changed the standard state for aqueous nitric acid, because now the activity of the undissociated species is given by the equation

$$a'_{HNO_3} = m_u f_u = a_u,$$ (22.74)

where the subscript u refers to the undissociated species. Equation (22.73) is therefore no longer applicable, and in its place we have

$$\frac{a_{H^+} a_{NO_3^-}}{a_u} = K = 21.$$ (22.75)

With these considerations clearly in mind, we are in a position to calculate $\Delta F f^\circ$ of undissociated HNO_3. For this purpose we may add the following equation to equation (22.72):

$$H^+ (aq, a_+ = 1) + NO_3^- (aq, a_- = 1) = HNO_3 (aq, a_u = 1).$$ (22.76)

For this reaction

$$\Delta F^\circ = -RT \ln \frac{1}{K} = 1800 \text{ cal mole}^{-1}.$$ (22.77)

Hence the standard free energy of formation of molecular, undissociated aqueous HNO_3 is $-24,450$ cal mole^{-1}, the sum of the ΔF°'s for reactions (22.72) and (22.76):

$$\tfrac{1}{2}H_2 (g) + \tfrac{1}{2}N_2 (g) + \tfrac{3}{2}O_2 (g) = HNO_3 (aq, a_u = 1);$$
$$\Delta F f^\circ = -24,450 \text{ cal mole}^{-1}.$$ (22.78)

F. General comments. These problems illustrate some of the methods which can be used to combine data for the free energies of pure phases with information on the behavior of constituents of a solution, in order to calculate fundamental thermodynamic properties for the compound in solution. Not all possible situations can be anticipated in a limited number of examples. Nevertheless, the solutions to the problems illustrated, combined with those of some of the exercises at the end of this chapter, should emphasize the general approach: break down the reaction into a series of consecutive steps for each one of which the free-energy change can be calculated from suitable experimental data.

IV. ENTROPIES OF IONS

In dealing with solutions we encounter the same challenge in connection with free-energy data that led to the development of the third law for pure substances. It may be necessary frequently to obtain values of ΔF for a process in solution for which only thermal data are available. If absolute entropy data could be obtained for solutions also, then it would be possible again to calculate ΔF° from a calorimetric determination of ΔH° and the equation

$$\Delta F^\circ = \Delta H^\circ - T \Delta S^\circ.$$ (22.79)

For aqueous solutions of electrolytes a concise method of tabulating such entropy data is in terms of the individual ions, since entropies for these species can be combined to give information on a wide variety of salts. The initial assembling of the ionic entropies is carried out generally by a reverse application of equation (22.79), that is, $\Delta Sf°$ of a salt is calculated from known values of $\Delta Ff°$ and $\Delta Hf°$ for that salt. The entropy of formation of the cation and anion together can then be separated, after a suitable convention has been adopted, into the entropies for the individual ions.

A. The entropy of an aqueous solution of a salt. The method of calculation of the absolute entropy of a salt may be illustrated with an example, NaCl, for which we have already solved part of the problem, the calculation of $\Delta F°$. From equation (22.58) we have

$$Na\ (s) + \tfrac{1}{2}Cl_2\ (g) = Na^+\ (a_+ = 1) + Cl^-\ (a_- = 1);$$
$$\Delta F^°_{298} = -93,933\ cal\cdot mole^{-1}. \quad (22.58)$$

The value of $\Delta H°$ for the same reaction can be obtained in a fashion analogous to that used for $\Delta F°$. The heat of formation of NaCl (s) is[16] $-98,330$ cal mole^{-1}. Hence we may write

$$Na\ (s) + \tfrac{1}{2}Cl_2\ (g) = NaCl\ (s); \quad \Delta H^2_{298} = -98,330\ cal\ mole^{-1}. \quad (22.80)$$

To this we wish to add the enthalpy change for the reaction

$$NaCl\ (s) = [Na^+ + Cl^-]\ (aq,\ standard\ state). \quad (22.81)$$

This quantity can be obtained readily from available data if we recollect the meaning of standard state in connection with partial molal enthalpies. As was pointed out in Chapter 19, a solute in its standard state has an activity of unity but a partial molal heat content which is the same as that of the solute in an infinitely dilute solution. Thus, the NaCl on the right-hand side of equation (22.81), in its standard state, the "*hypothetical* one-molal solution," has the partial molal enthalpy $\bar{H}^°_2$, that is, the same $\bar{H}_2$ as in the infinitely dilute solution. Consequently, for reaction (22.81) we may write

$$\Delta H° = \bar{H}^°_2 - \bar{H}^\bullet_{2(s)} = -\bar{L}^\bullet_{2(s)}. \quad (22.82)$$

Hence all we need to obtain $\Delta H°$ is the relative partial molal enthalpy of solid NaCl, which is -923 cal/mole.[17] Therefore we can now write

$$Na\ (s) + \tfrac{1}{2}Cl_2\ (g) = Na^+\ (aq) + Cl^-\ (aq);$$
$$\Delta H° = -97,407\ cal\ mole^{-1},$$
$$\Delta F° = -93,933\ cal\ mole^{-1},$$
$$\Delta S° = -11.65\ cal\ mole^{-1}\ deg^{-1}. \quad (22.83)$$

---

[16] F. R. Bichowsky and F. D. Rossini, *Thermochemistry of Chemical Substances*, Reinhold Publishing Corporation, New York, 1936.

[17] T. F. Young and O. G. Vogel, *J. Am. Chem. Soc.*, **54**, 3030 (1932).

By this procedure we have obtained the standard entropy of formation of aqueous NaCl.

B. Calculation of entropy of formation of individual ions. As in the case of free-energy changes, we may also divide the entropy change for a reaction such as (22.83) into two parts and assign one portion to each ion. Since *actual* values of individual ion entropies cannot be determined, we must establish some convention for apportioning the molecular entropy among the constituent ions.

In the treatment of the free energies of individual ions, we adopted the convention that $\Delta Ff°$ of the hydrogen ion shall be set equal to zero for all temperatures; that is,

$$\tfrac{1}{2}H_2 \text{ (g)} = H^+ \text{ (aq)} + e; \quad \Delta Ff° = 0. \tag{22.48}$$

We have also shown previously [equation (8.30)] that

$$\left(\frac{\partial \Delta F}{\partial T}\right)_P = -\Delta S. \tag{22.84}$$

If equation (22.48) is valid at all temperatures, it follows that the entropy change in the formation of hydrogen ion from gaseous hydrogen must be zero; that is,

$$\tfrac{1}{2}H_2 \text{ (g)} = H^+ \text{ (aq)} + e; \quad \Delta Sf° = -\left(\frac{\partial \Delta Ff°}{\partial T}\right)_P = 0. \tag{22.85}$$

A consistent convention therefore would set the entropy of formation of aqueous H^+ ion equal to that of one-half mole of H_2 gas; that is,

$$\bar{s}^{\circ}_{H^+} = \tfrac{1}{2} s^{\circ}_{H_2} = 15.61 \text{ cal mole}^{-1} \text{ deg}^{-1}. \tag{22.86}$$

On this basis we can obtain $\bar{s}^{\circ}_{298}$ for Cl^- ion from any one of several reactions, for example, equation (22.47) for the formation of aqueous H^+ and Cl^-. Using values of $-31,373$ and $-39,940$ cal mole^{-1} for $\Delta F°$ and $\Delta H°$, respectively, Latimer[18] has calculated -28.75 E.U. for $\Delta S°$ at 298.1°K. If we adopt the convention stated in equation (22.86), and if s°_{298} for Cl_2 (g) is taken as 53.3 E.U., it follows that for reaction (22.47)

$$\Delta Sf^{\circ}_{298} = \bar{s}^{\circ}_{H^+} + \bar{s}^{\circ}_{Cl^-} - \tfrac{1}{2} s^{\circ}_{H_2} - \tfrac{1}{2} s^{\circ}_{Cl_2} = -28.75 \text{ E.U.}; \tag{22.87}$$

hence
$$\bar{s}^{\circ}_{Cl^-} = -2.1 \text{ cal mole}^{-1} \text{ deg}^{-1}. \tag{22.88}$$

Having a value for $\bar{s}^{\circ}_{Cl^-}$, we can proceed immediately to obtain the entropy of formation of Na^+ (aq) from $\Delta S°$ for reaction (22.83):

$$\Delta Sf^{\circ}_{298} = \bar{s}^{\circ}_{Na^+} + \bar{s}^{\circ}_{Cl^-} - s^{\circ}_{Na} - \tfrac{1}{2} s^{\circ}_{Cl_2} = -11.65 \text{ cal mole}^{-1} \text{ deg}^{-1}. \tag{22.89}$$

Consequently,
$$\bar{s}^{\circ}_{Na^+} = 29.3 \text{ cal mole}^{-1} \text{ deg}^{-1}. \tag{22.90}$$

---

[18] W. M. Latimer, *Chem. Rev.*, 18, 349 (1936).

By procedures analogous to those described in the preceding two examples, we can obtain entropies for many aqueous ions. A list of such values has been assembled in Table 1.

Also listed in Table 1 is a set of ionic entropies based on a convention different from the one just described. Historically, the usefulness of ionic entropies was first emphasized by Latimer and Buffington,[19] who established the convention of setting the standard entropy of hydrogen ion equal to zero, that is,

$$\bar{s}^{\circ}_{H^+} = 0. \tag{22.91}$$

With that convention, the entropy of formation of H^+ (aq) is not zero:

$$\tfrac{1}{2}H_2 \text{ (g)} = H^+ \text{ (aq)} + e; \quad \Delta Sf^{\circ} = \bar{s}^{\circ}_{H^+} - \tfrac{1}{2}\, s^{\circ}_{H_2} = -15.61 \text{ E.U.} \tag{22.92}$$

Consequently, equation (22.84) for the temperature coefficient of the hydrogen electrode cannot be used, since ΔFf° of the aqueous hydrogen ion is still retained as zero at all temperatures. Nevertheless, since this is the convention which has been followed in the literature, a column of ionic entropies on this basis has been included in Table 1. In practice the same values of ΔS° are obtained for any reaction no matter which basis for ion entropies is used, so long as the same column of $\bar{s}^{\circ}$'s is used throughout the calculation. The validity of this statement will be evident in the numerical example described in the next section.

C. Utilization of ion entropies in thermodynamic calculations. With tables of ion entropies available, it is possible to estimate the free-energy change which cannot be obtained at all, or not with any reasonable precision, by direct experiment. For example, an adequate calcium electrode has yet to be prepared; nevertheless it is possible to calculate the electrode potential, or more specifically the free-energy change in the reaction

$$Ca \text{ (s)} + 2H^+ \text{ (aq)} = Ca^{++} \text{ (aq)} + H_2 \text{ (g)}, \tag{22.93}$$

from the data in Table 1 plus a knowledge of the ΔH° of this reaction. Thus

$$\begin{aligned} \Delta S^{\circ}_{298} &= \bar{s}^{\circ}_{Ca^{++}} + s^{\circ}_{H_2} - s^{\circ}_{Ca} - 2\bar{s}^{\circ}_{H^+} \\ &= 18.5 + 31.21 - 9.95 - 31.21 = 8.55 \text{ E.U.} \end{aligned} \tag{22.94}$$

Since ΔH° is $-129{,}800$ cal mole^{-1}, we find a value of ΔF° of $-132{,}300$ cal mole^{-1}. Hence $\mathcal{E}^{\circ}$ for reaction (22.93) is 2.86 volts.

If we used the Latimer and Buffington entropies, instead of the entropies based on the convention of equation (22.86), we would still obtain the same answer for ΔS°_{298}. Thus, in equation (22.94) $\bar{s}^{\circ}_{Ca^{++}}$ and $2\bar{s}^{\circ}_{H^+}$ would have values of -12.7 and 0 entropy units, respectively, instead of the values

---

[19] W. M. Latimer and R. M. Buffington, *J. Am. Chem. Soc.*, **48**, 2297 (1926).

shown. Nevertheless, $\Delta S°$ turns out to be 8.55 E.U., the same answer we obtained with our convention.

With this example we conclude our discussion of absolute ion entropies. It is hardly necessary to point out that they can be used in free-energy calculations involving equation (22.79), in exactly the same manner as was outlined for the use of absolute entropies of elements and compounds. The procedures described in connection with the use of the third law of thermodynamics (Chapters 11 and 12) thus are applicable to solutions also.

TABLE 1

ENTROPIES OF AQUEOUS IONS AT 298.1 °K*

Ion	$\bar{s}°$, cal mole^{-1} deg^{-1}		Ion	$\bar{s}°$, cal mole^{-1} deg^{-1}	
	Convention of Equation (22.86)	Convention of Equation (22.91)		Convention of Equation (22.86)	Convention of Equation (22.91)
H$^+$	(15.61)	(0.00)	OH$^-$	-18.10	-2.49
Li$^+$	20.3	4.7	F$^-$	-17.9	-2.3
Na$^+$	29.6	14.0	Cl$^-$	-2.11	13.50
K$^+$	39.8	24.2	Br$^-$	-4.1	19.7
Rb$^+$	44.3	28.7	I$^-$	9.7	25.3
Cs$^+$	47.4	31.8	ClO$^-$	-5.6	10.0
NH$_4^+$	42.2	26.6	ClO$_2^-$	8.5	24.1
Ag$^+$	33.15	17.54	ClO$_3^-$	23.8	39.4
Ag(NH$_3$)$_2^+$	73.4	57.8	ClO$_4^-$	28.0	43.6
Tl$^+$	46.1	30.5	BrO$_3^-$	22.9	38.5
Mg^{++}	-0.4	-31.6	IO$_3^-$	12.4	28.0
Ca^{++}	18.5	-12.7	HS$^-$	-0.7	14.9
Sr^{++}	23.9	-7.3	HSO$_3^-$	17.0	32.6
Ba^{++}	33.5	2.3	SO$_3^{--}$	-28	3
Fe^{++}	5.3	-25.9	HSO$_4^-$	15.0	30.6
Cu^{++}	4.7	-26.5	SO$_4^{--}$	-26.8	4.4
Zn^{++}	5.5	-25.7	NO$_2^-$	14.3	29.9
Cd^{++}	14.8	-16.4	NO$_3^-$	19.4	35.0
Hg$_2^{++}$	48.9	17.7	H$_2$PO$_4^-$	6.0	21.6
Sn^{++}	26.3	-4.9	HPO$_4^{--}$	-39.9	-8.7
Pb^{++}	35.1	3.9	PO$_4^{---}$	-99	-52
Al^{+++}	-29	-76	HCO$_3^-$	6.6	22.2
Fe^{+++}	-14	-61	CO$_3^{--}$	-44.3	-13.0
			C$_2$O$_4^{--}$	-21.6	9.6
			CN$^-$	9	25
			MnO$_4^-$	31.1	46.7
			CrO$_4^{--}$	-20.7	10.5
			H$_2$AsO$_4^-$	12	28

* Based on values from W. M. Latimer, K. S. Pitzer, and W. V. Smith, *J. Am. Chem. Soc.*, **60**, 1829 (1938); C. C. Stephenson, *ibid.*, **66**, 1436 (1944).

Exercises

1. The free energy of formation of NH_4^+ (aq) may be obtained by the following procedure:

(a) $\Delta F f^\circ$ of NH_3 (g) is -3910 cal mole^{-1} at 298.1°K.   Therefore

$$\tfrac{1}{2}N_2 \text{ (g)} + \tfrac{3}{2}H_2 \text{ (g)} = NH_3 \text{ (g, } a = f = 1); \quad \Delta F f^\circ_{298} = -3910 \text{ cal mole}^{-1}. \tag{22.95}$$

(b)   A graph of the partial pressure of NH_3 (g) vs. the molality of undissociated ammonia dissolved in water has a limiting slope, p/m, of 0.01764 as m approaches zero.   If the fugacity may be taken as equal to the partial pressure, show that f_2° of the standard state *in solution*, that is, the "hypothetical one-molal solution," is 0.01764 atm.   For this calculation the quantity of ammonia which is dissociated in solution is negligible (that is, the limiting slope is reached at relatively high values of m, where the fraction of NH_4^+ ions is very small).

(c)   Show that ΔF°_{298} must be -2395 cal mole^{-1} for the reaction

$$NH_3 \text{ (g, } a = f = 1) = NH_3 \text{ (aq, } a_2 = 1). \tag{22.96}$$

Keep in mind that though a and a_2 are both unity, they have not been defined on the basis of the same standard state.

(d)   The equilibrium constant for the reaction

$$NH_3 \text{ (aq)} + H_2O \text{ (l)} = NH_4^+ \text{ (aq)} + OH^- \text{ (aq)} \tag{22.97}$$

may be taken as 1.8×10^{-5}, and the standard free energy of formation of H_2O (l) at 298.1° as $-56,560$ cal mole^{-1}.   Show that $\Delta F f^\circ_{298}$ for $NH_4^+ + OH^-$ ions must be $-56,395$ cal mole^{-1}.

(e)   If $\Delta F f^\circ_{298}$ for OH^- is $-37,455$ cal mole^{-1}, show that the corresponding quantity for NH_4^+ (aq) is $-18,940$ cal mole^{-1}.

2.   Given the following information on CO_2 and its aqueous solutions, and using standard sources of reference for any other necessary information, find the standard free energy of formation at 298.1°K for the CO_3^- ion.

Henry's law constant for solubility in H_2O = 29.5 for p in atmospheres and m in moles (kilogram $H_2O)^{-1}$.
Ionization constants of H_2CO_3: $K_1 = 3.5 \times 10^{-7}$,
$\qquad\qquad\qquad\qquad\qquad\qquad K_2 = 3.7 \times 10^{-11}$.

3.   The solubility of pure solid glycine at 25° in water is 3.33 moles (kilogram $H_2O)^{-1}$.   The activity coefficient of glycine in such a saturated solution is 0.729. Data on the relative partial enthalpies of glycine in aqueous solution are tabulated in Exercise 7 of Chapter 14.   Given ΔF° and ΔH° for solid glycine, complete the following table.

	Glycine (s)	Glycine (aq)
ΔF°	$-\ 88,520$ cal mole^{-1}	. . . . .
ΔH°	$-126,270$ cal mole^{-1}	. . . . .
ΔS°	. . . . .	. . . . .

Correct answers may be found in an article by F. T. Gucker, Jr., H. B. Pickard, and W. L. Ford, *J. Am. Chem. Soc.*, **62**, 2698 (1940).

4. The solubility of α-d-glucose in aqueous 80 per cent ethyl alcohol at 20°C is 20 grams liter^{-1}; that of β-d-glucose, 49 grams liter^{-1} [C. S. Hudson and E. Yanovsky, *J. Am. Chem. Soc.*, **39**, 1013 (1917)]. If an excess of solid α-d-glucose is allowed to remain in contact with its solution for sufficient time, some β-d-glucose is formed, and the total quantity of dissolved glucose increases to 45 grams liter^{-1}. If excess solid β-d-glucose remains in contact with its solution, some α-d-glucose is formed in the solution, and the total concentration rises to a limit, which we shall refer to as T_β.

(a) Assuming that the activity of each sugar is proportional to its concentration and that neither substance has an appreciable effect on the escaping tendency of the other, determine $\Delta F°$ for:

(1) α-d-glucose (solid) $\rightarrow$ β-d-glucose (solid); (2) α-d-glucose (solid) $\rightarrow$ α-d-glucose (solute); (3) α-d-glucose (solute) $\rightarrow$ β-d-glucose (solute).

(b) Compute the concentration T_β. Is such a solution stable? Explain.

5. The dissociation constant of acetic acid is 1.754×10^{-5}. Calculate the degree of dissociation of 0.01-molar acid in the presence of 0.01-molar NaCl. Use the Debye-Hückel limiting law for calculation of the activity coefficients of the ions. Take the activity coefficient of the undissociated acid as unity. In your calculation, neglect the concentration of H^+ in comparison with (Na^+).

6. When $\Delta F°$ is calculated from an equilibrium constant, K, what error would result at 25°C from an error of a factor of 2 in K?

7. (a) What will happen to the degree of dissociation of 0.02-molar acetic acid if NaCl is added to the solution? Explain.

(b) What will happen to the degree of hydrolysis of 0.02-molar sodium acetate if NaCl is added? Explain.

8. Two forms of solid A exist. For the transition $A' \rightarrow A''$, $\Delta F° = -1000$ cal mole^{-1}. Which is the more soluble? A' and A'' produce the same dissolved solute.

9. Henry's law is obeyed by a solute over a certain temperature range. Prove that $\bar{L}_2$, $\bar{L}_1$, and the integral heat of dilution are zero within this range. Do not assume that the constant of Henry's law is independent of temperature; it is not, in general.

10. The average value of $\bar{L}_2$ between 0° and 25°C of NaCl in 0.01-molal solution is about 45 cal mole^{-1}. According to the Debye-Hückel limiting law, $\gamma_\pm$ of NaCl in this solution at 25°C is 0.89.

(a) Calculate $\gamma_\pm$ at 0°C from the thermodynamic relation for the temperature coefficient of log $\gamma_\pm$ [equation (21.127)].

(b) Calculate $\gamma_\pm$ at 0°C from the Debye-Hückel limiting law at that temperature. Compare the result with that obtained in Part (a).

11. The activity coefficient of $CdCl_2$ in 6.62-molal solution is 0.025. The e.m.f. of the cell

$$Cd, CdCl_2 (6.62m), Cl_2$$

is 1.8111 volts at 25°C. The 6.62m solution is a saturated one; it can exist in equilibrium with solid $CdCl_2 \cdot 5\frac{1}{2}H_2O$ and water vapor at a pressure of 16.5 mm (Hg). Calculate $\Delta F°$ for the reaction

$$Cd (s) + Cl_2 (g) + 5\frac{1}{2}H_2O (l) \rightarrow CdCl_2 \cdot 5\frac{1}{2}H_2O (s).$$

For the solution of Problems 12−18, the Debye-Hückel limiting law is sufficiently accurate, and the difference between molality and molarity may be neglected. Temperature is to be taken as 25°C.

12. Calculate the e.m.f. of the cell pair

H_2; HCl (0.001), KNO_3 (0.009); AgCl, Ag—Ag, AgCl; HCl (0.01); H_2.

13. Calculate the e.m.f. of the cell

H_2; HCl (0.001), KCl (0.009); AgCl, Ag,

using the fact that the cell

H_2; HCl (0.01); AgCl, Ag

has an e.m.f. of 0.46395 volt.

14. Calculate the e.m.f. of the cell pair

Ag, AgCl; KCl (0.1), TlCl (satd), Tl—Tl, TlCl (satd), KCl (0.001); AgCl, Ag.

15. MacInnes and Shedlovsky have made conductance measurements which indicate that 0.02-molar acetic acid is 2.987 per cent ionized. Assuming that the activity coefficient of the undissociated acid is unity,

(a) Compute the ionization constant of acetic acid.

(b) Using this constant, calculate the degree of dissociation of 0.01-molar acetic acid.

16. Two solutions contain only hydrochloric acid, acetic acid, and water. In the first, the concentration of acetate ion is 0.0004-molar; in the second, 0.0001-molar. The total ionic strength in each solution is 0.01 (molar). Compute the ratio of the fugacities of acetic acid, $C_2H_4O_2$, in the two solutions.

17. Two solutions contain only sodium chloride, acetic acid, and water. In the first solution the concentration of acetate ion is 0.0004-molar; in the second it is 0.0001. The total ionic strength of each solution is 0.01. Compute the ratio of the fugacities of acetic acid, $C_2H_4O_2$, in the two solutions. What can be said about the relative (partial) vapor pressures of the monomeric form of acetic acid above the solutions? Of acetic acid dimer?

18. The solubility in 0.0005-molar KNO_3 of luteo ferricyanide, $[Co(NH_3)_6]$-$[Fe(CN)_6]$, was found by V. K. La Mer, C. V. King, and C. F. Mason [*J. Am. Chem. Soc.*, **49**, 363 (1927)] to be 3.251×10^{-5} mole per liter.

(a) In what concentration of $MgSO_4$ is its solubility the same?

(b) What is its solubility in pure water? Since the ionic strength is not known, the problem may be solved by a method of successive approximations (or by a graphical method).

(c) Calculate its solubility in 0.0025-molar $MgSO_4$. How much error is introduced by neglecting the contribution of the luteo salt itself to the ionic strength?

(d) The total molarity (C of luteo ferricyanide + C of NaCl) of a solution containing NaCl and saturated with the luteo salt is 0.0049. Calculate the molarities of each of the solutes.

CHAPTER 23

Concluding Remarks

WITH THE discussion of the calculation of thermodynamic quantities for reactions which involve solutions, we conclude our technical considerations of the theory and methods of chemical thermodynamics. We have achieved our primary objective as chemists, in that we have established the principles and procedures by which the chemical-thermodynamic properties associated with a given transformation may be determined; and we have learned how these quantities may be used to judge the feasibility of a particular reaction.

In emphasizing these aspects of the subject, however, we have neglected numerous broad fields within the realm of thermodynamics. Even within the restricted areas to which we have limited ourselves, we have omitted any discussion of topics such as surface reactions and higher-order phase transitions, and have paid only superficial attention to problems of phase equilibria and to electrochemical processes. Had we allowed ourselves to break out of our narrow confines, we might also have examined briefly some of the topics of more theoretical interest, such as equilibria involving radiation, thermoelectric and magnetic phenomena, and the effects of gravitational fields. Similarly, but with an emphasis on more practical applications, we could have considered flow processes and some problems concerned with combustion and power, to which a great deal of information has been contributed by engineers.

A textbook, and particularly an introductory one, however, cannot hope—in fact, must not aspire—to be comprehensive, lest the student become overwhelmed by the magnitude and the ramifications of the subject. For those who may wish to pursue some of these special topics, a series of appropriate references is listed at the end of this chapter.

It should be emphasized also, in concluding these general remarks, that the point of view adopted toward thermodynamics in this book is the classical or phenomenological one. This approach was inaugurated over a century ago and reached its fruition in theoretical formulation near the turn of the century. Established on firm foundations, it continues to grow, particularly in the elaboration of its details and in the extension of the range of its applications.

Parallel with this development, however, there has arisen an alternative point of view toward thermodynamics—a statistical approach to the subject. The stimulus for this new viewpoint came from two sources. On the theoretical side, the kinetic theory of gases led almost naturally to the statistical theory of matter as a whole. Perhaps even more important, however, was the discovery of experimental phenomena such as Brownian motion and density fluctuations giving rise to light scattering. These fluctuation phenomena showed that some of the principles of classical thermodynamics—in particular, the second law—cannot be extrapolated to systems of very small size.

Though statistical mechanics was initially formulated principally to account for these small-scale deviations, it has since been extended enormously in scope and applied to many problems in systems of macroscopic size. In the field of thermodynamics, statistical theory has also provided a mechanical model which permits an alternative method of formulation, on a more concrete, molecular basis, of some of the abstract concepts of classical thermodynamics. In addition, it has extended the range of thermodynamic reasoning to new kinds of experimental data—spectroscopic properties of matter—and has been fundamental to the building of a bridge between thermodynamics and kinetics of chemical reactions. For these reasons, a knowledge of statistical thermodynamics is essential as a companion to phenomenological thermodynamics for the effective solution of present problems and for the formulation of stimulating new questions.

References

Bridgman, P. W., *The Nature of Thermodynamics*, Harvard University Press, Cambridge, Mass., 1941.

Epstein, P. S., *Textbook of Thermodynamics*, John Wiley & Sons, Inc., New York, 1937.

Fowler, R. H., and Guggenheim, E. A., *Statistical Thermodynamics*, The Macmillan Company, New York, 1939.

Gurney, R. W., *Introduction to Statistical Mechanics*, McGraw-Hill Book Company, Inc., 1949.

Harkins, W. D., Young, T. F., and Boyd, E., "The Thermodynamics of Films," *J. Chem. Phys.*, **8**, 954 (1940).

Hougen, O. A. and Watson, K. M., *Chemical Process Principles; * Part II: *Thermodynamics*, John Wiley & Sons, Inc., New York, 1947.

Mayer, J. E., and Mayer, M. G., *Statistical Mechanics*, John Wiley & Sons, Inc., New York, 1940.

Glasstone, S., *Theoretical Chemistry*, D. Van Nostrand Company, Inc., New York, 1944.

Index

L

Least squares method, 21
LEWIS, G. N., 226
LEWIS and RANDALL:
  j function, 327–328
  rule for gas solutions, 243, 244
  statement of third law, 153–154
Limitations of thermodynamics, 2–3
Limiting law (*see also* Limiting slope), 279
  Debye-Hückel theory, 329–330
  dilute solutions, 275–276, 309, 329
    electrolytes, 265
    non-electrolytes, 257–265
Limiting slope (*see also* Limiting law):
  activities:
    electrolytes, 303–304, 316
    non-electrolytes, 272
  Debye-Hückel theory, 329–330
  electromotive force measurements, 314,
    315, 316
  fugacity:
    electrolytes, 300, 301, 303, 306, 307
    non-electrolytes, 275

M

MACH, 27
Maximum work, 111, 117—119, 130
Maximum yield, 2, 111
Mean ionic activity (*see also* Activity), 304
Mean ionic activity coefficient (*see also*
  Activity coefficient), 305
Mean ionic molality, 306, 308, 309
Mixed electrolytes, 308, 310
Molal activity coefficient (*see* Activity co-
  efficient)
Molal heat capacity, 4, 29, 39
Molal volume, 183–184
Molar activity coefficient, 267

N

NERNST:
  distribution law, 258–259
  heat theorem, 152, 153
NEWTON, 27

O

Objectives of thermodynamics (*see*
  Thermodynamics)
Open system, 224–225
Operational definitions, 28, 79
Osmotic pressure, 261–263

P

PARKS and HUFFMAN, 177–178
Partial molal enthalpy (*see also* Enthalpy),
  205–222, 232–233, 264, 352

activity calculations, 323–324
  definition, 205
  freezing-point calculations, 323–324, 335
  from apparent molal enthalpy, 217–222
  from heats of solution, 213–217
  infinite dilution, 280
  reference state (*see* Standard states)
  relative (*see* Relative partial molal en-
    thalpy)
  standard state, 277, 279–280, 281–284
Partial molal entropy (*see also* Entropy):
  standard state, 281–28
Partial molal free energy, 188, 201–202,
  224–233, 259
  activity, 266
  standard state, 282
Partial molal heat capacity, 202, 223
  freezing-point calculations, 323–324, 335
  standard state, 282, 284
Partial molal quantities, 13, 183–202
  calculation of, 193–201
  definition, 184–189
  equations of, 190–193
  pure phase, 189
Partial molal volume (*see also* Volume),
  184–188, 190
  calculation of, 193–199
  ideal solution, 247–248, 252
  standard state, 282
Partial pressure, 244
Perfect crystalline system, 152, 166
Perfect gas (*see* Ideal gas)
Perfect solution (*see* Ideal solution)
Perpetual-motion machine, 78
Phase changes (*see also* Equilibrium), 2,
  134–141
  entropy change, 94, 99–100
PLANCK, 152–153
POINCARE, 27
Potential (*see also* Electromotive force),
  132
Practical activity coefficient (*see* Activity
  coefficient)
Pressure, 114
  critical, 68–69, 161–162, 240
  ideal, 239
  osmotic, 261–263
  partial, 244
  reduced, 239–240
  temperature coefficient, 156
Pressure-volume diagram, 80, 81, 107, 235
Principles of impotence, 77–78

R

Radiation, 359, 360
Radii of ions (*see* Ionic diameters)
Raoult's law, 246, 248, 257–258, 259–261,
  262